EVIDENCE-BASED SOFTWARE ENGINEERING AND SYSTEMATIC REVIEWS

Chapman & Hall/CRC Innovations in Software Engineering and Software Development

Series Editor
Richard LeBlanc
Chair, Department of Computer Science and Software Engineering, Seattle University

AIMS AND SCOPE

This series covers all aspects of software engineering and software development. Books in the series will be innovative reference books, research monographs, and textbooks at the undergraduate and graduate level. Coverage will include traditional subject matter, cutting-edge research, and current industry practice, such as agile software development methods and service-oriented architectures. We also welcome proposals for books that capture the latest results on the domains and conditions in which practices are most effective.

PUBLISHED TITLES

Computer Games and Software Engineering
Kendra M. L. Cooper and Walt Scacchi

Software Essentials: Design and Construction
Adair Dingle

Software Metrics: A Rigorous and Practical Approach, Third Edition
Norman Fenton and James Bieman

Software Test Attacks to Break Mobile and Embedded Devices
Jon Duncan Hagar

Software Designers in Action: A Human-Centric Look at Design Work
André van der Hoek and Marian Petre

Evidence-Based Software Engineering and Systematic Reviews
Barbara Ann Kitchenham, David Budgen, and Pearl Brereton

Fundamentals of Dependable Computing for Software Engineers
John Knight

Introduction to Combinatorial Testing
D. Richard Kuhn, Raghu N. Kacker, and Yu Lei

Building Enterprise Systems with ODP: An Introduction to Open Distributed Processing
Peter F. Linington, Zoran Milosevic, Akira Tanaka, and Antonio Vallecillo

Software Engineering: The Current Practice
Václav Rajlich

Software Development: An Open Source Approach
Allen Tucker, Ralph Morelli, and Chamindra de Silva

CHAPMAN & HALL/CRC INNOVATIONS IN
SOFTWARE ENGINEERING AND SOFTWARE DEVELOPMENT

EVIDENCE-BASED SOFTWARE ENGINEERING AND SYSTEMATIC REVIEWS

Barbara Ann Kitchenham
Keele University, Staffordshire, UK

David Budgen
Durham University, UK

Pearl Brereton
Keele University, Staffordshire, UK

CRC Press
Taylor & Francis Group
Boca Raton   London   New York

CRC Press is an imprint of the
Taylor & Francis Group an **informa** business

A CHAPMAN & HALL BOOK

CRC Press
Taylor & Francis Group
6000 Broken Sound Parkway NW, Suite 300
Boca Raton, FL 33487-2742

First issued in paperback 2020

Version Date: 20150922

ISBN 13: 978-0-367-57533-5 (pbk)
ISBN 13: 978-1-4822-2865-6 (hbk)

Visit the Taylor & Francis Web site at
http://www.taylorandfrancis.com

and the CRC Press Web site at
http://www.crcpress.com

Contents

III Guidelines for Systematic Reviews 291

22 Systematic Review and Mapping Study Procedures 293

List of Figures

List of Tables

Preface

As a relatively young (and as we will later argue, still somewhat immature) discipline, *software engineering* is in an emergent[1] state for many purposes. Its foundations as a distinct sub-discipline of computing are widely considered to have been laid down at the 1968 NATO conference, although the term was probably in fairly regular use before that. Since then, ideas have ebbed and flowed, along with the incredibly rapid expansion and evolution of computing from an activity largely concerned with 'crunching numbers' in support of scientific research, to something that forms a pervasive element of everyday life. While this has helped to drive the development of software engineering as a discipline, the headlong pace has also meant that there has often been little opportunity to appraise and reflect upon our experiences of how software systems can be developed, how well the different approaches work, and under what conditions they are likely to be most effective.

The emergence of the concept of *evidence-based software engineering* (EBSE) can certainly be assigned a clear starting point, with the seminal paper being presented at the 2004 International Conference on Software Engineering (ICSE). In the decade that has followed, ideas about EBSE, and about its key tool, the systematic review, have evolved and matured; it has taken its place in the empirical software engineer's toolbox; and has helped to categorise and consolidate our knowledge about many aspects of software engineering research and practice. While few commercial software development activities can as yet even be described as 'evidence-informed', the philosophy of EBSE is beginning to be widely recognised and appreciated. As such then, this seems to be a suitable time to bring this knowledge together in a single volume, not least to help focus thinking about what we as a community might usefully do with that knowledge in the future.

Like Gaul, our book is divided into three parts[2]. In the first part we discuss the nature of evidence and the evidence-based practices centred around the systematic review, both in general and also as applying to software engineering. The second part examines the different elements that provide inputs to a systematic review (usually considered as forming a *secondary* study), especially the main forms of primary empirical study currently used in software

---

[1] An emergent process is one that is 'in a state of continual process change, never arriving, always in transition' (Truex, Baskerville & Klein 1999).

[2] Those with a classical education will remember that this was the first observation in Julius Caesar's *The Conquest of Gaul*, and quite possibly, that is the only thing that many of us remember from that work!

engineering. Lastly, the third part provides a practical guide to conducting systematic reviews (the *guidelines*), drawing together accumulated experiences to guide researchers and students when they are planning and conducting their own studies. In support of these we also include an extensive *glossary*, and an appendix that provides a *catalogue* of reviews that may be useful for practice and teaching.

This raises the question of who we perceive to be the audience for this book. We would like to think that almost anyone with any involvement in software engineering (in the broadest sense) can find something of use within it, given that our focus is upon seeking to identify what works in software engineering, when it works, and why. For the researcher, it provides guidance on how to make his or her own contribution to the corpus of knowledge, and how to determine where the research efforts might be directed to best effect. For practitioners, the book both explains the foundations of evidence-based knowledge related to software engineering practices, and also identifies useful examples of this. Finally, for teachers and students, it provides an introduction to the nature and role of empirical software engineering and explains what empirical studies can tell us about our subject.

So, how should the aspiring empiricist, or even the merely curious, approach all of this material, assuming that he or she might be reluctant to attempt to devour each chapter in turn, in the way that they would read a novel? We would suggest that the first few chapters provide a background to EBSE that should be relevant to anyone. These chapters explain the basic thinking about evidence-based studies and concepts, and show how they can be applied within a software engineering context.

The researcher, including of course, all PhD students, should additionally read the rest of Part I, so as to understand how to plan a secondary study. Armed with this understanding they can then turn to Part III, which provides essential practical guidance on the conduct of such a study, and which can then lead them through the steps of putting their plan into action. And, should any researcher determine that the ground is not yet solid enough for a secondary study, they can turn to Part II to learn something about how to conduct and report on a primary study in such a way as to make it a useful input to a future secondary study. Indeed, even when undertaking a secondary study, Part II should also be useful to the systematic reviewer when he or she is facing the tasks of data extraction and synthesis, by explaining something of the context behind the different forms of empirical study that provide the inputs to their analysis.

Practitioners and others who want to know more about EBSE and the use of secondary studies may find that Part I provides much of what they need in order to understand (and use) the outcomes from secondary studies. Likewise, teachers will, we hope, find much useful material in both Part I and Part II, in the latter case because an understanding of secondary studies is best founded upon a solid appreciation of the roles and forms of primary studies. Both of these groups should also find material that is of direct usefulness in the catalogue of reviews provided in the appendix.

We are teachers as well as researchers, and should observe here that teaching the practices used in performing secondary studies to advanced undergraduates can be beneficial too. Students usually need to undertake a literature review as part of their individual 'capstone' projects, and adopting a systematic and objective approach to this can add valuable rigour to the outcomes.

In writing this book, we have drawn upon our own experiences with conducting systematic reviews and primary studies, and so our material and its organisation build upon the lessons that we have learned through these. These experiences have included both designing our own studies and reviewing the designs of others, and with conducting both methodological studies as well as ones that examine some established software engineering practices. Wherever possible we have tried to illustrate our points by drawing upon these experiences, as well as learning from those of many others, whose contribution to EBSE and its development we gratefully acknowledge.

This leads to an issue that always presents something of a challenge for evidence-based researchers such as ourselves, namely that of how to handle *citation*. As evidence-based software engineering researchers we usually feel it necessary to justify everything we possibly can by pointing to relevant evidence—but equally as authors, we are aware that this risks present the reader with a solid wall of reference material, which itself can form a distraction from gaining an understanding of key concepts. We have therefore tried to find a balance, providing citations whenever we think that the reader may possibly wish to confirm or clarify ideas for themselves. At the same time we have tried to avoid a compulsive need for justification at every opportunity, and especially when this is not really essential to enjoying the text—and of course, a sense of interest and enjoyment is exactly what we sincerely hope others will be able to experience from learning about EBSE and how the use of systematic reviews can help to inform software engineering as a discipline.

Finally, as a related point, since all the chapters of Part I relate to different aspects of secondary studies, we have provided a single set of suggestions for *further reading* at the end of this part, in order to avoid undue repetition. In Part II, where we address different forms of primary study in each chapter, we have reverted to the more conventional approach of providing recommendations for further reading at the end of each chapter.

Glossary

The vocabulary used in this book has been derived from a variety of sources and disciplines, which is not unreasonable, as that is how the ideas of empirical software engineering have themselves been derived. Our glossary does not purport to be definitive, the aim is to convey the relevant concepts quickly, so that when consulting it, the reader does not have to stray far from the flow of what they are reading.

absolute (measurement scale): This is the most restrictive of the measurement scales and simply uses counts of the elements in a set of entities. The only operation that can be performed is a test for equality. (See also *measurement scales*.)

accuracy: The accuracy of a measurement is an assessment of the degree of conformity of a measured or calculated value to its actual or specified value.

accuracy range: The accuracy range tells us how close a sample is to the true population of interest, and is usually expressed as a plus/minus margin. (See also *confidence interval*.)

aggregation: The process of gathering together knowledge of a particular type and form (for example, in a table).

attribute: An attribute is a measurable (or at least, identifiable) characteristic of an entity, and as such provides a mapping between the abstract idea of a *property* of the entity and something that we can actually measure in some way.

between-subject: (Also known as *between-groups* or *parallel experiment*.) In this form of study, participants are assigned to different treatment (intervention) groups and each participant only receives one treatment.

bias: A tendency to produce results that depart systematically from the true results.

blinding: A process of concealing some aspect of an experiment from researchers and participants. In single-blind experiments, participants do not know which treatment they have been assigned to. In double-blind experiments, neither participants nor experimenters know which treatment the participants have been assigned to. In software engineering we sometimes use blind-marking, where the marker does not know which treatment the participants adopted to arrive at their answers or responses.

case study: A form of *primary study*, which is an investigation of some phenomenon in a real-life setting. Case studies are typically used for *explanatory*, *exploratory* and *descriptive* purposes. The main two forms are *single-case* studies which may be appropriate when studying a representative case or a special case, but will be less trustworthy than *multiple-case* forms, where replication is employed to see how far different cases predict the same outcomes. (Note that the term *case study* is sometimes used in other disciplines to mean a narrative describing an example of interest.) Case study research is covered in detail in Yin (2014) and for software engineering, in Runeson, Höst, Rainer & Regnell (2012).

causality: The link between a stimulus and a response, in that one *causes* the other to occur (also termed cause and effect). The notion of some form of causality usually underpins *hypotheses*.

central tendency: The 'typical value' for a probability distribution. The three most common measures used for this are the *mean*, the *median* and the *mode*. (See the separate definitions of these.)

closed question: (As used in a questionnaire.) Such a question constrains respondents by requiring them to select from a pre-determined list of answers. This list may optionally include 'other' or 'don't know' options. (See also *open question*.)

conclusion validity: (See *validity*.)

confidence interval: This is an assessment of how sure we are that the region within the stated interval around our measured mean does contain the true mean. This is expressed as a percentage, for example, a confidence interval of 95% (which corresponds to two standard deviations either side of the mean) means that there is a 95% likelihood that the true population mean lies within two standard deviations of our sample mean. So, for this value of the confidence interval, if we did many independent experiments and calculated confidence intervals for each of these, the true mean of the population being studied would be within the confidence limits in 95% of these.

confounding factor: An undesirable element in an experimental study that produces an effect that is indistinguishable from that of one of the treatments.

construct validity: (See *validity*.)

content validity: (As used in a survey.) Concerned with whether the questions are a well-balanced sample for the domain we are addressing.

control group: For laboratory experiments we can divide the participants into two groups—with the *treatment group* receiving the experimental treatment being investigated, and the experimental context of the *control group* involving no manipulation of the independent variable(s). It is then possible to attribute any differences between the outcomes for the two groups as arising from the treatment.

controlled experiment: (See *laboratory experiment, field experiment* and *quasi-experiment.*)

convenience (sample): A form of *non-probabilistic sampling* in which participants are selected simply because it is easy to get access to them or they are willing to help. (See *sampling technique.*)

cross-over: (See *within-subject.*)

dependent variable: (Also termed *response variable* or *outcome variable.*) This changes as a result of changes to the independent variable(s) and is associated with an *effect.* The outcomes of a study are based upon measurement of the dependent variable.

descriptive (survey): (See *survey.*)

direct measurement: Assignment of values to an attribute of an entity by some form of counting.

divergence: A divergence occurs when a study is not performed as specified in the *experimental protocol,* and all divergences should be both recorded during the study and reported at the end.

double blinding: (See *blinding.*)

dry run: For an experiment, this involves applying the experimental treatment to (usually) a single recipient, in order to test the experimental procedures (which may include training, study tasks, data collection and analysis). May sometimes be termed a *pilot experiment.* A similar activity may be performed for a survey instrument.

effect size: The effect size provides a measure of the strength of a phenomenon. There are many measures of effect size to cater to different types of treatment outcome measures, including the standardized mean differences, the log odds ratio, and the Pearson correlation coefficient.

empirical: Relying on observation and experiment rather than theory (*Collins English Dictionary*).

ethics: The study of standards of conduct and moral judgement (*Collins English Dictionary*). Codes of ethics for software engineering are published by the British Computer Society and the ACM/IEEE. Any empirical study that involves human participants should be vetted by the researcher's local *ethics committee* to ensure that it does not disadvantage any of the participants in any way.

ethnography: A form of observational study that is purely observational, and hence without any form of intervention or participation by the observer.

evidence-based: An approach to empirical studies by which the researcher seeks to identify and integrate the best available research evidence with domain expertise in order to inform practice and policy-making. The normal mechanism for identifying and aggregating research evidence is the *systematic review*.

exclusion criteria: After performing a search for papers (primary studies) when performing a systematic review, the exclusion criteria are used to help determine which ones will not be used in the study. (See also *inclusion criteria*.)

experiment: A study in which an intervention (i.e. a treatment) is deliberately controlled to observe its effects (Shadish, Cook & Campbell 2002).

external attribute: An external attribute is one that can be measured only with respect to how an element relates to other elements (such as reliability, productivity, etc.).

field experiment: An experiment or quasi-experiment performed in a natural setting. A field experiment usually has a more realistic setting than a laboratory experiment, and so has greater external validity.

field study: A generic term for an empirical study undertaken in real-life conditions.

hypothesis: A testable *prediction* of a cause–effect link. Associated with a hypothesis is a *null hypothesis* which states that there are no underlying trends or dependencies and that any differences observed are coincidental. A statistical test is normally used to determine the probability that the null hypothesis can or cannot be rejected.

inclusion criteria: After performing a search for papers (primary studies) when performing a systematic review, the inclusion criteria are used to help determine which ones contain relevant data and hence will be used in the study. (See also *exclusion criteria*.)

independent variable: An independent variable (also known as a *stimulus* variable or an *input* variable) is associated with *cause* and is changed as a result of the activities of the investigator and not of changes in any other variables.

indirect measurement: Assigning values to an attribute of an entity by measuring other attributes and using these with some form of 'measurement model' to obtain a value for the attribute of interest.

input variable: (See *independent variable.*)

instrument: The 'vehicle' or mechanism used in an empirical study as the means of data collection (for the example of a survey, the instrument might be a questionnaire).

internal attribute: A term used in software metrics to refer to a measurable attribute that can be extracted directly from a software document or program without reference to other software project or process attributes.

interpretivism: In information systems research and computing in general, interpretive research is 'concerned with understanding the social context of an information system: the social processes by which it is developed and construed by people and through which it influences, and is influenced by, its social setting' (Oates 2006). (See also *positivism.*)

interval scale: An interval scale is one whereby we have a well-defined ratio of intervals, but have no absolute zero point on the scale, so that we cannot speak of something being 'twice as large'. Operations on interval values include testing for equivalence, greater and less than, and for a known ratio. (See also *measurement scales.*)

interview: A mechanism used for collecting data from participants for surveys and other forms of empirical study. The forms usually encountered are *structured, semi-structured* and *unstructured.* The data collected are primarily subjective in form.

laboratory experiment: Sometimes referred to as a *controlled laboratory experiment,* this involves the identification of precise relationships between experimental variables by means of a study that takes place in a controlled environment (the 'laboratory') involving human participants and supported by quantitative techniques for data collection and analysis.

longitudinal: Refers to a form of study that involves repeated observations of the same items over long periods of time.

mapping study: A form of secondary study intended to identify and classify the set of publications on a topic. May be used to identify 'evidence gaps' where more primary studies are needed as well as 'evidence clusters' where it may be practical to perform a systematic review.

mean: Often referred to as the *average*, and one of the three most common measures of the *central tendency*. Computed by adding the data values and dividing by the number of elements in the dataset. It is only meaningful for data forms that have genuinely numerical values (as opposed to codes).

measurement: The process by which numbers or symbols are assigned to attributes of real-world entities using a well-defined set of rules. Measurement may be direct (for example, length) or indirect, whereby we measure one or more other attributes in order to obtain the value (such as measuring the length of a column of mercury on a thermometer in order to measure temperature).

measurement scales: The set of scales usually used by statisticians are absolute, nominal, ordinal, interval and ratio. (See the separate definitions of these for details). A good discussion of the scales and their applicability is provided in Fenton & Pfleeger 1997.

median: (Also known as the 50th percentile.) One of the three most common measures of the *central tendency*. This is the value that separates the upper half of a set of values from the lower half, and is computed by ordering the values and taking the middle one (or the average of two middle ones if there is an even number of elements). Then half of the elements have values above the median and half have values below.

meta-analysis: The process of statistical pooling of similar quantitative studies.

mode: One of the three most common measures of the *central tendency*. This is the value that occurs most frequently in a dataset.

nominal measurement scale: A nominal scale consists of a number of categories, with no sense of ordering. So the only operation that is meaningful is a test for equality (or inequality). An example of a nominal scale might be programming languages. (See also *measurement scales*.)

null hypothesis: (See *hypothesis*.)

objective: Objective measures are those that are independent of the observer's own views or opinions, and so are repeatable by others. Hence they tend to be quantitative in form.

observational scale: An observational scale seeks simply to record the actions and outcomes of a study, usually in terms of a pre-defined set of factors, and there is no attempt to use this to confirm or refute any form of hypothesis. Observational scales are commonly used for diagnosis or making comparison between subjects or between subjects and a benchmark. For research, they may be used to explore an issue and to determine whether more rigorous forms might then be employed.

open question: (As used in a questionnaire.) An open question is one that leaves the respondent free to provide whatever answer they wish, without any constraint on the number of possible answers. See also *closed question*.

ordinal scale: An ordinal scale is one that *ranks* the elements, but without there being any sense of a well-defined interval between the different elements. An example of such a scale might be *cohesion*, where we have the idea that particular forms are better than others, but no measure of how much. Operations are equality (inequality) and greater than/less than. (See also *measurement scales*.)

outcome variable: (See *dependent* variable.)

participant: Someone who takes part (participates) in a study, sometimes termed a *subject*. Participant is the better term in a software engineering context because involvement nearly always has an active element, whereas subject implies a passive recipient.

population: A group of individuals or items that share one or more characteristics from which data can be extracted and analysed. (See *sampling frame*.)

positivism: The philosophical paradigm that underlies what is usually termed the 'scientific method'. It assumes that the 'world' we are investigating is ordered and regular, rather than random, and that we can investigate it in an objective manner. It therefore forms the basis for hypothesis-driven research. For a fuller discussion, see (Oates 2006).

power: (See *statistical power*.)

precision: (See also *recall*.) In the context of information retrieval, the *precision* of the outcomes of a search is a measure of the proportion of studies found that are *relevant*. (Note that this makes no assumptions about whether or not all possible relevant documents were found.) If the number of relevant documents N_{rel} is defined as

$$N_{rel} = N_{retr} - \overline{N_{rel}}$$

where N_{retr} is the number retrieved and $\overline{N_{rel}}$ is the number that is classified as not relevant, then

$$precision = \frac{N_{rel}}{N_{retr}}$$

Hence if we retrieve 20 documents, of which 8 are not relevant, the value for precision will be $(20 - 8)/20$ or 0.6. So a value of 1.0 for precision indicates that all of the documents found were relevant, but says nothing about whether every relevant document was found.

primary study: This is an empirical study in which we directly make measurements about the objects of interest, whether by surveys, experiments, case studies, etc. (See also *secondary study*.)

proposition: (In the context of a case study.) This is a more detailed element derived from a *research question* and performs a role broadly similar to that of a *hypothesis* (and like a hypothesis can be derived from a theory). Propositions usually form the basis of a case study and help to guide the organisation of data collection (Yin 2014). However, an *exploratory* case study would not be expected to involve the use of any propositions.

protocol: In the context of empirical studies, this term is used in two similar (but different) ways.

- For empirical studies in general, the *experimental protocol* is a document that describes the way that a study is to be performed. It should be written before the study begins and evaluated and tested through a 'dry run'. During the actual study, any *divergences* from the protocol should be recorded. It is this interpretation that is used throughout this book.

- The practice of *protocol analysis* can be used for qualitative studies, forming a data analysis technique that is based upon the use of *think-aloud*. In this, the protocol provides a categorisation of possible utterances that can be used to analyse the particular sequence of words produced by a participant while performing a task, as well as to strip out irrelevant material (Ericsson & Simon 1993).

qualitative: A measurement form that (typically) involves some form of human judgement or assessment in assigning values to an attribute, and hence which may use an ordinal scale or a nominal scale. Qualitative data is also referred to as *subjective data*, but such data can be quantitative, such as responses to questions in survey instruments.

quantitative: A measurement form that involves assigning values to an attribute using an interval scale or (more typically) a ratio scale. Quantitative data is also referred to as *objective data*, however this is incorrect, since is it possible to have quantitative subjective data.

quasi-experiment: An experiment in which units are not assigned at random to the interventions (Shadish et al. 2002).

questionnaire: A data collection mechanism commonly used for surveys (but also in other forms of empirical study). It involves participants in answering a series of questions (which may be 'open' or 'closed').

randomised controlled trial (RCT): A form of large-scale controlled experiments performed in the field using a random sample from the population of interest and (ideally) *double blinding*. In clinical medicine this is regarded as the 'gold standard' in terms of experimental forms, but there is little scope to perform RCTs in disciplines (such as software engineering) where individual participant skill levels are involved in the treatment.

randomised experiment: An experiment in which units are assigned to receive the treatment or alternative condition by a random process such as a coin toss or a table of random numbers.

ratio scale: This is a scale with well-defined intervals and also an absolute zero point. Operations include equality, greater than, less than, and ratio—such as 'twice the size'. (See also *measurement scales*.)

reactivity: This refers to a change in the participant's behaviour arising from being tested as part of the study, or from trying to help the experimenter (hypothesis guessing). It may also arise because of the influence of the experimenter (forming a source of bias).

recall: (See also *precision*.) In the context of information retrieval, the *recall* of the outcomes of a search (also termed *sensitivity*) is a measure of the proportion of all relevant studies found in the search. If the number of relevant documents N_{rel} is defined as

$$N_{rel} = N_{retr} - \overline{N_{rel}}$$

where N_{retr} is the number retrieved and $\overline{N_{rel}}$ is the number that is classified as not relevant, then

$$recall = \frac{N_{rel}}{N_{rel}^{tot}}$$

where N_{rel}^{tot} is the total number of documents that are relevant (if you know it). Hence if we retrieve 20 documents of which 8 are not relevant, and we know that there are no other relevant ones, then the value for recall will be $(20 - 8)/12$ or 1.0. So while a value of 1.0 for recall indicates that all relevant documents were found, it does not indicate how many irrelevant ones were also found.

research question: The research question provides the rationale behind any primary or secondary empirical study, and states in broad terms the issue that the study is intended to investigate. For experiments this will be the basis of the *hypothesis* used, but the idea is equally valid when applied to a more observational form of study.

response rate: For a survey, the response rate is the proportion of surveys completed and returned, compared to those issued.

response variable: An alternative term for the *dependent variable*.

sample: This is the set (usually) of people who act as participants in a study (for example, a survey or a controlled laboratory experiment). Equally, it can be a sample set of documents or other entities as appropriate. An important aspect of a sample is the extent to which this is representative of the larger population of interest.

sample size: This is the size of the sample needed to achieve a particular *confidence interval* (with a 95% confidence interval as a common goal). As a rule of thumb, if any statistical analysis is to be employed, even at the level of calculating means and averages, a sample size of at least 30 is required.

sampling frame: This is the set of entities that could be included in a survey, for example, people who have been on a particular training course, or who live in a particular place.

sampling technique: This is the strategy used to select a sample from a sampling frame and takes two main forms:

non-probabilistic sampling Employed where it is impractical or unnecessary to have a representative sample. Includes purposive, snowball, self-selection and convenience sampling.

probabilistic sampling An approach that aims to obtain a sample that forms a representative cross-section of the sampling frame. Includes random, systematic, stratified and cluster sampling.

secondary study: A secondary study does not generate any data from direct measurements, instead it analyses a set of *primary studies* and usually seeks to aggregate the results from these in order to provide stronger forms of *evidence* about a particular phenomenon.

statistical power: The ability of a statistical test to reveal a true pattern in the data (Wohlin, Runeson, Höst, Ohlsson, Regnell & Wesslen 2012). If the power is low, then there is a high risk of drawing an erroneous conclusion. For a detailed discussion of statistical power in software engineering studies, see (Dybå, Kampenes & Sjøberg 2006).

stimulus variable: (See *independent variable*.)

subjective: Subjective measures are those that depend upon a value judgement made by the observer, such as a ranking ('A is more significant than B). May be expressed as a qualitative value ('better') or in a quantitative form by using an ordinal scale.

survey: A comprehensive research method for collecting information to describe, compare or explain knowledge, attitudes and behaviour. The purpose of a survey is to collect information from a large group of people in a standard and systematic manner and then to seek *patterns* in the resulting data that can be generalised to the wider population. Surveys can be broadly classified as being

- *experimental* when used to assess the impact of some intervention
- *descriptive* if used to enable assertions to be made about some phenomenon of interest and the distribution of particular attributes— where the concern is not *why* the distribution exists, but *what* form it has

synthesis: The process of systematically combining different sources of data (evidence) in order to create new knowledge.

systematic (literature) review: This is a particular form of *secondary study* and aims to provide an objective and unbiased approach to finding relevant primary studies, and for extracting, aggregating and synthesising the data from these.

tertiary study: This is a secondary study that uses the outputs of secondary studies as its inputs, perhaps by examining the secondary studies performed in a complete discipline or a part of it.

test–retest: Conventionally, this forms a measure of the *reliability* and *stability* of a survey instrument. Respondents are 'tested' at two well-separated points in time, and the responses are compared for consistency by means of a correlation test, with correlation values of 0.7–0.8 usually being considered satisfactory. Use of test–retest is only appropriate in situations where 'learning' effects are unlikely to occur within the intervening time period. In the context where a single researcher is performing a systematic review, the use of test–retest can be interpreted as being for the researcher to perform such tasks as *selection* and *data extraction* twice, with these being separated by a suitable time interval, and to check for consistency between the two sets of outcomes. Where possible, these tasks should be performed using different orderings of the data items, in order to reduce possible bias.

treatment: This is the 'intervention' element of an experiment (the term is really more appropriate to *randomised controlled trials* where the participants are recipients). In software engineering it may take the form of a task (or tasks) that participants are asked to perform such as writing code, testing code, reading documents.

triangulation: Refers to the use of multiple elements that reinforce one another in terms of providing evidence, where no single source would be adequately convincing. The 'sources' may be different forms of data, or the outcomes from different research methods.

validity: This is concerned with the degree to which we can 'trust' the outcomes of an empirical study, usually assessed in terms of four commonly encountered forms of *threat to validity*. The following definitions are based upon those provided in Shadish et al. (2002).

- *internal:* Relating to inferences that the observed relationship between treatment and outcome reflects a cause–effect relationship.
- *external:* Relating to whether a cause–effect relationship holds over other conditions, including persons, settings, treatment variables and measurement variables.
- *construct:* Relating to the way in which concepts are operationalised as experimental measures.
- *conclusion:* Relating inferences about the relationship between treatment and outcome variables.

within-subject: Refers to one of the possible design forms for a quasi-experiment. In this form, participants receive a number of different treatments, with the order in which these are received being randomised. A commonly encountered design (two treatments) is the *A/B–B/A crossover* whereby some participants receive treatment *A* and then treatment *B*, while others receive them in reverse order. A weaker version is a *before–after* design, whereby all participants perform a task, are then given some training (the treatment), and are then asked to undertake another task. (Also known as a *sequential* or *repeated-measures* experiment.)

Part I

Evidence-Based Practices in Software Engineering

Chapter 1

The Evidence-Based Paradigm

Since this is a book that is about the use of evidence-based research practices, we feel that it is only appropriate to begin it by considering what is meant by *evidence* in the general sense. However, because this is also a book that describes how we acquire evidence about software engineering practices, we then need to consider some of the ways in which ideas about evidence are interpreted within the rather narrower confines of science and technology.

Evidence is often associated with *knowledge*. This is because we would usually like to think that our knowledge about the world around us is based upon some form of evidence, and not simply upon wishful thinking. If we go to catch a train, it might be useful to have evidence in the form of a timetable that shows the intention of the railway company to provide a train at the given time that will take us to our destination. Or, rather differently, if we think that some factor might have caused a 'population drift' away from the place where we live, we might look at past census data to see if such a drift really has occurred, and also whether some groups have been affected more than others. Of course the link between evidence and knowledge is rarely well-defined, as in our second example, where any changes in population we observe might arise from many different factors. Indeed, it is not unusual, in the wider world at least, for the same evidence to be interpreted differently (just think about global warming).

In this chapter we examine what is meant by evidence and knowledge, and the processes by which we interpret the first to add to or create the second. We also consider some limitations of these processes, both those that are intrinsic, such as those that arise from the nature of the things being studied, and of data itself, and also those that arise from the inevitable imperfections of research practice. In doing so, we prepare the ground for Chapter 2, where we look at how the discipline of software engineering interprets these concepts, and review the characteristics of software engineering that influence the nature of our evidence—and hence the nature of our knowledge too.

1.1   What do we mean by evidence?

As noted above, evidence can be considered as being something that underpins *knowledge*, and we usually expect that knowledge will be derived from evidence through some process of *interpretation*. The nature of that interpretation can take many forms. For example, it might draw upon other forms of knowledge, as when the fictional detective Sherlock Holmes draws upon his knowledge about different varieties of tobacco ash, or about the types of earth to be found in different parts of London, in order to turn a clue into evidence. Interpretation might also be based upon mathematical or statistical procedures, such as when a scientist gathers together different forms of experimental and observational data—for example, using past medical records to demonstrate that smoking is a cause of lung cancer. Yet another, less scientific, illustration of the concept is when the jury at a criminal trial has to consider the evidence of a set of witnesses in order to derive reasonable knowledge about what actually happened. Clearly these differ in terms of when they arise, the form of knowledge derived, and the rigour of the process used for its derivation (and hence the *quality* of the resulting knowledge). What they do have in common though, is that our confidence about the knowledge will be increased if there is more than one source (and possibly form) of evidence. For the fictional detective, this may be multiple clues; for the clinical analysis it might involve using records made in many places and on patients who have different medical histories; for the jury, it may be that there are several independent witnesses whose statements corroborate each other. This process of *triangulation* between sources (a term derived from navigation techniques) is also an important means of testing the *validity* of the knowledge acquired.

Science in its many forms makes extensive use of these concepts, although not always expressed using this vocabulary. Over the years, particular scientific disciplines have evolved their own accepted set of empirical practices that are intended to give confidence in the validity and quality of the knowledge created from the forms of evidence considered to be appropriate to that discipline, and also to assess how strong that confidence is. Since this book is extensively concerned with different forms of *empirical* study, this is a good point to note that such studies are ones that are based upon *observation* and *measurement*. Indeed, this is a reminder that, strictly speaking, scientific processes never 'prove' anything (mathematics apart), they only 'demonstrate' that some relationship exists between two or more factors of interest. Even physicists, who are generally in the best position to isolate factors, and to exclude the effect of the observation process, are confronted with this issue. The charge on an electron, or the universal gravitational constant, may well be known to a very high level of precision, and with high confidence, but even so, some residual uncertainty always remains. For disciplines where it can be harder to separate out the key experimental characteristics and where (horrors), humans are involved in roles other than as observers, so the element of variability will inevitably increase. This is of course the situation that occurs for many software engineering research studies, and we will look at some of the consequences in the next chapter.

When faced with evidence for which the values and quality may vary, the approach generally adopted is to use repeated observations, as indicated above, and even better, to gather observations made by different people in different locations. By pooling these, it becomes easier to identify where we can recognise repeated occurrences of patterns in the evidence that can be used to provide knowledge. This repetition also helps to give us confidence that we are not just seeing something that has happened by chance.

The assumption that it is meaningful to aggregate the observations from different studies and to seek patterns in these is termed a *positivist* philosophy. Positivism is the philosophy that underpins the 'scientific method' in general, as well as almost all of the different forms of empirical study that are described in this book.

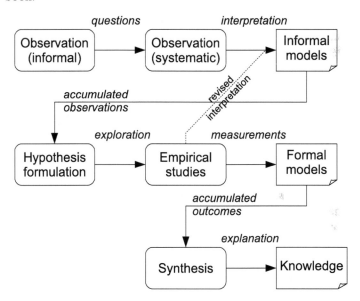

FIGURE 1.1: A simple model of knowledge acquisition.

Figure 1.1 shows a simple model that describes how these concepts relate to one another in a rather general sense. The top row represents how, having noticed the possible presence of some effect, we might begin gathering observations to create a rather informal model to describe some phenomenon. This model might well identify more than one possible cause. If this looks promising, then we might formulate a hypothesis (along the lines that "factor X causes outcome Y to occur") and perform some more systematically organised studies to explore and test this model, during which process, we may discard or revise our ideas about some of the possible causes. Finally, to confirm that our experimental findings are reliable, we encourage others to repeat them, so that our knowledge is now accumulated from many sources and gathered together by a process that we refer to as *synthesis*, so that the risk of bias is reduced. Many well-known scientific discoveries have followed this path in some way, such as the discovery of X-rays and that of penicillin.

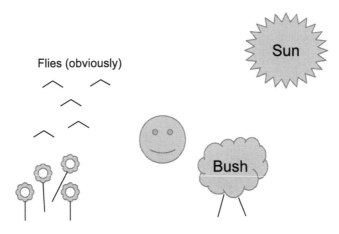

FIGURE 1.2: Does the bush keep the flies off?

Since this is rather abstract, let's consider a simple (slightly contrived but not unrealistic) example. This is illustrated (very crudely) in Figure 1.2. If we imagine that, while sitting out in a garden one day in order to enjoy the summer sunshine, we notice that we are far less bothered by flies when sitting near a particular bush, then this provides an example of informal observation. If we get enough good weather (we did say this example was contrived), we might try repeating the observation, perhaps by sitting near other bushes of that variety. If we continue to notice the effect, then this now constitutes an informal model. Encouraged by visions of the royalties that could arise from discovering a natural insecticide, we might then go on to pursue this rather more systematically, and of course, in so doing we will probably find all sorts of other possible explanations, or indeed, that it is not really an effect at all. But of course, we might also just end up with some systematically gathered knowledge about the insect-repellent nature of this plant (or perhaps, of this plant in conjunction with other factors).

This book is mainly concerned with the bottom two layers of the model shown in Figure 1.1. In Part I and Part III we are concerned with how knowledge from different sources can be 'pooled', while in Part II we provide a subject-specific interpretation of what is meant by the activities in the middle layer. In particular, we will be looking at ways of gathering evidence that go beyond just the use of formal experiments.

In the next section we examine how the concepts of *evidence-based* knowledge and of *evidence-informed* decision-making, have been interpreted in the 20th and 21st centuries. In particular, we will discuss the procedures that have been adopted to produce evidence that is of the best possible quality.

1.2 Emergence of the evidence-based movement

It is difficult to discuss the idea of evidence-based thinking without first providing a description of how it emerged in clinical medicine. And in turn, it is difficult to categorise this as other than a movement that has influenced the practice and teaching of medicine (and beyond). At the heart of this lies the *Cochrane Collaboration*[1], named after one of the major figures in its development. This is a not-for-profit body that provides both independent guardianship of evidence-based practices for clinical medicine, and also custodianship of the resulting knowledge.

So, who was Cochrane? Well, Archie Cochrane was a leading clinician, who became increasingly concerned throughout his career about how to know what was the best treatment for his patients. His resulting challenge to the medical profession was to find the most effective and fairest way to evaluate available medical evidence, and he was particularly keen to put value upon evidence that was obtained from randomised controlled trials (RCTs). Cochrane's highly influential 1971 monograph "Effectiveness and Efficiency: Random Reflections on Health Services" (Cochrane 1971) particularly championed the extensive use of randomisation in RCTs, in order to minimise the influence of different sources of potential bias (such as trial design, experimenter conduct, allocation of subjects to groups, etc.). Indeed, he is quoted as saying that "you should randomise until it hurts", in order to emphasise the critical importance of conducting fair and unbiased trials.

Cochrane also realised that even when performed well, individual RCTs could not be relied upon to provide unequivocal results, and indeed, that where RCTs on a given topic were conducted by different groups and in different places, they might well produce apparently conflicting outcomes. From this, he concluded in 1979 that "it is surely a great criticism of our profession that we have not organised a critical summary by speciality or subspeciality, adapted periodically, of all relevant randomised controlled trials".

Conceptually, this statement was at complete variance with accepted scientific practice (not just that in clinical medicine). In particular, the role of the *review paper* has long been well established across much of academia, with

---

[1]www.cochrane.org

specialist journals dedicated to publishing reviews, and with an invitation to write a review on a given topic often being regarded as a prestigious acknowledgement of the author's academic standing. However, a problem with this practice was (and still is) that two people who are both experts on a given topic might well write reviews that draw contrasting conclusions—and with each of them selecting a quite different set of sources in support of their conclusion.

While this does not mean that an expert review is necessarily of little value, it does raise the question of how far the reviewer's own opinions may have influenced the conclusions. In particular, where the subject-matter of the review requires interpretation of empirical data, then how this is selected is obviously a critical parameter. A widely-quoted example of this is the review by Linus Pauling in his 1970 publication on the benefits of Vitamin C for combatting the common cold. His 'cherry-picking' of those studies that supported his theory, and dismissal of those that did not as being flawed, produced what is now regarded as an invalid conclusion. (This is discussed in rather more depth in Ben Goldacre's book, *Bad Science* (2009), although Goldacre does observe that in fairness, cherry-picking of studies was the norm for such reviews at the time when Pauling was writing—and he also observes that this remains the approach that is still apt to be favoured by the purveyors of 'alternative' therapies.)

Finding the most relevant sources of data is, however, only one element in producing reviews that are objective and unbiased. The process by which the outcomes (findings) from those studies are *synthesised* is also a key parameter to be considered. Ideas about synthesis have quite deep roots—in their book on literature reviews, Booth, Papaioannou and Sutton (2012) trace many of the ideas back to the work of the surgeon James Lind and his studies of how to treat scurvy on ships—including his recognition of the need to discard 'weaker evidence', and to do so by using an objective procedure. However, the widespread synthesis of data from RCTs only really became commonplace in the 1970s, when the term *meta-analysis* also came into common use[2].

Meta-analysis is a statistical procedure used to pool the results from a number of studies, usually RCTs or controlled experiments (we discuss this later in Chapters 9–11). By identifying where individual studies show consistent outcomes, a meta-analysis can provide much greater statistical authority for its outcomes than is possible for individual studies.

Meta-analysis provided one of the key elements in persuading the medical profession to pay attention. In particular, what Goldacre describes as a "landmark meta-analysis" looking at the effectiveness of an intervention given to mothers-to-be who risked premature birth, attracted serious attention. Seven

---

[2]One of us (DB) can claim to have had relatively early experience of the benefits of synthesis, when analysing scattering data in the field of elementary particle physics (Budgen 1971). Some experiments had suggested the possible presence of a very short-lived Σ particle, but this was conclusively rejected by the analysis based upon the composite dataset from multiple experiments.

trials of this treatment were conducted between 1972 and 1981, two finding positive effects, while the other five were inconclusive. However, in 1989 (a decade later) a meta-analysis that pooled the data from these trials demonstrated very strong evidence in favour of the treatment, and it is a "Forest Plot" of these results that now forms a central part of the logo of the *Cochrane Collaboration*, as shown in Figure 1.3[3]. With analyses such as this, supported by the strong advocacy of evidence-based decision making from David Sackett and his colleagues (Sackett, Straus, Richardson, Rosenberg & Haynes 2000), clinicians became more widely persuaded that such pooling of data could provide significant benefits. And linking all this back with the ideas about evidence, Sackett et al. (2000) defined *Evidence-Based Medicine* (EBM) as "the conscientious, explicit and judicious use of the current best evidence in making decisions about the care of individual patients".

FIGURE 1.3: The logo of the Cochrane Collaboration featuring a forest plot (reproduced by permission of the Cochrane Collaboration).

The concept has subsequently been taken up widely within healthcare, although, as we note in Section 1.4, not always without some opposing arguments being raised. It has also been adopted in other disciplines where empirical data is valued and important, with education providing a good example of a discipline where the outcomes have been used to help determine policy as well as practice. A mirror organisation to that of the Cochrane Collaboration is the Campbell Collaboration[4], that "produces systematic reviews of the effects of social interventions in Crime & Justice, Education, International Development, and Social Welfare". And of course, in the following chapters, we will explore how evidence-based ideas have been adopted in software engineering.

So, having identified two key parameters for producing sound evidence from an objective review process as being:

- objective selection of relevant studies

- systematic synthesis of the outcomes from those studies

---

[3] We provide a fuller explanation of the form of Forest Plots in Chapter 11. The horizontal bars represent the results from individual trials, with any that are to the left of the centre line favouring the experimental treatment, although only being statistically significant if they do not touch the line. The results of the meta-analysis is shown by the diamond at the bottom.

[4] www.campbellcollaboration.org

we can now move on to discuss the way that this is commonly organised through the procedures of a *systematic review*.

1.3   The systematic review

At this point, we need to clarify a point about the terminology we use in this book. What this section describes is something that is commonly described as a process of *systematic review* (SR). However, in software engineering, a commonly-adopted convention has been to use the term *systematic literature review* (SLR). This was because when secondary studies were first introduced into software engineering, there was concern that they would be confused with code inspection practices (also termed reviews) and so the use of 'literature' was inserted to emphasise that it was published studies that were being reviewed, not code.

Now that secondary studies as a key element of evidence-based software engineering (EBSE) are part of the empirical software engineer's toolbox, the likelihood of confusion seems much less. So we feel that it is more appropriate to use the more conventional term 'systematic review' throughout this book. However, we do mention it here just to emphasise that when reading software engineering papers, including many of our own, a systematic literature review is the same thing as a systematic review.

The goal of a *systematic review* is to search for and identify all relevant material related to a given topic (with the nature of this material being determined by the underlying question and the nature of the stakeholders who have an interest in it). Knowledge about that topic is then used to assist with drawing together the material in order to produce a collective result. The aim is for the procedures followed in performing the review to be as objective, analytical, and repeatable as possible—and that this process should, in the ideal, be such that if the review were repeated by others, it would select the same input studies and come to the same conclusions. We often refer to a systematic review as being a *secondary study*, because it generates its outcomes by aggregating the material from a set of *primary studies*.

Not surprisingly, conducting such a review is quite a large task, not least because the 'contextual knowledge' required means that much of it needs to be done by people with some knowledge of the topic being reviewed. We will encounter a number of factors that limit the extent to which we can meet these goals for a review as we progress through the rest of this part of the book. However, the procedures followed in a systematic review are intended to minimise the effects of these factors and so even when we don't quite meet the aim as fully as we would like, the result should still be a good quality review. (This is not to say that expert reviews are not necessarily of good quality, but they are apt to lack the means of demonstrating that this is so, in contrast to a systematic review.)

So, a key characteristic of a systematic review is that it is just that, *systematic*, and that it is conducted by following a set of well-defined procedures. These are usually specified as part of the *Review Protocol*, which we will be discussing in more detail later, in Chapter 4. For this section, we are concerned simply with identifying what it is that these procedures need to address. Figure 1.4 illustrates how the main elements of a systematic review are related once a sensible question has been chosen. Each of the ovals represents one of the processes that needs to be performed by following a pre-defined procedure. Each process also involves making a number of decisions, as outlined below.

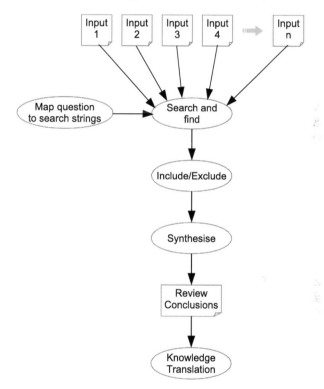

FIGURE 1.4: The systematic review process.

What searching strategy will be used? An important element of the review is to make clear *where* we will search, and *how* we will search for appropriate review material. In addition, we need to ensure that we have included all the different keywords and concepts that might be relevant. We address this in detail in Chapter 5.

What material is eligible for inclusion? This relates to both the different *forms* in which material (usually in the form of the outcomes of empirical studies) might occur, and also any characteristics that might affect its *quality*. Indeed, we often have more detailed specifications for what is to be *excluded* than for what is to be included, since we want to ensure that we don't miss anything that could be in a form that we didn't anticipate, or expect to encounter. Again, these issues will be considered more fully in Chapters 6 and 7.

How is the material to be synthesised? This addresses the analytical procedures that are to be followed. These may be fairly simple, as we explain below, or quite complex. Chapters 9, 10 and 11 consider the relevant issues for a software engineering context.

How to interpret the outcomes of the review? This is not necessarily a single process, since the outcomes might need to be interpreted differently when used in specific contexts. The processes involved are termed *Knowledge Translation* (KT), and are still the topic of extensive discussion in domains where evidence-based practices are much more established than they are in software engineering. However, in Chapter 14, we do examine how KT can be applied in a software engineering context.

The point to emphasise though, is that all of these activities involve *procedures* that need to be applied and interpreted by human beings, with many of them also needing knowledge about the topic of the review. While tools can help with managing the process, the individual decisions still need to be made by a human analyst. In particular, because there will almost certainly be a wide variation of potential inputs to a review, it is possible that some of these will be interpreted differently by different people. To minimise the effects of this, systematic reviews are often conducted by two (or even more) people, who compare results at each stage, and then seek to resolve any differences (again in a systematic manner).

As indicated, because systematic reviews have different forms, the process of synthesis can also take many forms. (A very good categorisation of the wide range of forms of synthesis used across those disciplines that employ systematic reviews is provided in the book by Booth et al. (2012).) At its most simple, synthesis can consist mainly of classification of the material found, identifying where there are groups of studies addressing a particular issue, or equally, where there is a lack of studies. We term this a *mapping study*, and software engineering research has made quite extensive use of this form. A value of a mapping study lies partly in identifying where there is scope to perform a fuller review (the groups of related studies), and also where there is a need for more primary studies (the gaps). At the other extreme, where the material consists mostly of RCTs, or good quality experiments, synthesis may be organised in the form of a statistical meta-analysis. Meta-analyses do exist in the software engineering literature, but only in small numbers. Most software engineering studies use less rigorous forms (and sometimes forms that are less rigorous than could actually be used), and again, we will examine this in much more detail in Chapters 9, 10 and 11.

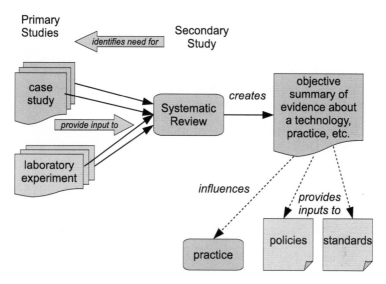

FIGURE 1.5: The context for a systematic review.

Figure 1.5 illustrates the wider context for a systematic review. So far we have mainly described the things that affect a review, but as we can see, the review itself also has some quite important roles. One of these is in providing a context for primary studies. Until the adoption of the evidence-based paradigm, these were mostly viewed as essentially being isolated studies that formed 'islands' of knowledge. When primary studies are viewed in terms of their role as inputs to a systematic review, there are two new factors that may influence the way that they are organised. One is the choice of topic—perhaps because a review has identified the need for further studies. The other is the way that primary studies report their results—one of the frequent complaints from analysts who conduct a systematic review is that important information is apt to be omitted from papers and reports. So designing and reporting of primary studies now needs to be more influenced by this role as an input to a secondary study than was the case in the past. Reviews also influence policies, standards and decisions about practice—and while this is still less likely to be the case in software engineering than in disciplines such as education and clinical medicine, consideration of these aspects should increasingly be a goal when performing systematic reviews.

The systematic review is the main instrument used for evidence-based studies and so will be discussed in depth through most of this book, and certainly in the rest of Part I. So, to conclude this introductory chapter, we need to consider some of its limitations too. This is because an appreciation of these is really needed when designing and conducting reviews as well as when seeking to understand what the outcomes of a review might mean to us.

1.4   Some limitations of an evidence-based view of the world

Not surprisingly, there has been a growing tendency for researchers, at least, to consider that knowledge that has been derived from an evidence-based process must inevitably be better than 'expert' knowledge that has been derived, albeit less systematically, from experience. And as the preceding sections indicate, we would to some degree support such a view, although replacing "inevitably" with the caveat "depending upon circumstances".

In clinical medicine and in wider healthcare, it has been argued that evidence-based research practices have become the "new orthodoxy", and that there are dangers in blind acceptance of the outcomes from this. Some of the arguments for this position are set out in a paper by Hammersley (2005). In particular, he questions whether professional practice can be wholly based on research evidence, as opposed to informed by it, noting that research findings do themselves rely upon judgement and interpretation. While many of the arguments focus upon how to interpret outcomes for practice, rather than upon the research method itself, the appropriateness of this form of research for specific topics does need to be considered. Even for systematic reviews, the two well known adages of "to a person with a hammer everything looks like a nail" and "garbage in–garbage out" may sometimes be apt.

So here we suggest some factors that need to be kept in mind when reading the following chapters. They are in every sense 'limitations', in that they do not necessarily invalidate specific evidence-based studies, but they might well limit the extent to which we can place full confidence in the outcomes of a systematic review.

A systematic review is conducted by people. There is inevitably an element of *interpretation* in the main activities of a systematic review: performing searches; deciding about inclusion and exclusion; and making various decisions during synthesis. All of these contain some potential for introducing *bias* into the outcomes. The practice of using more than one analyst can help with constraining the degree of variability that might arise when performing these tasks, but even then, two analysts who have the same sort of background might arrive at a set of joint decisions about which primary studies to include that would be different from those that would be made by two analysts who come from different backgrounds. Both the selection of studies, and also the decisions made in synthesis, can affect the outcomes of a review.

The outcomes depend upon the primary studies. The *quality* of the primary studies that underpin a systematic review can vary quite considerably. A review based upon a few relatively weak primary studies is hardly likely to be definitive.

Not all topics lend themselves well to empirical studies. To be more specific, the type of empirical study that is appropriate to some topics may well offer poorer scope for using strong forms of synthesis than occur (say) when using randomised controlled experiments. We will examine this more fully in Part II.

All of these are factors that we also need to consider when planning to perform a systematic review. And in the same way that a report of a primary study will usually make an assessment of the limitations upon its conclusions imposed by the relevant "threats to validity" (we discuss this concept further later), so a report of the outcomes from a systematic review needs to do the same. Such an assessment can then help the reader to determine how fully they can depend upon the outcomes and also how limited or otherwise the scope of these is likely to be.

In the next chapter we go on to look at the way that systematic reviews are performed in software engineering, and so we also look at some of these issues in rather more detail and within a computing context.

Chapter 2

Evidence-Based Software Engineering (EBSE)

Although this chapter is mainly about how evidence-based ideas can be used in software engineering, we actually begin by examining some prior activities that helped pave the way for an acceptance of evidence-based thinking. To do so, we first examine some of the 'challenges' that empirical software engineer researchers were already posing, as well as some of the other factors that helped to make it the right time to introduce evidence-based thinking to software engineering in the years after 2004. We then describe a few examples of how evidence-based research has contradicted some widely-held beliefs about software engineering practices, after which we discuss what the concept of EBSE implies for software engineering and what the use of a systematic review might expect to achieve within a software engineering context. Finally, we examine some limitations that apply to evidence-based practices as used in software engineering research.

2.1   Empirical knowledge before EBSE

From around the mid-1990s there was a perceptible growth in the use of empirical studies to assess software engineering practices. In particular, some of these studies looked more widely at what was happening in software engineering research, and so we first look briefly at three such studies that have been quite widely cited, and at what they found.

- Zelkowitz & Wallace (1998) developed a classification of empirical validation forms, and to test this, they used it to categorise 612 papers published in the three years: 1985, 1990 and 1995. These were taken from a major conference, ICSE (International Conference on Software Engineering); an archival journal, (IEEE Transactions on Software Engineering); and a 'current practices' magazine (IEEE Software). After removing 50 papers because they addressed topics for which a validation was not appropriate, they then classified the remaining 562. They observed that about a third of the papers had no validation at all (although the percentage of these dropped from 36% in 1985 to 19% in 1995), and that a third relied upon informal 'assertions' (in effect, "we tried it out on a sample and it worked"). They also noted that "experimentation terminology is sloppy".

- At around the same time Walter Tichy raised the question "should computer scientists experiment more?" (Tichy 1998), and addressed many of the fallacies that were apt to be raised in opposition whenever the use of empirical studies was advocated. He particularly argued that computing in general was sufficiently well established to justify wider use of empirical validation than was being observed, and also observed that "experimentation can build a reliable base of knowledge, and thus reduce uncertainty about which theories, methods and tools are adequate".

- Somewhat later, Glass, Vessey and Ramesh conducted a series of classification studies of the ways that research was being conducted in the three major branches of computing: computer science, information systems and software engineering. These were based on papers published in a range of journals over the period 1995–1999. Their consolidated overview was published as (Glass, Ramesh & Vessey 2004), and showed that each branch had quite distinct characteristics. Once again though, based upon a sample of 369 papers, software engineering research methods were predominantly non-empirical, with 44% of the papers being classified as being "(non-mathematical) concept analysis" and 17% being "concept implementation" (loosely interpreted as "we built it and it worked").

So, when the idea of employing the evidence-based paradigm in software engineering research was proposed in 2004 by Kitchenham, Dybå & Jørgensen, this created considerable interest among researchers. We can suggest several reasons why this was well-timed.

- Firstly, there was the influence of the concerns raised in the studies of practice described above. These played an important role in widening awareness of the poor evidential basis available for software engineering techniques and practices.

- Secondly, empirical software engineering had also been making an increasing impact upon the academic software engineering community over the previous decade or so. How can we tell this was so? Two good indicators are:

- The establishment of a specialist journal (*Empirical Software Engineering*) in 1996, together with the publication of increasing numbers of empirical papers in many other journals.
- The establishment of two successful conference series. The first of these was the IEEE-sponsored ISESE (*International Symposium on Empirical Software Engineering*)—which began in 2002, and in 2006 merged with the *Metrics* conference to form the ESEM (*Empirical Software Engineering & Measurement*) series. The second was the smaller and more informal EASE (*Evaluation & Assessment in Software Engineering*) series of conferences, which began in 1996.

Taken together, these helped to promote an interest in, and better understanding of, empirical studies among researchers, as well as providing a useful corpus of material for secondary studies.

- A further factor in favour of the acceptance of the concepts of EBSE has been the growing recognition that the results from individual empirical studies are often inconclusive, and that such studies are difficult to replicate successfully (Sjøberg, Hannay, Hansen, Kampenes, Karahasanović, Liborg & Rekdal 2005, Juristo & Vegas 2011). Since software engineering researchers are partly motivated by the goal of providing input to both software engineering practitioners and also policy-makers, an approach that offers the potential for creating more convincing demonstrations to these audiences is likely to be favourably received.

The rest of this chapter examines some of the ways in which evidence-based thinking has begun to influence software engineering research. We begin by looking at a number of examples of where evidence-based studies have contradicted 'expert' opinion and established practice to explain some of the challenges this approach has created. We then look at how EBSE is organised; consider some aspects of software and software engineering practices that influence its effectiveness; and finally look at some examples of how evidence-based studies can provide guidelines for using some specific software engineering practices.

2.2 From opinion to evidence

Expert opinion and experience are often linked in software engineering. Techniques that have proved effective in one context are apt to be extrapolated to others, without this necessarily being appropriate. Expert opinion can also easily become linked to what might loosely be termed 'academic dogma', such as the belief that something that uses mathematically based formalisms or algorithms will be 'better' in some way. For some situations, it is certainly true that mathematical forms of reasoning are appropriate of course (the design of compilers is a good example). However, given that software engineering can be characterised as a 'design discipline', the associated non-deterministic nature of many software engineering activities (and the corresponding absence of 'right' or 'wrong' solutions) means that we need to be careful of overly emphasising any assumptions about rigour that the use of mathematical formalisms can confer. Indeed, and in contrast, one of the strengths of evidence-based studies is the rigour with which they can be conducted, although this is not conventionally 'mathematical' in its form. Their use of systematic and well-defined procedures provides an appropriate means for both linking experience to knowledge and also addressing the non-deterministic nature of software engineering activities.

One consequence of the formulation of ideas about EBSE has been the proliferation of published secondary studies over the following decade. A series of three broad 'tertiary' studies (a *tertiary* study is a secondary study that performs a mapping study of other secondary studies) identified over 100 published systematic reviews in the period up to 2009 (Kitchenham, Brereton, Budgen, Turner, Bailey & Linkman 2009, Kitchenham, Pretorius, Budgen, Brereton, Turner, Niazi & Linkman 2010, da Silva, Santos, Soares, França, Monteiro & Maciel 2011). Keeping up with this proliferation of secondary studies and indexing them has proved to be quite a challenge[1], but we can estimate that there have been over 200 secondary studies published in the first decade of EBSE. Inevitably, some of these have contradicted expert opinion (or "common wisdom" if you prefer) based on experience and expertise. Here we briefly examine three examples that highlight particular aspects of the clashes that can occur between evidence and opinion.

Estimating software development effort. Project planning for software projects, like all planning, is a challenging exercise. Over the years, algorithmic cost modelling approaches, such as that employed by the well-known COCOMO model (Boehm 1981) has often been viewed as the 'right' approach to predicting project costs. In part, this may well be because it is much more tractable to teach about using models than about using experience when teaching students about software engineering, and so greater emphasis has been placed upon the former. Anyway, whatever the reason, this belief is clearly challenged by the findings of Jørgensen (2004), who, from a set of 15 primary studies comparing models with expert judgement, found that:

- For one third of them, a formal cost model worked best;

---

[1] We do maintain a database on our website at www.ebse.org.uk, but this is inevitably always well behind the 'current' position.

- In another third, expert cost estimation was most effective;
- The remaining third identified no difference between expert judgement and model-based forms.

From this, and from examining similar studies in other disciplines, Jørgensen observed that "there is no substantial evidence supporting the superiority of model estimates over expert estimates". He noted that there were "situations where expert estimates are more likely to be more accurate, e.g. situations where experts have important domain knowledge not included in the models". And conversely, that "there are situations where the use of models may reduce large situational or human biases, e.g. when the estimators have a strong personal interest in the outcome".

So, here we see an example of how an evidence-based approach can be used to resolve the different outcomes from a range of studies with outcomes that may appear to be contradictory, and can synthesise the results in order to provide useful guidelines on how to use such techniques.

Our next example again highlights the point that the benefits claimed for well-known software engineering techniques are not always found to occur upon closer inspection.

Pair-Programming. The emergence of agile methods for software development, and of *extreme programming* in particular, has popularised the use of *pair programming*, with this often being used as a technique outside of an agile context. In pair programming, two programmers work together with a single keyboard, mouse and screen, taking it in turns to be the 'driver' and the 'observer' or 'navigator'. The perceived benefits include roles such as training of novices, producing better quality code, and speeding up the development process.

Pair programming does lend itself to experimentation. However, the range of experiments that have been performed is quite wide, and making any form of comparison with 'solo programming' is something of a challenge (it is easier to specify what pair programming involves, but not quite so easy to do so for solo programming). The meta-analysis of the outcomes from 18 primary studies reported in Hannay, Dybå, Arisholm & Sjøberg (2009) demonstrates this very clearly—although with some caveats about the possible existence of reporting bias[2]. After looking at the effects of pair programming upon measures of quality, duration and effort, the authors advise that:

> "If you do not know the seniority or skill levels of your programmers, but do have a feeling for task complexity, then

---

[2] *Reporting bias* occurs when we find the outcomes of studies with inconclusive or negative results do not get published, either because the authors do not think them of interest, or referees reject the submitted papers because they do not show significant results.

> employ pair programming either when task complexity is low and time is of the essence, or when task complexity is high and correctness is important."

So, this example also shows that while there may be benefits to using a particular technique, they are unlikely to be universal, nor will they necessarily be consistent with every claim made for it.

Finally, we look at an example that shows that the benefits claimed for a technique on the basis of early studies may not be supported when a fuller set of studies is taken into account.

Inspections. The practice of performing inspections has long been accepted as being a useful technique for validating software and related documents. So not surprisingly, efforts have been made to optimise the benefits, usually by structuring the reading technique, with one of these being *perspective-based reading* or PBR (Basili, Green, Laitenberger, Lanubile, Shull, Sorumgard & Zelkowitz 1996). Early studies of its use suggested that, when compared to other forms, it was possible to achieve a 35% improvement when using this approach to inspection.

However, the systematic review performed by Ciolkowski (2009) found that, when used with requirements documents, there was no significant difference between PBR and any of the ad-hoc code inspection techniques in terms of their effectiveness. Further, PBR was less effective as a way of structuring inspections than the use of checklists.

One concern was that many of the studies were effectively replications of the original study that used the same dataset as the original study. However, one of the independent studies did also find positive results for PBR, and there are other factors that might explain some of the variation in results.

In many ways this third example shows that initial claims for the benefits of new software engineering techniques need to be treated carefully, and that the developers of a technique may not be the most appropriate people to conduct such studies, however carefully they try to avoid being biased.

We should add an important caveat here, that systematic reviews (in any discipline) provide evidence that represents the best knowledge available at a given point in time. If and when more (and hopefully better quality) primary studies become available, a later extended systematic review may well be able to refine and revise the original findings. So we should always view the outcomes of any systematic review as representing the "best knowledge" that is currently available, and indeed, one of the tasks in reporting a review is to assess the quality of the primary studies used, and also the effect this may have upon any conclusions (see Chapter 7).

Two key aims of evidence-based studies are to avoid bias and encourage objectivity, and in the next section we examine how the procedures of a systematic review can be organised for use with software engineering topics.

2.3   Organising evidence-based software engineering practices

The previous section looked at what an evidence-based approach can tell us about software engineering practices. In this section we discuss how this is done, and why this should be able to give us confidence in its objectivity.

In proposing the adaptation of evidence-based practices for use in software engineering, Kitchenham, Dybå & Jørgensen (2004) suggested that this could be structured as a five-step process.

1. Convert the need for information into an answerable question.

2. Find the best evidence with which to answer the question.

3. Critically appraise the evidence for its validity (how close it comes to the truth), its impact (the 'size' of the effects observed), and its applicability (how useful it is likely to be).

4. Integrate the critical appraisal with software engineering expertise and stakeholders' values.

5. Evaluate the effectiveness and efficiency in the previous steps 1–4, and seek ways to improve them.

The first three steps are essentially the role of the *systematic review*, while the fourth is that of *Knowledge Translation* (we explained what was meant by KT in the previous chapter, and discuss KT later in Chapter 14). The fifth is one of ensuring that the research procedures are themselves subject to constant scrutiny. (An illustration of this is the use of a systematic process to produce the revised *Guidelines* that form Part III of this book.)

Our concern here is with the first three steps. These tasks can be structured as a set of nine *activities*, grouped as three *phases* (note that the phases do not map directly on to the steps, although the number is the same). This is illustrated in Figure 2.1—although we should note that this shows a somewhat idealised model, and that in practice, there is likely to be some iteration between the different activities. We will not discuss each activity in detail here, since all of them are described more fully in the following chapters. So the main task here is to identify what each one involves. Practical guidance on performing these activities is also provided in Part III.

Phase 1: Plan the review. The first phase addresses the task of designing how the study is to be performed, with this being documented through the *review protocol*. Planning a review involves three important activities.

  1. *Specify the research question.*

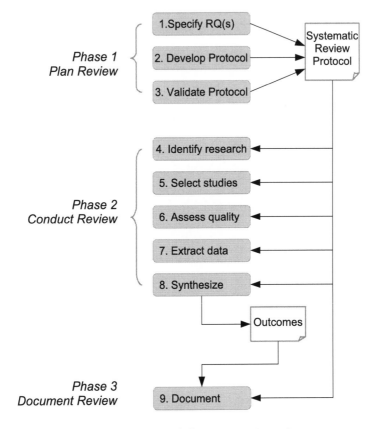

FIGURE 2.1: Overview of the systematic review process.

2. *Develop the review protocol.*

3. *Validate the review protocol.*

We discuss the activities of this phase in more detail in Chapter 4.

Phase 2: Conduct the review. In this phase we put the plan into action. Phase 2 is very much driven by the research protocol, and any *divergences* that occur, requiring that we change the plan to reflect unexpected or other circumstances, need to be carefully documented.

4. *Identify relevant research.* We discuss this activity in more depth in Chapter 5.

5. *Select primary studies.* A fuller explanation of this activity is provided in Chapter 6.

While all of the different forms of systematic review that we discuss in the next chapter should involve performing the first two activities of this phase, not all will necessarily undertake the remaining three in full detail.

6. *Assess study quality.* We discuss these issues further in Chapter 7.

7. *Extract required data.* This is discussed in more detail in Chapter 8.

8. *Synthesize the data.* This is a challenging task that we examine in much greater detail in Chapters 9, 10 and 11.

Phase 3: Document the review. Reporting about the processes and outcomes of a review is discussed in Chapter 12.

We should note here that applying evidence-based ideas in software engineering is not necessarily confined to conducting systematic reviews, although this is the form that has largely been taken so far, and that we focus upon in this book. In the *Further Reading* section provided at the end of Part I, we discuss the study reported by Kasoju, Peterson & Mäntylä (2013) which used EBSE practices as the means of investigating a particular industry problem (related to the software testing processes used in the automotive industry). The important distinction here is that whereas a systematic review forms a *topic*-specific application of evidence-based ideas, the approach used in that study was *problem*-specific, and employed a multi-stage process involving a mix of empirical forms (including a systematic review).

2.4 Software engineering characteristics

At this point, it is useful to consider how the review process outlined in the previous section is influenced by some characteristics of the software engineering discipline—and sometimes by the characteristics of its practitioners too.

The challenges posed by the characteristics of software were outlined by Fred Brooks Jr. in one of software engineering's seminal papers (Brooks Jr. 1987). These are obviously important to any researcher conducting a primary study, and so clearly do have influence upon a secondary study in terms of the likely spread of results they create. Here we look at some factors that are in part consequences of the main characteristics identified by Brooks (invisibility, changeability, mix of static and dynamic properties) and in part consequences of the way that the discipline has evolved.

- *Primary studies involve active participation.* In software engineering it is common to refer to the people who take part in primary studies as *participants* rather than *subjects*. This is because they perform active tasks (coding, reviewing, classifying, etc.) rather than simply receiving some form of treatment (as occurs in much of clinical medicine). Not only does this make it impractical to conduct Randomized Controlled Trials (RCTs) in software engineering, since 'blinding' of participants and experimenters is virtually impossible, it also means that the outcomes of primary studies may well be quite strongly influenced by the characteristics of the particular set of participants involved, by the skills that they have, and by their previous experiences. Like some of the other characteristics we consider here, this one complicates the task of *synthesis*. We examine some aspects of this further in Part II when we look at how primary studies are organized.

- *Software engineering lacks strong taxonomies.* The terms that we use are often imprecise, and software engineers are rather prone to create new terms to describe ideas that may well be closely related to existing ones. This can complicate *searching* since we need to consider all possible forms of terminology that might have been used in the titles and abstracts of papers. Snowballing may help with this, but essentially it stems from the constraint of studying procedures and artifacts.

- *Primary studies lack statistical power.* Because software engineering studies usually need specialist skills and knowledge, it is often difficult for experimenters to recruit enough participants to provide what is generally regarded as an acceptable level of statistical power (Dybå et al. 2006). This in turn reduces the strength of the *synthesis* that can be achieved in a systematic review.

- *There are too few replicated studies.* There may be many reasons for this, not least the problem of getting a paper describing a replicated study published, particularly one that is considered to be a *close* replication (Lindsay & Ehrenberg 1993). Although this view may be inaccurate, if researchers think it is so, then they will be reluctant to conduct replicated studies. There is also debate about what exactly constitutes a 'satisfactory' replication study (we will examine this issue in Chapter 21). Again, this presents a problem for *synthesis* in particular.

- *Reporting standards are often poor.* Many primary studies are reported in a manner that effectively ignores the likelihood that, at some time in the future, a systematic reviewer will attempt to extract data from the paper. While this might have been more excusable in the past, that really is not the case now. Another form of reporting problem is related to our culture of refereed conferences, which can lead to researchers publishing more than one paper that uses the same set of results—requiring the systematic reviewer to take care not to count such studies more than once. Similarly, some papers also describe more than one experiment, complicating separation of individual studies when the analyst conducting a systematic review is performing *data extraction*. We address reporting needs for primary studies in Part II.

2.5 Limitations of evidence-based practices in software engineering

We touched on some factors that constrained the use of the evidence-based paradigm in Chapter 1. In this section we discuss these in the context of EBSE, and look at how they may be affected by the characteristics of software and software engineering. We also introduce the concept of *threats to validity* in rather more detail.

2.5.1 Constraints from software engineering

We begin by considering how these factors are influenced by the nature of our discipline, as characterised in the preceding section.

A systematic review is conducted by people. As we identified earlier, a major risk arising from this aspect is that of *bias*. This can arise in various stages, for example when searching using electronic forms, our choice of search engines and of search terms may favour our finding some studies and perhaps missing others. (As we mentioned in the preceding section, software engineering does lack strong taxonomies.) Equally, when searching manually, our choice of journals and conferences may influence the outcomes. Similarly, our inclusion/exclusion criteria might lead to bias — for example, in software engineering, replicated studies are probably less likely to be published than original ones, but may be available as technical reports. Equally, the analyst needs to be aware that the common practice of expanding conference papers into journal papers can easily lead to a study being counted twice. The issue of bias is discussed quite extensively by Booth et al. (2012) in their discussion of analysis, where they also discuss some strategies that might be used to cope with this.

The outcomes depend upon the primary studies. Even when it is systematic, the main contribution of any review will arise from its *synthesis* of the outcomes from the primary studies. For software engineering these primary studies typically:

- exhibit poor statistical power arising from having small numbers of participants;
- address a wide variety of research questions;

- employ a range of empirical forms;
- have a tendency to employ student participants for tasks that might actually be performed rather differently by more experienced practitioners.

These are all factors that impede the production of reliable outcomes from a secondary study, or at least, constrain the scope of any outcomes.

Not all topics lend themselves well to empirical studies. In software engineering research we are concerned with the study of *artefacts*, which we create, rather than of 'physical' entities. Glass et al. (2004) identified a wide range of forms for evaluation that were used in software engineering, and while we might feel that our discipline could make more use of empirical evaluation, we need to also recognise that forms such as 'concept implementation' are valid approaches to research, and may sometimes be more appropriate than empirical studies.

Not only do we create and study *artefacts*, these are also often being subject to continuous change and evolution. This turn means that different studies that make use of a given artefact in some way may actually all be based upon different versions. This can also apply to our conceptual tools—for example, the UML (Unified Modeling Language) has gone through a number of versions, adding new diagrammatical forms as it evolves. So different studies based on using the UML may not always be directly comparable.

2.5.2   Threats to validity

The concept of limitations upon the rigour of an empirical study, expressed as *threats to validity*, is well established for primary studies, and we discuss them in that context in Part II. However, the concept does apply to secondary studies too, and indeed, they are often discussed when reporting a systematic review. The factors that influence the validity of a study are largely those discussed above, but cast into a slightly different perspective within the structure of a systematic review. Shadish et al. (2002) identify four major forms of threat arising for primary studies, and here we briefly discuss how each of these might be interpreted in the context of a secondary study.

Construct Validity is concerned with how well the *design* of the study is able to address the research question. Essentially this relates to the consistency and comparability of the operationalisation of the outcome measures as used in the primary studies.

Internal Validity is concerned with the *conduct* of the study, particularly related to data extraction and synthesis, and whether there are factors that might have caused some degree of bias in the overall process.

Conclusion Validity is concerned with how reliably we can draw conclusions about the link between a treatment and the outcomes of an empirical study (particularly experiments). For a secondary study, we can therefore relate this to the *synthesis* element of a systematic review, and how well this supports the conclusions of the review. Hence for secondary studies there is little distinction between internal and conclusion validity.

External Validity is concerned with how widely a cause-effect relationship holds, given variations in conditions. For a secondary study this should be based upon an assessment of the range covered by the primary studies in terms of their settings, materials and participants.

Taken together, these provide a framework that can be employed to assess the possible limitations that apply to the outcomes of a review. They need to be reported by the systematic review team, mainly because they are the people who are in the best position to assess whether these factors are likely to have had any effect upon the outcomes, and if so, how significant this might be. In turn, this knowledge may be important to anyone wanting to make use of the outcomes in making decisions about practice or policy, as well as to anyone who might in the future wish to extend and update the review.

But do note that...

> None of these issues are likely to make it impossible to conduct a useful systematic review, although they may well limit its scope and usefulness, since we are often studying differences in practice that have fairly small effects upon the outcomes. However, where possible, it is important to anticipate the influence of the likely 'threats' when writing the research protocol.

Chapter 3

Using Systematic Reviews in Software Engineering

We conduct secondary studies in order to answer a variety of research questions, so it is not surprising that we need to adapt the way that such a study is organised according to the question being addressed. Some reviews seek to answer questions about software engineering practices (for example, "in what situations is pair programming likely to be a good strategy to adopt?"); others may examine research trends (such as "what have been the 'hot topics' in cloud technology research and how have they changed with time?"); while a 'broad' form of review might be used to help determine whether a more focused review of a topic is feasible—or whether it is first necessary to perform more primary studies on that topic.

The way that we conduct a secondary study, and in particular, the way that searching is organised and the choice of procedures used to synthesize the results, will therefore vary substantially. Indeed, the very concept of 'synthesis' is apt to have many interpretations, especially when conducting studies of research trends, where the reviewers may well choose to include non-empirical forms of input.

Booth et al. (2012) catalogue a range of forms that are used for organising secondary studies across a range of disciplines. Since not all of these are very relevant to software engineering, in this chapter we briefly examine those forms that software engineers do use, and what they use them for. (Each of these forms will also be discussed in much greater depth in the following chapters.) The key forms are as follows.

- *Systematic reviews* (using both qualitative and quantitative inputs). In some cases, it is also possible to perform a meta-analysis for a quantitative review.

- *Mapping studies* (used both for secondary and }tertiary studies).

Note that our use of this terminology does differ a little from that used in Booth et al., largely reflecting the way that the use of secondary studies has evolved in software engineering.

Figure 3.1 shows a simple summary of the roles of, and relationships between, the different forms. In the rest of this chapter we say a little more about each of these, and in particular, about the way that each form is used in software engineering.

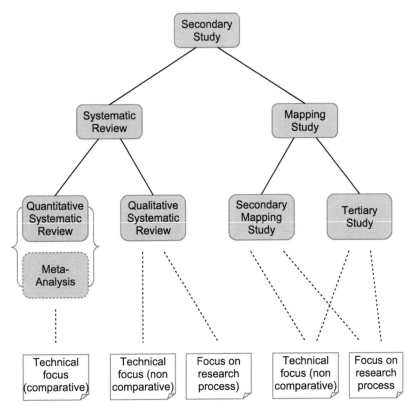

FIGURE 3.1: The hierarchy of study forms.

3.1   Systematic reviews

While 'systematic review' is often used as a generic term for all types of review conducted using evidence-based practices, a systematic review is also a well-defined form of study used to answer a specific research question. Systematic reviews can be further sub-classified according to whether they involve synthesizing qualitative or quantitative forms of data. In turn, this will determine how both data extraction and synthesis need to be organised. In particular, for a quantitative systematic review it may well make it possible to perform a statistical meta-analysis as the means of synthesis, providing greater confidence in both the statistical significance and also the statistical power of the outcomes. We discuss this a bit more in Section 3.3.

Since the organisation and use of systematic reviews is covered extensively in the rest of the book, this section will be confined to discussing how systematic reviews are commonly used in software engineering.

A matter of classification

> We should observe here that throughout Part I of this book, our classification of different publications as being mapping studies or systematic reviews may not always agree with those used by the original authors. This is because we use the way that synthesis is performed in a study as our key criteria for differentiating between these forms, and so consider some studies that have been described as "systematic reviews" when published, to be more correctly classified as mapping studies for that reason.

One role of a systematic review in software engineering is to establish whether particular techniques or practices work better than others, and if so, under what conditions this will be true. Existing systematic reviews span many activities and forms (agile methods, project estimation, design patterns, requirements elicitation techniques, regression testing techniques, just to name a few). They are also used for other purposes, such as to evaluate how far particular techniques have been adopted by industry and commerce, or to identify the benefits of using tools in a particular context.

The purpose of a review will determine the type of input that is expected, and hence the way that the inputs from different studies can be synthesized. For example, a systematic review addressing topics such as pair programming, inspection techniques or the use of software design patterns might be expected to involve synthesizing the results from experiments and quasi-experiments. Studies looking at the adoption of tools in industry, or the take-up of agile methods, are more likely to be synthesizing the outcomes from observational studies and case studies. In this book we are mainly concerned with the following two classes of systematic review, as identified in Figure 3.1.

Quantitative reviews Inputs for these are likely to come from experiments or quasi-experiments, or from data mining using existing repositories, and the studies themselves may well be performing comparisons or producing estimates based on past profiles. Associated research questions are likely to address comparative aspects such as "does technique X perform better than technique Y?". Synthesis may take a range of forms ranging from tabulation of the different outcomes through to a statistical meta-analysis, depending on how much the primary studies vary in terms of topics and measures used. A good example of a quantitative review is that of Dieste & Juristo (2011), comparing the effectiveness of different requirements elicitation techniques.

Qualitative reviews These usually address questions about the specific use of a technology, and so are unlikely to involve making comparisons (and hence less likely to address questions that involve any sense of something being 'better'). In a software engineering context they may well be used for such tasks as studying adoption issues, or perhaps more likely, the barriers to adoption, employing procedures for synthesis that can help to identify patterns in the data. This class of review also includes studies that look at research methodologies, not just practice, such as Kitchenham & Brereton (2013).

Systematic reviews, along with mapping studies are the two forms that have been most widely employed in software engineering to date. However, a few examples of the use of meta-analyses for software engineering topics do exist, and so we also discuss the role of meta-analysis in the final section.

3.2   Mapping studies

The goal of a mapping study is to survey the available knowledge about a topic. It is then possible to synthesise this by categorisation in order to identify where there are 'clusters' of studies that could perhaps form the basis of a fuller review, and also where there are 'gaps' indicating the need for more primary studies. Mapping studies may also be 'tertiary' studies, for which the inputs are secondary studies, so providing a higher level of categorisation of knowledge.

Mapping studies have found wide acceptance in software engineering, although this form of study appears to be less widely used in other disciplines. This may reflect the nature of software engineering and its vocabulary in particular. Software engineering is still not a truly 'empirical' discipline, even if it is slowly moving that way, and so we may often have both very limited knowledge about how widely a topic has been studied, and also relatively few studies that are empirical in form. Related to this, empirical studies may well be reported in many different venues, meaning that we need to search widely to find all relevant material. In addition, whereas the vocabulary of clinical medicine is based upon terms used for the parts of the human body and its conditions, which have been standardised over a long period of time, software engineering deals with artifacts, and so invents new terms to describe these. Unfortunately, software engineers are also apt to "reinvent the wheel" when doing this, sometimes using new terms to describe a concept that was developed earlier, but may not have been realised at the time because of (say) a lack of computational power.

If we return to the model of the systematic review process described in the previous chapter, we can describe a mapping study as a form that involves relatively little synthesis. A mapping study may also include an element of quality assessment, depending upon its purpose.

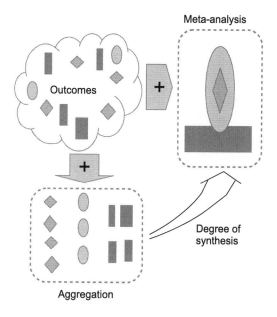

FIGURE 3.2: The spectrum of synthesis.

Figure 3.2 illustrates this issue. At one extreme, simple *aggregation* alone just groups together any data occurring within a given category, and while providing knowledge about the count for each category (such as number of papers published each year), it creates no new derived knowledge. Extending from that we have a spectrum of *synthesis*, extending through to a statistical meta-analysis, whereby new knowledge is derived about such aspects as patterns and associations in the data. The issues related to performing synthesis across the spectrum of study forms are discussed more fully in Section 10.2.

Searching may well use quite a broad set of terms in order to ensure that we do not overlook any particular group of studies. It also needs to be as thorough as possible in order to obtain a clear and useful picture of the topic.

The activity of categorisation may employ a number of different schemes, intended to reflect specific characteristics of the primary studies. It might use an existing scheme—for example, we might want to classify the studies found in terms of the research method employed for each primary study, using a set of categories such as experiment, quasi-experiment, survey, observational, and case study. Employing an existing categorisation scheme also provides a useful basis for identifying where there are 'gaps'. We might also derive the categories for a given characteristic by looking at the set of studies found, and grouping these to reflect the patterns observed.

So, in what context do we find it useful to conduct a mapping study? Below, we briefly examine two examples of situations where mapping studies may be particularly relevant.

Studying research trends. A mapping study may be useful as a means of analysing how research in a given topic has evolved over a period of time (so one of the categories used for the studies will need to be publication date). Such a study may focus upon identifying the "hot issues", or the techniques used, or even the countries where the research has been performed.

One example of its use in this role, mentioned earlier, is that of the *tertiary study*. To recap, a tertiary study is a form of systematic review for which the inputs are secondary studies. A *broad tertiary study* is organised as a mapping study where the purpose is to categorise these and to observe trends. The earliest tertiary study conducted in software engineering was that reported in (Kitchenham, Brereton, Budgen, Turner, Bailey & Linkman 2009). This study identified a set of secondary studies and categorised them by type (such as research trends) and topic. Later broad tertiary studies such as (Kitchenham, Pretorius, Budgen, Brereton, Turner, Niazi & Linkman 2010) and (da Silva et al. 2011) also included a quality assessment of the secondary studies. Viewed as a series, the value of these studies was therefore to index the emerging field of evidence-based studies, identifying those areas of software engineering where most activity was taking place. This is also an example of where aggregation is an appropriate form of analysis.

PhD literature review. Preparation for PhD study almost always requires a candidate to undertake a 'literature review' of the state of the art related to the intended topic. Traditionally this is conducted using informal searching, with expert guidance provided by the supervisor. However, for PhD projects involving empirical studies in particular, the use of a mapping study may well provide a very useful initial stage for a study, as examined in (Kitchenham, Budgen & Brereton 2011). Using this approach is not just appropriate for empirical topics of course, definitions and research trends may usefully be studied in this way too, as demonstrated in the review of different definitions of 'service oriented architecture' or SOA, provided in (Anjum & Budgen 2012). Here, it is appropriate to employ a degree of synthesis in analysing the outcome data.

While these are by no means the only roles that can be performed by a mapping study, they are fairly representative of the ways that this form has been used in software engineering.

3.3   Meta-analysis

For clinical medicine, where a secondary study may well be drawing together the results from a number of Randomized Controlled Trials (RCTs), the use of a statistical meta-analysis (which we discuss more fully in Chapter 11) is often appropriate. This is because primary studies of a new clinical treatment are likely to use the same baseline(s) for their comparisons and ask similar research questions about the effective use of a treatment. For software engineering however, even when a review is drawing together the outcomes of a set of experiments, these may well vary widely in form, as well as in their research questions.

Where the use of meta-analysis is an option, this is usually because there is a reasonable number of primary studies with similar forms, and that these also use comparable response variables. This was the case for the meta-analysis of pair programming studies performed by Hannay et al. (2009), although as here, the analysts may still have to cope with wide variation in the characteristics of the primary studies. In addition, many primary studies in software engineering have poor statistical power, as we observed earlier. However, one benefit of being able to use meta-analysis is that any outcomes can then be assigned a quantitative measure of confidence based on the use of inferential statistics.

One question we might well ask is whether this position is likely to change in the future so that we will see more use of meta-analysis. One of the aims of this book is to provide guidance for the use of evidence-based practices in software engineering, and in doing so, to encourage the use of more rigorous forms of synthesis. As we explain in Chapter 11, meta-analysis is possible in a variety of circumstances and there seem to be no reasons why it should not be employed more widely in software engineering. Certainly, if EBSE is to make greater impact upon the software engineering profession, then an increased use of meta-analysis is likely to be an important way of helping with providing potential users with appropriate levels of confidence in the findings from our studies.

Chapter 4

Planning a Systematic Review

We and other researchers have found that undertaking a systematic review or mapping study is an extremely time-consuming activity requiring a great deal of attention to detail (Babar & Zhang 2009). As with any complex project, planning is a key factor in achieving a successful outcome.

In this chapter we look at the tasks that need to be performed before and during Phase 1 of a systematic review or mapping study (see Figure 4.1 which highlights the planning phase of a review). The focus here is particularly on the development of a *review protocol*. The protocol plays a key role in planning a review, providing a framework within which to make and document the necessary study design decisions. The aim is to minimise bias by defining in advance the steps that will be followed and the criteria against which decisions will be made during the conduct of a review. We note though, that although it is important to agree and document a review design in advance, it is sometimes necessary to modify that design, and hence the protocol, during the conduct phase.

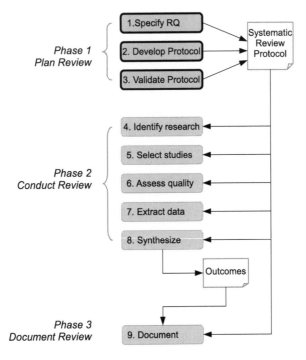

FIGURE 4.1: Planning phase of the systematic review process.

Even before developing and validating a protocol, reviewers should ensure that a review is both needed and feasible. We briefly consider these issues as well as aspects of managing the review process, before addressing the three main planning tasks:

1. Specifying the research questions,

2. Developing the protocol,

3. Validating the protocol.

4.1   Establishing the need for a review

To date, systematic reviews and mapping studies in software engineering have been largely motivated by the requirements of researchers (that is, to achieve academic goals) rather than by real problems from practice. Researchers undertake reviews to summarise evidence about some particular phenomenon in a thorough and unbiased manner. A recent survey by Santos & da Silva (2013) found that the four main factors that have motivated systematic reviewers in software engineering are:

- To gather knowledge about a particular field of study,

- To identify recommendations for further research,

- To establish the context of a research topic or problem,

- To identify the main methodologies and research techniques used in a particular research topic or field.

The results of the survey largely support the outcomes of a study by Zhang & Babar (2013) which found the most important motivators for performing systematic reviews and mapping studies to be (1) obtaining new research findings and (2) describing and organising the state-of-the-art in a particular area.

Whatever the motivation, before investing the substantial time and effort needed to carry out a thorough systematic review or mapping study, it is important to consider:

- whether it is likely to contribute to our knowledge about the topic,

- whether it is feasible, given the resources available within a review team.

Whether a review is needed and is feasible depends on a range of factors. For example, it may not be *needed* if a good quality review addressing the same or a similar topic already exists. The problem of multiple systematic reviews addressing the same topic is handled in other disciplines by researchers registering their intention to undertake a review with a central authority. For example, the Cochrane Collaboration provides a facility for such registration[1]. However, at present there is no such central authority for software engineering reviews. In fact, there are at least two examples of (pairs of) reviews addressing the same software engineering topic (Kitchenham, Mendes & Travassos 2007, MacDonell, Shepperd, Kitchenham & Mendes 2010, Verner, Brereton, Kitchenham, Turner & Niazi 2012, Marques, Rodrigues & Conte 2012). It may not be *feasible* to undertake a review if, for example, there are too many primary studies to analyse with the available resources or if there are too few good quality studies to make the synthesis or aggregation of their results meaningful.

In the examples below, we summarise the motivations for some published reviews and note that in each case the authors had previously undertaken research in the topic area and had first hand knowledge of the research issues. Further discussion about establishing the need for a systematic review or a mapping study can be found in Part III.

---

[1] http://www.cochrane.org/cochrane-reviews/proposing-new-reviews

Examples of justifications for systematic reviews

Hall, Beecham, Bowes, Gray & Counsell (2012) state that fault prediction modelling is an important area of research which has been the subject of many studies. They note that published fault prediction models are both complex and disparate and that before their review there was no up-to-date comprehensive picture of the state of fault prediction. They indicate that their results will enable researchers to develop models based on best knowledge and will enable practitioners to make effective decisions about which models are best suited to their context.

Kitchenham et al. (2007) argue that accurate cost estimation is important for the software industry, that accurate cost estimation models rely on past project data and that many companies cannot collect enough data to construct their own models. Thus, it is important to know whether models developed from data repositories can be used to predict costs in a specific company. A number of studies had addressed this issue but had come to different conclusions. They indicate that it is necessary to determine whether, or under what conditions, models derived from data repositories can support estimation in a specific company.

Examples of justifications for mapping studies

Zhang & Budgen (2012) recognised that the concept of design patterns for developing object oriented systems is valued by experienced developers. However, during preliminary investigations they found that much of the literature on patterns was in the form of advocacy or experience reports rather than empirical studies about effectiveness. They carried out the mapping study to try to identify studies that evaluate aspects of design patterns.

The mapping study by Penzenstadler, Raturi, Richardson, Calero, Femmer & Franch (2014), which focuses on software engineering for sustainability, updates an earlier mapping study on the same topic. The authors indicate that the updated study was motivated by:

- The wide range of journals, conferences and workshops which address this topic,

- A high level of research activity in recent years,

- A desire to broaden the scope of the review.

In a tertiary study, Cruzes & Dybå (2011*b*) review the methods used in systematic reviews to synthesise the outcomes of the primary studies that they include. The authors point out that "comparing and contrasting evidence is necessary to build knowledge and reach conclusions about the empirical support for a phenomenon". The motivation for the study therefore stems from the needs of systematic reviewers to address the challenges associated with integrating evidence from multiple sources, especially where there is a high degree of heterogeneity in the research methods used for the contributing studies.

4.2 Managing the review project

At the start of a review, it is important to consider how the review project as a whole will be managed. This is distinct from planning and specifying the technical aspects of the review process. During the planning phase, management activities include:

- Organising the development, validation and signing off of the review protocol,

- Specifying the time scales for the review,

- Assigning the tasks specified in the protocol to team members,

- Deciding what tools to use for managing data and for supporting collaboration (see Chapter 13).

Generally, reviews are performed by two or more reviewers who constitute the review team. One of the reviewers acts as the team leader, taking responsibility for ensuring the management activities are planned, monitored and refined when necessary. If a review forms part of PhD, ideally, the student will take the lead role.

4.3 Specifying the research questions

Specifying research questions is a critical part of planning a systematic review or mapping study and the factors that motivate the questions should be fully explained. The questions drive the entire review process providing the basis for:

- Deciding which primary studies to include in a review, and hence driving the search strategy,

- Deciding what data must be extracted and how the data is synthesised or aggregated in order to answer the questions.

The nature of the research questions depends very much on the type of review being carried out.

For *systematic reviews*, questions are about evaluating a particular software engineering technology or research process. The term 'technology' is used in a broad sense here to encompass software engineering methods or processes, or particular management-related characteristics such as the attributes of software engineers or of software engineering teams. The research questions are formulated in one of two ways:

- As a quantitative comparison of two (or more) technologies to determine which one is more effective or efficient (or is in some other way 'better') than the others within some specific context.

- As a qualitative evaluation of a specific software engineering technology (including management-related characteristics) or an approach or procedure used in software engineering research, with respect to benefits, risks, value, impact or some other aspect of adoption.

In both cases, the questions will be driven by some underlying model of the topic, involving, for example, a comparison of a new model (or technology) with a traditional (control) model or the identification of consequences of adopting a new model.

For *mapping studies*, research questions are broader and concerned with classifying the literature in some way. The research questions for mapping studies are the most likely to change as a review progresses and new categories emerge (that is, the underlying model evolves).

We note also that mapping studies and qualitative systematic reviews are usually less focused than quantitative systematic reviews and hence tend to have a greater number of research questions.

It is important in any review to ask the right question(s). For systematic reviews, ideally, this should be one that:

- Is meaningful and important to practitioners as well as researchers. For example, researchers might be interested in whether a specific analysis technique leads to a significantly more accurate estimate of remaining defects after design inspections. However, a practitioner might want to know whether adopting a specific analysis technique to predict remaining defects is more effective than expert opinion at identifying design documents that require re-inspection.

- Will lead either to changes in current software engineering practice or to increased confidence in the value of current practice. For example, researchers and practitioners would like to know under what conditions a project can safely adopt agile technologies and under what conditions it should not do so.

- Will identify discrepancies between commonly held beliefs and reality.

Nonetheless, as indicated earlier, many systematic reviewers ask questions that are primarily of interest to researchers. This is particularly the case for mapping studies which often ask questions that lead to the identification of opportunities for future research activities. For mapping studies, research questions should be ones that:

- Enable the literature on a particular software engineering topic to be classified in ways that are interesting and useful to researchers. For example, a mapping study undertaken as part of a PhD can provide the basis for the research student's work by enabling the student to show how the proposed research fits into the current body of knowledge.

- Are likely to identify clusters of research as well as gaps in the literature. Clusters can provide researchers with some indication of where there is a sufficient body of work to warrant a more focused systematic review. Gaps in the literature can indicate that further primary studies may be usefully performed in order to fill the gaps.

As described in Section 3.2, a tertiary study is a special form of mapping study that classifies or maps reviews relating to some aspect of software engineering. Research questions for tertiary studies are aimed at identifying trends in systematic reviews focusing, for example, on:

- identifying the topics addressed by the reviews,

- the specific review procedures or approaches used by researchers.

Further details and advice about specifying research questions can be found in Part III.

Examples of research questions from quantitative systematic reviews

The review by Mitchell & Seaman (2009) covers studies that compare the cost, duration and product quality for two approaches to software development. These are (1) the 'waterfall' approach and (2) iterative and incremental development (IID). The research questions posed in this review are:

"What is the development cost of software produced using waterfall or its variations versus using IID?"

"What is the development duration for software produced using waterfall or its variations versus using IID?"

"What is the quality of software produced using waterfall or its variations versus using IID?"

Jørgensen (2007) reports a review of evidence about the use of expert judgement, formal models and a combination of these to estimate software development effort. The research questions for the review are:

"Should we expect more accurate effort estimates when applying expert judgement or models?"

"When should software development effort estimates be based on expert judgement, on models, or on a combination of expert judgement and models?"

MacDonell & Shepperd (2007) review studies that compare the use of cross-company and within-company data within effort estimation models. Their research question is:

"What evidence is there that cross-company estimation models are at least as good as within-company estimation models for predicting effort for software projects?"

Examples of research questions from qualitative systematic reviews

The technology focused study by Beecham, Baddoo, Hall, Robinson & Sharp (2008) reviews studies on motivation in software engineering. Research questions are:

"What are the characteristics of Software Engineers?"

"What (de)motivates Software Engineers to be more (less) productive?"

"What are the external signs or outcomes of (de)motivated Software Engineers?"

"What aspect of Software Engineering (de)motivate Software Engineers?"

"What models of motivation exist in Software Engineering?"

The research-oriented review by Kitchenham & Brereton (2013) focuses on primary studies that address aspects of the systematic review process in software engineering. the research questions addressed are:

"What papers report experiences of using the systematic review methodology and/or investigate the systematic review process in software engineering between the years 2005 and 2012 (to June)"?

"To what extent has research confirmed the claims of the systematic review methodology?"

"What problems have been observed by software engineering researchers when undertaking systematic reviews?"

"What advice and/or techniques related to performing systematic review tasks have been proposed and what is the strength of evidence supporting them?"

Examples of research questions from mapping studies

Walia & Carver (2009) report a technology focused mapping study about the sources of requirements faults. The high level research question addressed by this review is:

> "What types of requirements errors can be identified from the literature and how can they be classified?"

This is decomposed into four more specific questions (some with sub-questions). The four specific questions are:

> "Is there any evidence that using error information can improve software quality?"

> "What types of requirement errors have been identified in the software engineering literature?"

> "Is there any research from human cognition or psychology that can propose requirement errors?"

> "How can the information gathered in response to the above questions be organized into an error taxonomy?"

Another technology focused mapping study by Marshall & Brereton (2013) identifies and classifies tools developed to support the systematic review process in software engineering. The research questions for this mapping study are:

> "What tools to support the systematic review process in software engineering have been reported?"

> "Which stages of the systematic review process do the tools address?"

> "To what extent have the tools been evaluated?"

The study by Ampatzoglou & Stamelos (2010) maps research relating to software engineering for games development. It addresses the following research questions:

> "What is the intensity of the research activity on software engineering methods for game development?"

> "What software engineering research topics are being addressed in the domain of computer games?"

> "What research approaches do software engineering researchers use in the domain of computer games?"

> "What empirical research methods do software engineering researchers use in the domain of computer games?"

Examples of research questions from tertiary studies

The tertiary study by Marques et al. (2012) maps reviews about distributed software development (DSD). The research questions are:

> "How many systematic literature reviews have been published in the DSD context?"

> "What research topics are being addressed?"

> "What research questions are being investigated?"

> "Which individuals and organizations are involved in systematic literature review-based DSD research?"

> "What are the limitations of systematic literature reviews in DSD?"

The study by Cruzes & Dybå (2011*b*) focuses on the synthesis stage of the systematic review process and addresses three questions:

> "In terms of primary study types and evidence that is included, what is the basis of software engineering systematic reviews?"

> "How, and according to which methods, are the findings of systematic reviews in software engineering synthesized?"

> "How are the syntheses of the findings presented?"

Kitchenham, Pretorius, Budgen, Brereton, Turner, Niazi & Linkman (2010) report a broad research-focused tertiary study of systematic reviews and mapping studies in software engineering. Research questions are:

> "How many systematic reviews were published between 1st January 2004 and 30th June 2008?"

> "What research topics are being addressed by systematic reviews in software engineering?"

> "Which individuals and organisations are most active in research on systematic reviews?"

4.4   Developing the protocol

A systematic review or mapping study protocol is a documented plan describing, as far as possible, all of the details about how a review will be conducted. A protocol is particularly valuable because: (1) it can help to reduce the probability of researcher bias by limiting the influence of researcher expectations on, for example, the selection of individual (primary) studies or the synthesis of results; (2) it can be evaluated by other researchers who can provide feedback about the design of a review in advance of its conduct; and (3) it can form the basis of the introduction and method sections of a report of a review.

It is important that a protocol is structured in such a way that it can be easily used as a reference document by a review team and can be updated as necessary during the conduct of a review. We stress that a review protocol is a living document that is likely to be updated during the conduct of a review. An example template for systematic review and mapping study protocols is shown in Figure 22.5.

As well as covering all of the technical elements of a review, a protocol can provide information about the management of a review project. This can include the allocation of roles, mechanisms for resolving disagreements and the project schedule.

The following subsections summarise the main components of a protocol.

4.4.1 Background

The background section of a protocol provides a summary of related reviews and the justification for a review. Establishing the need for a review is discussed in Section 4.1.

4.4.2 Research questions(s)

This is a critical component of a protocol because the research questions drive the later stages of the review process. Specifying the research questions is discussed in Section 4.3.

4.4.3 Search strategy

The strategy for finding appropriate studies will describe and justify the way in which specific searching methods, such as automated searching, manual searching, snowballing and contacting key researchers, are combined. If an automated search is planned, this component will include a description of the search strings and resources, such as digital libraries or indexing services, that will be used. For a manual search, suitable journals and conference proceedings should be specified and their selection justified. This part of a protocol will also include a description of the mechanism for validating the search process. Chapter 5 and Part III provide further details and advice about search strategies and approaches to validation. Management decisions that are specific to the search process, such as the allocation of members of the review team to the searching tasks and the approach to resolving disagreements can also be recorded here.

4.4.4   Study selection

In this component, reviewers specify (1) the study selection criteria for determining whether a primary study is included in or excluded from a review and (2) procedures that will be followed to apply the criteria. The inclusion and exclusion criteria relate closely to the research questions and hence will be formulated to ensure the inclusion of those studies that can contribute to answering these questions.

The likelihood is that criteria are applied in a number of stages. For example, initial decisions can be based on the title or the title, abstract and keywords of a paper in order to exclude those that are clearly irrelevant. In later stages, reviewers will read candidate papers in full. Marginal papers, or those for which inclusion/exclusion is uncertain, can be kept in the inclusion set with the final decision being made during data extraction. This situation is most likely to arise for qualitative systematic reviews. For quantitative systematic reviews the criteria are usually easier to apply and for mapping studies leaving out a few papers, or including a few extra papers is not usually critical. Plans might also address the allocation of team members to the stages of study selection and the resolution of disagreements. There is further discussion of study selection in Chapter 6 and in Part III.

4.4.5   Assessing the quality of the primary studies

This is a particularly challenging task relying on two key decisions. One is to decide on the criteria against which quality will be assessed and the other is to establish the procedures for applying the criteria. The *criteria* will usually be expressed as one or more checklists depending, at least in part, on the range of evaluation methods used in the primary studies. Evaluation methods may include experiments, surveys, case studies and experience reports. One approach is to use separate checklists for each study type. The alternative is to use a generic checklist across all study types. Each of these approaches has limitations. For mapping studies, where the goal is to map out a domain of interest, assessing the quality of the individual studies may not be needed.

Procedures for applying the quality criteria are specified in a way that aims, as far as is possible, to ensure the reliability of the assessment. Mechanisms that can be used for this include having all, or a sample of, assessments checked by another person or having two reviewers perform the assessment independently. As well as describing who will undertake the quality assessment and the mechanism for resolving disagreements, a protocol can record decisions about the use of forms or tools to manage both individual scores and the outcomes of the resolution process. Whatever the type of review, it is important to consider the purpose of assessing the quality of the primary studies and to justify the approaches taken. There is further discussion of quality checklists, their limitations and assessment procedures in Chapter 7 and Part III.

4.4.6 Data extraction

This part of a protocol defines the data that will be extracted and the procedures for performing the extraction and for validating the data. The data will include publication details for each paper plus the information that is needed to answer the research questions. Extracting qualitative information presents a particular challenge since specific pieces of text need to be extracted and linked to specific research questions. Where qualitative synthesis is planned, data extraction and synthesis can be combined within a single process. For mapping studies in particular, data extraction may be iterative since important trends and ways of categorising papers may only become evident as individual papers are read. These challenges have led to an interest in the use of textual analysis tools to support data extraction and other aspects of the systematic review process (see Chapter 13).

A protocol should also define how data will be recorded (for example, using a review support tool or spreadsheet), who will perform the data extraction and how disagreements will be resolved. One approach is for a review leader to extract standard publication data and for two reviewers to extract data that is specific to a review. Strategies for resolving disagreements include discussion and using a third reviewer. The data extraction strategy (i.e. the selected data items and the procedures) should be justified. There is further discussion and advice about data extraction in Chapter 8 and in Part III.

4.4.7 Data synthesis and aggregation strategy

This section of a protocol defines the strategy for summarising, integrating, combining and comparing the findings from the primary studies included in a review. For quantitative data there is usually little opportunity to undertake a formal meta-analysis for software engineering studies. However, where meta-analysis is planned, details of the techniques to be used should be included. More commonly, for systematic reviews in software engineering, primary studies are too heterogeneous for statistical analysis and a qualitative approach (such as vote counting) has to be used.

The studies included in a review are often qualitative in nature and use a wide range of empirical methods. For textual data, synthesis is generally an iterative process because authors use different terminology to describe the same concepts (and sometimes use the same terminology to describe different concepts). Also, if the text is to be coded, the codes will be derived after reading the papers and need to be agreed on by all of the members of the review team who are performing the coding. Combining findings across multiple methods is especially challenging ((Cruzes & Dybå 2011 b), (Kitchenham & Brereton 2013)). Common approaches to synthesis include narrative and thematic synthesis where data is tabulated in a way that is consistent with the research questions. For mapping studies, the goal should be to classify the findings in interesting ways and to present summaries using a variety of tabular and graphical forms. Further information and advice about a range of approaches to data synthesis and aggregation is provided in Chapters 9, 10 and 11, and in Part III.

4.4.8   Limitations

This section can be used to document the limitations of a review that are inherent to its context. Essentially these are limitations that have not or cannot be addressed by the review design. One example of this type of validity problem is where the data is extracted from papers that were written by the reviewers. The data could be based on the reviewers' understanding of their own research rather than the information actually reported in the papers.

4.4.9   Reporting

It is useful to consider, in advance, the approach that will be taken to disseminating the findings of a review. Usually a review is reported as a detailed technical report, as a conference paper (or papers) and/or as a journal paper. A technical report (or a chapter in a PhD thesis) and a journal paper can include all, or at least most, of the information that is needed to provide traceability from individual primary studies to the results and conclusions of a review and to demonstrate rigour in applying the review process. Conference papers, however, are usually limited in size and hence may need to provide links to additional information. A protocol should record agreements about the list of authors for each publication and about the target audience. Further details about reporting can be found in Chapter 12 and Part III.

4.4.10   Review management

This section covers management decisions, for example relating to scheduling and to tool support, that have not been recorded in other parts of the protocol. Further details about tool support can be found in Chapter 13.

4.5   Validating the protocol

In this component, reviewers specify the steps that will be taken, both internally and externally, to validate the protocol. Internal validation will include trialling specific aspects of the review plan such as the search strings and the data extraction forms to be used as well as the processes to be followed for data synthesis and/or aggregation. Also, we have emphasised the key role played by a protocol so it is important that it is evaluated by researchers who are external to a review team. PhD students should at least have their protocol evaluated by members of their supervisory team and might also call upon independent researchers, particularly if their supervisors have limited experience of the process. Evaluators can check a protocol against review guidelines, looking to confirm that the main elements are covered, that the decisions made are justified, that validation is adequately addressed and that a protocol is internally consistent. Authors of a protocol should provide evaluators with a checklist or set of questions addressing each of the elements of a protocol. Table 4.1 lists some examples of questions about each of the elements.

TABLE 4.1: Example Questions for Validating a Protocol

Components	Example questions
Background	Is the motivation for the review clearly stated and reasonable? Are related reviews summarised?
Research questions	Do these address a topic of interest to practitioners and/or researchers? Are they clearly stated?
Search strategy	Is the strategy justified and is it likely to find the right primary studies without the reviewers having to check or read a large number of irrelevant papers? For automated searches, is there likely to be a substantial level of duplication of papers found across the set of electronic resources used? Has the strategy been validated? Is it clear which members of the review team will perform the searching?
Study selection	Are inclusion/exclusion criteria clearly defined and related to research questions? Is a staged process defined? Is a validation process specified? Are the roles of the team members defined for each stage of the process and is the mechanism for resolving disagreements specified? Is there a process for handling marginal and uncertain papers (especially for qualitative reviews), and for managing multiple reports of individual studies?
Quality of primary studies	If quality is to be assessed, is it clear that the outcomes will be used in the later stages of the review? Are criteria for assessing quality provided and justified (given the range of primary study types anticipated in the review)? Is a validation process specified? Are the roles of team members and the process for resolving disagreements specified?

TABLE 4.1: Example Questions for Validating a Protocol

Data extraction	Does the data to be extracted properly address the research questions? Are the methods of recording the data appropriate for the types of data to be extracted (e.g. using forms, tables, spreadsheets or more advanced tools)? Have these been adequately piloted? Are there mechanisms for iteration where data is qualitative and categories are not (or cannot be) fully defined in advance of the extraction? Is a validation process specified? Are roles and strategies for resolving disagreements specified? If textual analysis tools are to be used, is their use justified? Will the data extracted by each reviewer, and any agreed values where reviewers differ, be appropriately stored for later analysis?
Data aggregation and synthesis	Will the process enable the research questions to be answered? Are the methods proposed for qualitative and quantitative data appropriate? Have they been piloted? Has consideration been given to combining results across multiple study types? Is the approach to aggregation and synthesis justified with reference to appropriate literature?
Reporting	Has this been considered? If the aim is to publish the review (or even if it is not!), has sufficient attention been paid to completeness, general interest, validation, traceability and the limitations of the review? Has the authorship of reports been considered?
Review management	Is the proposed schedule realistic? Have roles and responsibilities been defined for the stages in review? Are the tools that will be used for managing papers, studies and data specified and appropriate (and available)? Is the management of the many-to-many relationship between papers and studies addressed?

Chapter 5

Searching for Primary Studies

The focus of this chapter is on the identification of relevant primary studies. This process forms the first step of the conduct phase of the systematic review process, as highlighted in Figure 5.1.

An important element of any systematic review or mapping study is to devise a search strategy that will find as many primary studies as possible that are relevant to the research questions. The likelihood is that the strategy will involve a combination of search methods. One widely used method is automated searching of the literature using resources such as digital libraries and indexing systems. Other methods include manual searching of selected journals and conference proceedings, checking papers that *are cited* in the papers included in a review (backwards snowballing) and checking papers that *cite* the papers included in a review (forwards snowballing). The search strategy will aim to achieve an acceptable level of *completeness* (see Section 5.1) within the review's constraints of time and human resources. The level of completeness that might be targeted will depend on the type of review being undertaken. Generally, for a quantitative systematic review, that is, one which compares software engineering technologies, a high level of completeness

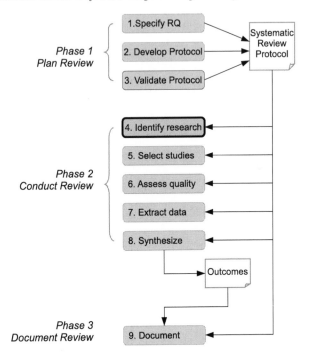

FIGURE 5.1: Searching stage of the systematic review process.

is essential. For other types of review, such as qualitative systematic reviews which assess risks, benefits, motivating factors, etc. or which review research processes, and for mapping studies, a lower level of completeness may be acceptable. This point is discussed in more detail in the following section.

Once a set of candidate papers have been found, and duplicate copies of the same paper have been removed, references can be managed in, for example, a spreadsheet or a database. Tools to support the systematic review process are discussed in Chapter 13.

Within the following sections we look at assessing the completeness of the set of papers found by the search process and discuss a range of searching methods that can be used as part of a strategy. Also, examples of strategies are presented for different types of systematic reviews and mapping studies.

5.1   Completeness

We consider two aspects of the completeness of the set of papers found by following a search strategy. The first relates to *completeness target*: how 'complete' should the set of papers be? The second relates to *completeness assessment*: once we have a target, how will we know whether we have achieved it?

How complete?

A big question facing many reviewers is when to stop searching; and of course the answer is 'it depends'.

For *quantitative systematic reviews*, completeness is crucial. If we look at the examples described in Chapter 3, which relate to methods of estimating software development effort (Jørgensen 2004), pair programming versus solo programming (Hannay et al. 2009) and perspective-based reading (PBR) compared to other approaches to reading (Basili et al. 1996) we see that 15 studies are included in the cost estimation review, 18 in the pair programming review and 21 in the PBR review (with many of the primary studies in the last of these considered to be replications rather than independent studies). Given the highly focused nature of these reviews and the small numbers of included studies, missing only a few of these could substantially affect the outcomes of the reviews.

For other types of review, a lower level of completeness may be acceptable. For example, the *qualitative systematic review* by Beecham et al. (2008) aimed to 'plot the landscape of current reported knowledge in terms of what motivates developers'. The review includes 92 papers which report motivators, many of which are common across many of the papers (most of which report some form of survey). Failing to include some of these 92 papers would not have substantially affected the 'landscape of knowledge'.

Another situation where completeness may not be critical is where *mapping studies* are performed during the early stages of a research project (such as a PhD project). The value of the mapping study may come from acquiring a broad understanding of the topic and from identifying clusters of studies rather than from achieving completeness (Kitchenham, Brereton & Budgen 2012). However, a point to note here is that if a mapping study provides the basis for a more detailed and focused analysis (for example, where the presence of a cluster indicates that quantitative analysis may be feasible and valuable) it should not be assumed that the set of papers identified is complete. In this case a more focused search should be performed unless it can be demonstrated that the mapping study is of high quality in terms of completeness and rigour (Kitchenham, Budgen & Brereton 2011).

It can be argued that in some cases the level of completeness of *tertiary studies* should be high. Where a tertiary study aims to provide a catalogue and detailed analysis of systematic reviews across the software engineering domain (or across a broad sub-domain such as global software development), it can provide a key reference document for the community and as such should be as complete as possible. The argument for a high level of completeness may not be quite so compelling where a tertiary review is performed as a preliminary study, for example, to identify related reviews in advance of a more focused mapping study or systematic review. In the end, knowing when to stop searching depends on what level of completeness is needed in order to provide satisfactory answers to the research questions being addressed by a review.

Completeness assessment

There are two fundamental ways of assessing the completeness of the set of studies found by searching the literature. One is to use personal judgement. This may be the judgement of members of the review team, especially if they are experienced researchers on the topic being reviewed or it may involve external researchers whose views are sought by a review team at some point in the process. Whatever the source of personal knowledge, it is difficult to quantify the level of completeness achieved using this subjective approach. The alternative is to use some objective measure of the level of completeness.

Two key criteria for assessing the completeness of an automated search are *recall* (also termed sensitivity) and *precision* (Dieste, Grimán & Juristo 2009, Zhang, Babar & Tell 2011).

The *recall* of a search (using particular search strings and digital libraries/indexing systems) is the proportion (or percentage) of all the relevant studies that are found by the search.

The *precision* of a search is the proportion (or percentage) of the studies found that are relevant to the research questions being addressed by a review. These can be calculated as follows:

$$Recall = \frac{R_{found}}{R_{total}} \qquad (5.1)$$

$$Precision = \frac{R_{found}}{N_{total}} \qquad (5.2)$$

where:
R_{total} is the total number of relevant studies
N_{total} is the total number of studies found
R_{found} is the number of relevant studies found

Of course the practical problem in calculating recall is that the denominator, that is, the total number of relevant studies (R_{total}), is not known. Ideally, a search should have high recall, that is, it should find most (if not all) of the relevant studies. Precision is also important and high precision is desirable. High precision means that the burden on reviewers to check papers that turn out not to be relevant is low. If precision is reduced, for example as a consequence of efforts to improve recall, the reading load on reviewers will increase.

In the following section we look at how these measures can be used to validate a search strategy by assessing the completeness of the set of studies found.

5.2 Validating the search strategy

Developing a search strategy is an iterative process, involving refinement based on some determination of the level of completeness achieved. An essential basis for the subjective assessment of completeness is having a set of papers which are known to report relevant studies. This *known set* can be obtained in a number of ways:

- Through an informal automated search using a small set of digital libraries or indexing systems, or a manual search of a small set of relevant conferences and journals,

- Using the personal knowledge of researchers who have experience in the topic of the review,

- Using a previous systematic or traditional literature review which addresses a similar or overlapping topic,

- Through the construction of a quasi-gold standard. The use of a quasi-gold standard to assess completeness is discussed later in this section.

If the number of studies in the known set is considered to be large (although of course it is not easy to decide what constitutes a large known set) then a search process that finds all of these may be judged adequate. The argument here is that if these are found then it is likely that most of the other relevant studies have also been found. Personal judgement, based on knowledge of the topic of a review, has to be used to decide whether the number of known papers can be considered large enough. To give an idea of the numbers of studies that might be included in a review, we note that the numbers included in the reviews catalogued by the third broad tertiary study (da Silva et al. 2011) range from 4 to 299, although well over half fall in the range 15–80.

When reviewers are not confident that the number of known studies can be considered large, a quasi-gold standard can be constructed and used to assess completeness (Zhang et al. 2011). The quasi-gold standard is determined by performing a manual search across a limited set of topic-specific journals and conference proceedings over a restricted time period. The set of relevant papers found is then used to assess the completeness of an automated search. The approach has been evaluated through two participant-observer case studies with promising results (Zhang et al. 2011). The approach, shown in Figure 5.2, has the following steps:

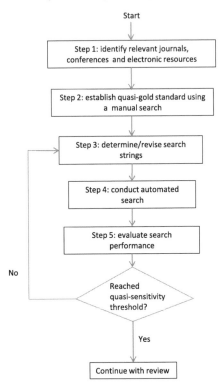

FIGURE 5.2: A process for assessing search completeness using a quasi-gold standard.

Step 1: Identify relevant journals, conferences and electronic resources

In this step, reviewers decide which journals and conference proceedings will be searched manually (in Step 2) and which digital libraries and indexing services to use for the automated search (Step 4). Manual searching is quite time consuming so the aim is to chose those outlets that are most likely to publish relevant papers. Selecting electronic sources for the automated search is discussed later in this chapter (see Section 5.3).

Step 2: Establish quasi-gold standard using a manual search

This step involves performing a manual search of the selected journals and conference proceedings over the chosen (and limited) time period. Essentially, the review team screen all of the papers in the selected sources and apply the inclusion and exclusion criteria, which should be defined in advance. Screening can be applied initially to the title and abstract of a paper (keywords could be considered too) and then, if a decision cannot be made, other parts of a paper, possibly the whole paper, can be read. The development and use of inclusion and exclusion criteria are discussed in more detail in Chapter 6.

Step 3: Determine/revise search strings

Zhang et al. suggest two ways of defining the strings to used to search the selected electronic resources. These are:

1. Subjective search string definition based on domain knowledge and past experience,

2. Objective elicitation of terms from the quasi-gold standard using a text analysis tool.

Search strings can also be derived from the research questions being addressed by a review (see Part III, Section 22.5.2.2 for practical advice about constructing search strings).

Step 4: Conduct automated search

Here, the selected electronic resources (digital libraries or indexing systems) are searched using the strings determined in Step 3 and for the chosen time period. Automated search is discussed in more detail in Section 5.3 and in part III, Section 22.5.2.

Step 5: Evaluate search performance

In this step, the results of the automated search are compared to the results of the manual search (the quasi-gold standard) and quasi-sensitivity is calculated. For the calculation, using equation 5.1, R_{found}, is the number of relevant studies found by the automated search (step 4) that are published in the venues used in Step 2 (the manual search) during the time period covered by the manual search. R_{total}, the total number of relevant studies for the selected venues and time period, is the number of papers found by the manual search (Step 2). Similarly quasi-precision can be calculated using equation 5.2 where N_{total} is the total number of papers found by the automated search.

Zhang et al. suggest that a sensitivity (recall) threshold (i.e. a completeness target) of between 70% and 80% might be used to decide whether to go back to Step 3 (and to refine the search terms) or whether to proceed to the next stage of the review. These percentages are based on the scales developed by Dieste & Padua (2007) who in turn based their scales on research in the medical domain. Clearly this is a judgement that must be made on a case by case basis and will depend on a number of factors such as the completeness target and the available human resources.

A refinement of the quasi-gold standard approach has been proposed by Kitchenham, Li & Burn (2011) who suggest that the set of known papers is divided into two sets with one being used to construct the search strings and the other to evaluate the effectiveness of the search process.

5.3 Methods of searching

As we have indicated in the introduction to this chapter, there are a number of ways of searching for relevant primary studies. In practice, methods are often combined in some way to achieve good coverage. In this section we describe the most commonly used methods and in the following section illustrate their use across a range of systematic reviews and mapping studies. In addition to the methods described here, reviewers can consider contacting researchers directly where they are known to be actively engaged in research in the specific topic area being addressed by the systematic review or mapping study.

We also note that it can be hard to find papers when the topic of a review is secondary to that of many of the relevant primary studies. This might arise, for example, where a review is about tool usage or about research methodology. In these circumstances the best method to choose for searching might be a manual search (looking at particular sections of a paper) or alternatively an automated search where the searching process accesses the complete text of a candidate paper (as opposed to just the title and abstract).

Automated search

This approach has been widely adopted by software engineering reviewers and involves the use of electronic resources such as digital libraries and indexing systems to search for relevant papers. In order to perform an automated search reviewers have to address two elements of the process. They have to decide which electronic resources to use and they have to specify the search strings that will drive the search.

Two key publisher-specific resources are the IEEE Digital Library (IEEEXplore) and the ACM Digital Library which together cover the most important software engineering (and more general computing) journals and conference proceedings. A tertiary review focusing on the period mid-2007 to end of 2008 found that IEEEXplore was used in 92% of the 38 reviews that were included and 81% used the ACM Digital Library (Zhang et al. 2011). ScienceDirect (Elsevier) was also quite extensively used for the systematic reviews included in the tertiary study.

General indexing services such as Scopus and Web of Science will find papers published by IEEE, ACM and Elsevier (although not necessarily the most recent conference proceedings). They also index papers published by Wiley and Springer and hence such services reduce the need for searching some publishers' sites.

Although some publishers provide open access to some papers, many require a payment, or a subscription, to obtain copies of full papers. Many universities now subscribe to publishers' packages of journals and conference proceedings, and there is also a growth in open access journals. Also, at some academic institutions, authors are required to put pre-publication versions of their papers into the University's open access catalogue. Additionally, pre-publication versions can sometimes be found by looking at an author's website. The publishing landscape for academic journals and conference proceedings is changing quite rapidly at the moment so we suggest that reviewers check with their library services and with publishers' websites to get an up-to-date-picture of their best route to acquiring access to full papers.

Generally, digital libraries and indexing systems provide mechanisms for exporting the bibliographic details of papers in a range of formats such as BibTeX, EndNote and Refworks.

Defining and refining search strings is an iterative process as illustrated in the quasi-gold standard approach described in the previous section. An initial set of keywords can be determined in a number of ways, such as:

- Extracting software engineering concepts and terms from the research questions,

- Reviewing terms used in the known papers,

- Identifying synonyms of the key terms.

As indicated earlier, it is a tricky balance between a search which finds most of the relevant papers (that is, having a high recall/sensitivity) and one which achieves a good level of precision (that is, not generating a large number of irrelevant papers). Even if the quasi-gold standard approach is not used, some iteration will be needed to ensure that all known papers that can be found by an automated search (that is, those that are indexed by the electronic systems being used) are included in the list of papers generated by the search.

Manual search

Manual searching of software engineering journals and conference proceedings can be very time consuming and onerous especially if the topic of a review is broad (so that the papers are not limited to a few specialist's outlets) or where the topic is quite mature (so that a large time span needs to be covered). The key decisions here are identifying the most appropriate journals and conferences and determining the date from which to start the search. Manual search can be particularly valuable for multidisciplinary reviews (see for example the mapping study by Jorgensen & Shepperd, summarised in Section 5.4, which addresses the topic of cost estimation). In general it is useful to have team members from the different domains covered by a multi-disciplinary review.

If the search validation mechanism is strong, for example where an independent search is performed by two or more reviewers and the agreement between them is high, a manual search can provide what is effectively a gold standard set of relevant papers. Achieving a gold standard set of papers in this way may not be practical except perhaps where the topic is highly focused, reviewers are experts in the subject area, and the time span for the search is not large.

Advice about selecting appropriate sources to use for a manual search can be found in Part III, Section 22.5.3.

Snowballing

Snowballing, also referred to as citation analysis, can take one of two forms. *Backwards snowballing* is where a search is based on the reference lists of papers that are known to be relevant (the included set). It is usually used as a secondary method to support automated search. *Forwards snowballing* is the process of finding all papers that cite a known paper or set of known papers. This approach is particularly useful where there are a small number of seminal papers that are likely to be cited by most of the subsequent papers on the topic. Skoglund & Runeson (2009) compare the recall (sensitivity) and precision of two snowballing approaches based on citation analysis with those of three historic reviews, where two had used automated searching and the other had used a manual search. The outcomes were quite varied across the three example reviews and no general conclusions were reached. A study by Jalali & Wohlin (2012) compared automated search and backwards snowballing for a review on Agile practices in global software engineering. They found that:

- Precision was better when using the snowballing approach,

- Although different papers were found by the two approaches there was a substantial degree of overlap,

- Conclusions drawn using each of the approaches were very similar.

5.4   Examples of search strategies

We summarise a range of strategies reported by researchers who have performed systematic reviews and mapping studies.

Examples of search strategies for systematic reviews

Kitchenham et al. (2007) report a quantitative systematic review of studies that compare cross-company and within-company cost estimation. The authors carried out their search in two stages. Initially, an automated search of six electronic databases and seven individual journals and conference proceedings, chosen because they had published known relevant papers, was performed. The set of known papers was also used to validate the automated search. For the second stage, the authors:

1. carried out backwards snowballing for the papers included after the initial search,

2. contacted researchers who have either authored relevant papers found by the initial search or who they believed to be working on the topic.

The qualitative systematic review by Beecham et al. (2008) focused on the motivation of software engineers. Following a piloting exercise, eight electronic resources were used in an automated search and a manual search was undertaken "directly on key conference proceedings, journals and authors". Additionally, for included papers, backwards snowballing was performed and the corresponding authors of the papers were asked whether they had any relevant material 'in press'.

Examples of search strategies for mapping studies

Jorgensen & Shepperd (2007) describe a mapping study which addresses a set of eight research questions about research on software cost estimation. The authors report a manual search of all volumes (up to their search date) of more than 100 peer-reviewed journals. Journals were identified by reading through the reference lists of known papers (important because it is a multidisciplinary topic), by searching for relevant journals and using their own experience. The reviewers constructed independent lists of potential journals and merged their lists.

da Silva et al. (2011) performed a research-focused broad tertiary study of systematic reviews and mapping studies in software engineering published between 1st July 2008 and 31st December 2009. The authors used a search strategy that combined automated search, manual search and backwards snowballing. The automated search was performed by two of the authors using six search engines and indexing systems (ACM Digital Library, IEEEXplore Digital Library, ScienceDirect, CiteSeerX, ISI Web of Science and Scopus). All of the searches except for the ISI Web of Science were based on the full texts of the published papers. The search process was validated using a set of known papers found by two earlier broad tertiary studies (Kitchenham, Brereton, Budgen, Turner, Bailey & Linkman 2009, Kitchenham, Pretorius,

Budgen, Brereton, Turner, Niazi & Linkman 2010). In parallel with the automated search, three of the authors performed a manual search of 13 journals and conference proceedings, selected because they had been used by the earlier tertiary studies (except where two of the conferences had been merged). The reviewers checked titles and abstracts. The two sets of candidate papers were merged and duplicates removed. Backwards snowballing was applied to the papers remaining after the study selection stage.

Zhang et al. (2011) replicated a published tertiary study (Kitchenham, Brereton, Budgen, Turner, Bailey & Linkman 2009) to evaluate a search strategy based on a quasi-gold standard (as described in Section 5.2). In Step 1 (identify relevant journals, conferences and electronic resources), the authors used personal experience and published journal and conference rankings to inform their selection of nine outlets for the manual search and four digital libraries for the automate search. In Step 2 (establish quasi-gold standard using a manual search), two of the authors performed independent manual searches of the selected outlets to establish a quasi-gold standard (after resolving disagreements). In Step 3 (determine/revise search strings) and Step 4 (conduct automated search), the search string was based on the authors' knowledge and on the papers in the quasi-gold standard and was coded to fit the syntax requirements of each of the search engines. In Step 5 (evaluate search performance), the quasi-sensitivity was calculated to be 65% which was considered to be below the required threshold (70–80%). The search string was reviewed and revised to include additional terms. This increased the quasi-sensitivity to 85% which was deemed acceptable.

Chapter 6

Study Selection

Once candidate papers have been identified through the search process, these need to be checked for relevance to the research questions being addressed by a review. The focus of this chapter is on this selection process which forms the second step of the conduct phase of the systematic review process, as highlighted in Figure 6.1.

Study selection is a multi-stage process which can overlap to some extent with the searching process. It is multi-stage because, ideally, many candidates that are clearly irrelevant can be quickly excluded, at an early stage, without the overheads of reading more than their titles and abstracts. In later stages candidate papers have to be read 'in full'. Study selection can overlap with the searching process when searching involves backwards snowballing or contacting authors of relevant papers (or studies). In this situation, relevance needs to be established before these searching methods are used.

In this chapter we discuss three aspects of study selection:

- The selection criteria,

- The selection process,

- The relationship between papers and studies.

The chapter concludes with some examples of the selection criteria used and the procedures followed by software engineering reviewers.

6.1   Selection criteria

The criteria for selecting studies to include in a review are formulated in order to identify those studies that are able to provide evidence that is of

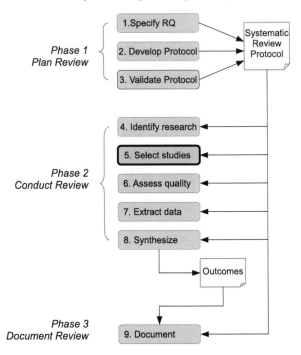

FIGURE 6.1: Study selection stage of the systematic review process.

relevance to the research questions. The criteria are generally (although not universally) expressed as two sets: one for the inclusion criteria and one for the exclusion criteria.

Some selection criteria are quite generic and fairly easily interpreted. For example, criteria relating to publication date are reasonably straightforward, although even this can be complicated by:

- The practice followed by some publishers of providing online access to draft papers before they are incorporated into a specific issue of a journal,

- Some studies being reported in more than one paper, particularly if not all of the papers fall within the scope (especially the time period) of a review.

It is also often the case that studies are included only if they are published in English and in 'full' peer-reviewed papers (as opposed, for example, to being reported in extended abstracts or in 'grey literature' such as technical reports and PhD theses).

For the more technical elements of a review, scoping the literature can be quite challenging and sometimes the criteria need to be revised as reviewers become more familiar with the topic and its boundaries. A point to note here is that it is important to be explicit about the scope of a review in any resulting publications, by fully reporting:

- The criteria used in the selection process,

- Details of the papers that are included in the review,

- The rationale for excluding marginal or 'near-miss' papers.

When planning a review, the study selection criteria can be piloted to ensure that they can be sensibly and consistently interpreted by members of a review team and that for the known papers they lead to the desired outcome. Even then, some refinement of the criteria (and hence of the protocol) may be needed as a review progresses.

6.2   Selection process

Study selection is usually carried out in a number of stages. Initially, once a set of candidate papers has been identified, those that are clearly irrelevant can be excluded on the basis of their title or their title and abstract. After this early screening, papers have to be looked at in more detail. For example, reviewers might decide to exclude a paper after reading some of the sections (such as the introduction, a methods section or the conclusions), however, the likelihood is that many papers will have to be read 'in full' before the decision to exclude (or not) can be made. Sometimes, the decision to include a study is overturned later in the review process. This may arise, for example, if the required data cannot be extracted or if a study fails to reach a quality threshold. In the end, there may well be marginal studies and the best that reviewers can do is report and explain their decisions in such cases.

Reviewers may find that they have a very large number of candidate papers (and what constitutes 'very large' will depend to some extent on the size of the review team). Possible strategies for dealing with this problem are:

- Refining the search strings to improve recall and precision,

- Reducing the scope of the review (through refinement of research questions),

- Use of a text mining tool to support the selection process,

- Increasing the size of the review team.

Also, if the selection process results in a large number of papers being included in a review then reviewers may choose to complete the process using only a sample of the papers.

Where study selection is being performed by a team of reviewers, there is the opportunity to validate the outcomes of the selection process by two (or more) members of the team independently applying the inclusion/exclusion criteria and checking their level of agreement (for example, by performing a *kappa* analysis (Cohen 1960)). Although tools are available to calculate the kappa coefficient, this is briefly explained below. As well as calculating the level of agreement, a mechanism is needed for resolving any differences that arise. Common approaches are to do this through discussion or by using a third reviewer to act as mediator.

A kappa (κ) coefficient is calculated using the following equation[1]:

$$\kappa = \frac{\text{actual agreement} \; - \; \text{agreement expected by chance}}{\text{scope for doing better than by chance}} \qquad (6.1)$$

TABLE 6.1: Example Data for Study Selection by Two Reviewers

		Reviewer B Included	Reviewer B Excluded	Total
Reviewer A	Included	10	3	13
Reviewer A	Excluded	4	25	29
	Total	14	28	42

Consider the data shown in Table 6.1. Reviewer B has classified 14 of 42 studies as 'included' while Reviewer A has included 13 of the 42 studies. The number of studies for which there is actual agreement is 10 plus 25 giving a total of 35 out of 42 which equals 0.8333 (83.33%) of the studies. By chance alone, the probability of an 'include' from Reviewer A is $13/42 = 0.3095$ and for Reviewer B is $14/42 = 0.3333$. The chances of agreement by chance are these two probabilities multiplied together, that is, $0.3095 \times 0.3333 = 0.1032$. Using a similar calculation, the chances of agreement to exclude by chance is 0.4604. Adding together these two probabilities of agreement by chance gives a total of 0.5636. That is, 56.36% agreement would be expected by chance. This gives a kappa score as shown below:

$$\kappa = (0.8333 - 0.5636)/(1 - 0.5635) = 0.618 \qquad (6.2)$$

Kappa scores are generally interpreted as shown in Table 6.2. We see, therefore, in our example, that agreement between the two reviewers is *Good/Substantial*.

One approach to sharing the workload associated with study selection is for the lead reviewer to perform the early screening stage(s), excluding papers on the basis of titles or on titles and abstracts, which is usually quite straightforward, with the later, more difficult, stages being performed independently by two members of the review team. See, for example, the process followed

---

[1]Further details can be found at: http://www.ganfyd.org/index.php?title=Statistical tests for agreement

TABLE 6.2: Interpretation of Kappa

Value of kappa	Strength of agreement
0 - 0.29	Poor
0.21 - 0.40	Fair
0.41 - 0.60	Moderate
0.61 - 0.80	Good/Substantial
0.81 - 1.00	Very good/Almost perfect

by Marshall & Brereton (2013), described in the next section, which adopted this approach.

For PhD students, it is not always possible for selection to be performed independently by two reviewers. Where this is the case there are a number of ways that confidence in the decisions made can be enhanced. For example, a member of the supervisory team can check a random sample of papers (or those papers that are considered marginal or about which the student is uncertain). Alternatively, PhD students or other lone researchers can use a test-retest approach which entails repeating (after a suitable time delay) some or all of the study selection actions and comparing the outcomes. For each of these approaches to study selection validation, if agreement is good, then the review can proceed with some confidence in its reliability, if it is not, then the criteria and their interpretation need to be reconsidered.

Another means of checking the decisions made (whether by one or by multiple reviewers) is to carry out some form of text analysis (also referred to as text mining) to help determine whether papers that are 'similar' in some way have been either all included or all excluded during the study selection process. The general approach is to use a text mining tool to identify and count the frequency of important words or phrases in each paper. A visual display tool can then be used to show clustering with respect to these, highlighting where papers in the same cluster (that is, papers that seem to be 'similar') have been treated differently in the selection process. A number of small studies have demonstrated the feasibility of using text mining to support study selection (Felizardo, Andery, Paulovich, Minghim & Maldonado 2012). Some text mining and visualisation tools that have been used to support the systematic review process are listed in Chapter 13.

6.3 The relationship between papers and studies

The relationship between research papers (or other dissemination forms) and the studies that they report is important for systematic reviews. Researchers undertaking systematic reviews or mapping studies are usually (although not exclusively) looking for empirical studies that provide some sort of evidence about a topic of interest. They will find, however, that

- Papers can report more than one study,

- Studies can be reported in more than one paper.

Where a paper reports multiple studies, these can generally be considered as separate studies for the purposes of a systematic review. The study selection process may result in some of the studies being included in a review and some being excluded. Although this seems quite straightforward, this is not always the case. Sometimes, one or more studies are preliminary or pilot studies undertaken in advance of the 'main' study. Also, sometimes, several case studies are reported which could be treated separately or as a single multi-case study. These issues are discussed further in Part III, Section 22.6.4.

A study may be reported in more than one paper. This is not unusual in software engineering. A conference paper may be followed by a more detailed or enhanced journal paper. Also, a large study may be reported in many papers which focus on different aspects of the research. It is important that such multiple publications of a (single) study are identified, so that the results are not counted more than once, and, where the quality of the study is being assessed, all of the published information about the study can be used for making that assessment. It is not always straightforward to establish that multiple publications report a single study. Of course there may be some cross-citation and it may be that titles and author sets are similar across a set of papers. In the absence of these fairly obvious indicators, reviewers should pay particular attention to sets of papers where the same number of participants are recorded for 'similar' studies reported by similar sets of authors. Again this issue is discussed further in Part III, Section 22.6.4.

6.4   Examples of selection criteria and process

In this section we describe the criteria used and the processes followed for some of the published reviews. We can see in these examples that quite a wide range of approaches to applying the criteria is taken. Sometimes, however, only limited information about the process and specifically about the roles taken by members of a review team is available in published papers.

Examples of study selection for quantitative systematic reviews

We look at two reviews in this category. They compare:

- Two approaches to software effort estimation (MacDonell & Shepperd 2007),

- Two development life cycle models (Mitchell & Seaman 2009).

The review by MacDonell & Shepperd (2007) compares the effectiveness of software effort estimation models that use within-company (that is, local) data with models that use cross-company (that is, global) data. The reviewers only included studies where the experimental design met the following (inclusion) criteria:

- Data was from five or more projects per company and for at least two companies,

- There was a comparison between within-company and cross-company models,

- The projects covered were substantially software projects (that is, not hardware or co-design),

- Projects were commercial (that is, not student projects),

- Publications were demonstrably peer-reviewed, written in English and published between 1995 and 2005.

Abstracts of all papers retrieved by the search process were read by the reviewer who had performed the search to determine whether the paper should be included. If the decision could not be made, the reviewer read the whole paper and then applied the inclusion/exclusion criteria. The second reviewer provided comments on a small number of borderline papers.

Mitchell & Seaman (2009) performed a review of studies that compare the cost, development duration and quality of software produced using a traditional waterfall approach with those of software produced using iterative and incremental development (IID). Their search process found 30 candidate papers, nine pairs of which were found to be duplicates, leaving 21 unique papers. At this stage, the reviewers applied the following inclusion criteria, requiring that papers should:

- Be written in English,

- Be peer-reviewed,

- Report a primary study,

- Report empirical results,

- Compare waterfall and IID processes,

- Present results concerning development cost and/or development duration and/or resulting product quality.

This process reduced the number of candidates to 11. The subsequent identification of duplicate reports of the same study, the realisation that the waterfall process for one of the studies included iteration, plus the application of a quality threshold reduced the final count of studies to five. It is interesting to note here, that the first two of these additional 'criteria' (relating to duplicate reports and details about the processes being compared) are essentially exclusion criteria although in the paper they are not labelled in this way. There is no indication in the paper about whether both or only one of the authors performed the study selection.

Examples of study selection for qualitative systematic reviews

Here we summarise the criteria and the process for study selection in a management focused review relating to motivation in software engineering (Beecham et al. 2008) and in a research-oriented review of studies about the systematic review process (Kitchenham & Brereton 2013).

The study by Beecham et al. (2008) reviewed knowledge about what motivates developers, what de-motivates them and how existing models address motivation. Before the authors applied the inclusion and exclusion criteria, they checked for duplicate publications of individual studies and only included one of the reports (either the most comprehensive or the most recent). The reviewers stated that they included 'texts' that:

- Directly answer one or more of the research questions,

- Were published from 1980 to June 2006,

- Relate to any practitioner directly producing software,

- Focus on de-motivation as well as motivation,

- Use students to study motivation to develop software,

- Focus on culture (in terms of different countries and different software environments),

- Focus on 'satisfaction' in software engineering.

They excluded texts:

- In the form of books or presentations,

- Relating to cognitive behaviour,

- Not relating to software engineering,

- Focusing on company structures and hierarchies unless expressly linked to motivations,

- In the form of opinion pieces, viewpoints or purely anecdotal,

- That focus on software managers (who do not develop software),on group dynamics or on gender differences.

Beecham et al. retrieved over 2000 references through the search process and eliminated approximately 1500 of these on the basis of titles and abstracts. This left 519 papers. These (except for 9 papers which could not be obtained) were looked at in full by 'a group of primary researchers' who accepted 95 papers. An independent researcher looked at 58 of the 519 papers, which were randomly selected by taking (approximately) every 10th paper from an alphabetic list, and re-applied the inclusion/exclusion criteria. The inter-rater reliability was 99.4% indicating a high level of agreement and giving confidence in the decisions made. A further validation exercise was carried out on the 95 included papers by an independent expert on motivation in software engineering who checked how each paper addressed the research questions. There was a high level of agreement (99.8%) and the three papers where the decision differed were considered by a third independent researcher. Once the disagreements were resolved, 92 papers remained in the set of included papers.

Kitchenham & Brereton (2013) performed a qualitative systematic review to identify and analysis research about using and improving the systematic review process. As well as reporting selection criteria, these researchers also explain the rationale for each criterion. For conciseness, the rationale is omitted from the following descriptions of the criteria used and the process followed. The inclusion criteria used were:

- the main objective of the paper is to discuss or investigate a methodological issue relating to systematic reviews.

- The paper addresses the construction and/or evaluation of quality instruments,

- There must be a software engineering context,

- The paper must be written in English,

- The paper may be a short paper.

Papers were excluded if:

- Their main objective was to report a systematic review or mapping study,

- They discussed evidence-based software engineering principles,

- They were methodological studies with a general (that is, a non-software engineering) focus,

- In form of PowerPoint presentations or extended abstracts,

- They produced guidelines for performing or reporting primary studies.

The search strategy for this review involved an initial informal search followed by a 3-stage process which included both a manual search and an automated search. Here we summarise the selection aspects of the search and selection process.

Stage 1 A manual search was performed by both authors who independently applied the inclusion and exclusion criteria, with an emphasis on inclusion unless a paper was clearly irrelevant. Disagreements were discussed and where agreement was not reached, the paper was included. Following an automated search, both reviewers applied the selection criteria, using the title and abstract of the papers found. Again the main emphasis was on including papers unless they were clearly irrelevant. Disagreements were discussed and where agreement could not be reached, the papers were provisionally included.

Stage 2 Papers included from the manual search, from the automated search and from the known set (determined through the informal search and using personal knowledge) were collated into a set of candidate papers. Where papers were treated differently across these inclusion sets, they were discussed and if no decision could be reached the paper remained a candidate. The final inclusion/exclusion decisions were made when the full papers were read during data extraction and quality assessment; again disagreements were discussed until agreement was reached.

Stage 3 At this stage, additional searching methods were used (snowballing and approaching individual researchers), and search and selection validation was carried out. Validation was based on the kappa agreement achieved between the authors for the decisions made during manual selection and for the selection from the candidates identified by the automated search.

Examples of study selection for mapping studies

The following examples report details of the processes followed as well as the criteria used. These mapping studies aimed to:

- Find out how extensively, and by what means, the Gang of Four (GoF) design patterns have been evaluated (Zhang & Budgen 2012),

- Identify and classify tools to support the systematic review process (Marshall & Brereton 2013).

Zhang & Budgen (2012) aimed to identify which Gang of Four (GoF) design patterns had been evaluated, what lessons had been learned from the evaluation studies and what further research might be needed to address 'gaps' in the evidence. The reviewers applied the following inclusion criteria:

- Papers describe software design patterns, although only empirical papers were used for the analysis,

- If several papers report the same study only the most comprehensive would be included,

- Where several studies were reported in a paper, each study would be treated independently.

Studies were excluded if they were:

- Reported in the form of abstracts or PowerPoint presentations,

- Documented in technical reports or papers submitted for publication.

The authors followed a 3-step process for excluding irrelevant papers or studies.

1. exclude on the basis of title,

2. exclude after reading the abstract,

3. exclude after reading the full paper.

The authors performed each of these steps independently, and then produced an agreed-upon list. They took a conservative approach for steps 1 and 2. A kappa score was calculated for an inter-rater agreement for each step.

The study by Marshall & Brereton (2013), which looked at the use of tools to support systematic reviews and mapping studies, included papers that:

- Report on a tool to support any stage of a systematic review or mapping study in software engineering,

- Report on a tool that is at any stage of development (e.g. proposal, prototype, functional, etc.)

Exclusion criteria were:

- Papers not written in English,

- Abstract or PowerPoint presentations.

The inclusion and exclusion criteria were applied in two stages. Initially, the first author checked the titles and abstracts of candidate papers and those that were clearly not relevant were excluded. After this stage, 21 papers were included. In the second stage, both authors checked the full texts, resulting in 16 papers remaining in the inclusion set. Subsequently two further papers were excluded during data extraction.

Chapter 7

Assessing Study Quality

As well as defining and applying inclusion and exclusion criteria to select *relevant* studies from a set of candidates, it is also important for many types of review to define and apply criteria for assessing the *quality* of the selected primary studies. This stage of the process is highlighted in Figure 7.1.

In this chapter we discuss three aspects of quality assessment:

- *Why* (and when) it is important to assess quality

- Defining the *criteria* to use for quality assessment

- Establishing and applying *procedures* for performing quality assessment

We note also, that although quality assessment is quite distinct from data extraction, which is covered in Chapter 8, these two stages can be performed sequentially or together (performing data extraction and quality assessment on a study-by-study basis).

Examples of quality assessment criteria and of procedures for applying them are described to illustrate some of the approaches taken by systematic reviewers in software engineering.

7.1 Why assess quality?

Quality assessment is about determining the extent to which the results of an empirical study are valid and free from bias. For systematic reviews

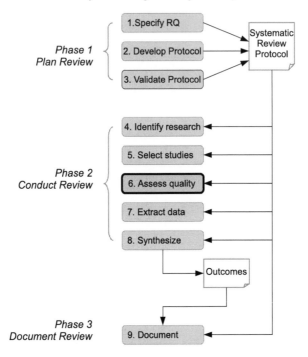

FIGURE 7.1: Quality assessment stage of the systematic review process.

and for some types of mapping study, evaluating the quality of the primary studies contributing to a review can enhance its value in a number of ways. For example:

- Differences in the quality of primary studies may explain differences in the results of those studies

- Quality scores can be used to weight the importance of individual studies when determining the overall outcomes of a systematic review or mapping study

- Quality scores can guide the interpretation of the findings of a review

For quantitative systematic reviews in particular, it is essential to assess the quality of the primary studies included in the review because if their results are invalid (or there is doubt about their validity) or if they are biased in some way, then this should be taken into account during the synthesis process. Reviewers might (simply) choose to exclude low quality primary studies from the synthesis process or they may choose to check whether their exclusion has a significant effect on the overall outcomes of a review. There have been a number of reports in the medical domain that have shown that if low-quality studies are omitted from the synthesis process of a systematic review (or from a meta-analysis) then the results of the review (or analysis) change. One example is a systematic review of homoeopathy which suggested that it performs well if low-quality studies are included, whereas high-quality studies found no significant effect (Shang, Huwiler-Müntener, Nartey, Jüni, Dörig, Sterne, Pewsner & Egger 2005). If reviewers intend to exclude low quality studies then effort can be saved by assessing quality in advance of data extraction.

Quality assessment can be of less importance when undertaking a mapping study since the focus for these is usually on classifying information or knowledge about a topic. However, it can be important for some mapping studies, especially tertiary studies, if for example their research questions relate to changes in quality over time.

Assessing the quality of a primary study is a particularly challenging task as there is no agreed, standard definition of study 'quality'. Many of the guidelines and criteria for assessing study quality, some of which are described later in this chapter, indicate that quality relates to the extent to which the design and execution of a study minimises bias and maximises validity. These concepts are summarised in Table 7.1 (see also Section 2.5.2). Further discussions and pointers to relevant literature about quality, bias and validity, can be found in Dybå & Dingsøyr (2008*b*).

TABLE 7.1: Quality Concepts

Term	Synonyms	Definition
Bias	Systematic error	A tendency to produce results that depart systematically from the 'true' results. Unbiased results are internally valid.
Internal validity	Validity	The extent to which the design and conduct of a study are likely to prevent systematic error. Internal validity is a prerequisite for external validity.
External validity	Generalisability, Applicability	The extent to which the effects observed in a study are applicable outside of the study.

As well as being an intrinsically difficult task to perform consistently, quality assessment is confounded by:

1. constraints imposed by the publication venue for papers reporting a primary study,

2. the range of primary study types included in a review.

The first of these factors, the constraints, usually relating to length, imposed by publishers can mean that researchers are not able to include all of the details of their study in a single paper. This is particularly problematic for conference proceedings where papers are often limited to 10 or 12 pages. This can result in the omission of important methodological (and other) information that would provide *evidence* of study quality. It can also lead to a study being reported in more than one paper adding to the difficulties of mapping papers to studies (as discussed in section 6.3). One approach to alleviating publishing constraints is to provide supplementary material on an associated web site. This facility is supported by a number of publishers. Another way of publishing more detailed information about a study than can be included in a single conference paper is to report different aspects of a study across multiple conference papers (which of course adds difficulties for the reviewer who has to extract and combine these) or in journal papers where length is usually less restricted.

The second of these factors which relates to the types of primary study, can cause problems where the studies included in a review are of diverse types. For example reviews can include quantitative studies, such as experiments and quasi-experiments, and qualitative studies, such as case studies and ethnographic studies. Here a dilemma arises — whether to use a generic set of quality criteria across all of the studies included in a review, regardless of their type, or whether to use specific sets of criteria tailored to each of the types of study that occurs in the set of primary studies. Not surprisingly each of these options has some strengths and some limitations. We look more closely at this issue in Section 7.2.2.

Once the decision has been made to assess the quality of the primary studies included in a systematic review or mapping study (and to use that assessment during the synthesis stage of the review) then a reviewer has two key questions to address. These are:

1. Against what criteria will the primary studies be assessed?

2. How will the assessment be performed and who should do it?

We also note that some reviews include non-empirical papers such as those reporting lessons learned or discussing some aspect of the topic of the review. The quality criteria discussed in this chapter are not appropriate for these types of papers.

7.2   Quality assessment criteria

A large number of quality assessment criteria and checklists for different types of empirical studies are published in the medical and social sciences literature. In addition to those indicated in the following section, the Support Unit for Research Evidence (SURE)[1], in the UK, provides a range of relevant links and resources including a set of critical appraisal checklists for quantitative studies, qualitative studies and systematic reviews. Work in the

---

[1]http://www.cardiff.ac.uk/insrv/libraries/sure/

medical and social science fields has provided the basis for many of the quality checklists proposed, used and/or evaluated for empirical studies in the software engineering field. We summarise here the checklists most widely used in software engineering reviews and also briefly discuss the problems associated with quality assessment across multiple study types.

7.2.1 Study quality checklists

A number of checklists that are tailored to specific study types have been proposed. For case studies, Runeson et al. (2012) present a checklist for both readers of case studies and for researchers who are performing case studies. These checklists are synthesised from a range of sources including literature in the social sciences and information systems fields and adapted to software engineering. The checklists for readers (and hence for reviewers) of case studies can be used to assess the quality of case studies included in a review. Primary studies of this type are commonly found in qualitative systematic reviews and mapping studies. The readers' checklist is shown in Table 7.2. Further details of the case study methodology and its use in systematic reviews can be found in Chapter 18.

TABLE 7.2: A Case Study Quality Checklist (Taken from Runeson, P., Höst, M., Rainer, A. & Regnell, B. (2012)). Reproduced with permission.

	Criteria
1.	Are the objectives, research questions, and hypotheses (if applicable) clear and relevant?
2.	Are the case and its units of analysis well defined?
3.	Is the suitability of the case to address the research questions clearly motivated?
4.	Is the case study based on theory or linked to existing literature?
5.	Are the data collection procedures sufficient for the purpose of the case study (data sources, collection, validation)?
6.	Is sufficient raw data presented to provide understanding of the case and the analysis?
7.	Are the analysis procedures sufficient for the purpose of the case study (repeatable, transparent)?
8.	Is a clear chain of evidence established from observations to conclusions?
9.	Are threats to validity analyses conducted in a systematic way and are countermeasures taken to reduce threats?
10.	Is triangulation applied (multiple collection and analysis methods, multiple authors, multiple theories)?
11.	Are ethical issues properly addressed (personal intentions, integrity, confidentiality, consent, review board approval)?
12.	Are conclusions, implications for practice and future research, suitably reported for its audience?

A quality checklist constructed for technology-intensive testing experiments is described by Kitchenham, Burn & Li (2009). The checklist focuses specifically on studies relating to testing; however, reviewers addressing other technology-intensive topics, such as cost estimation and performance, might find the approach to checklist construction and validation of interest. An adaptation of this checklist with suggestions for scoring each of the questions is shown in Figure 22.8.

A further quality checklist was developed and used for a qualitative, technology-focused systematic review on Agile methods (Dybå & Dingsøyr 2008*b*, Dybå & Dingsøyr 2008*a*). The 11 criteria making up the checklist were based on those proposed for the Critical Appraisal Skills Programme[2] and by the principles of good practice for empirical research in software engineering described by Kitchenham, Pfleeger, Pickard, Jones, Hoaglin, El Emam & Rosenberg (2002). The criteria, shown in Figure 7.3, cover four main areas of empirical research:

- *Reporting* - criteria 1-3 relate to the quality of reporting an empirical study,

- *Rigour* - criteria 4-8 address the details of the research design,

- *Credibility* - criteria 9 and 10 focus on whether the findings of the study are valid and meaningful,

- *Relevance* - criteria 11 concerns the relevance of the study to practice.

In the systematic review on Agile methods, the reviewers applied the checklist to 33 empirical studies, 24 of which were case studies, four were surveys, three were experiments and two used a mix of research methods. This checklist has been quite widely used by reviewers in software engineering as a basis for quality assessment. See for example, the reviews by Alves, Niu, Alves & Valença (2010), Chen & Babar (2011) and Steinmacher, Chaves & Gerosa (2013).

As discussed in Section 3.2, a tertiary study is a mapping study where systematic reviews and mapping studies constitute the 'primary' studies under review. Many researchers who undertake tertiary studies carry out quality assessment in order to identify trends in the quality of systematic reviews and/or mapping studies. To date, criteria to assess the quality of systematic reviews and mapping studies have not been developed specifically for software engineering reviews. However, one of the sets of criteria used in the medical domain, the DARE[3] criteria[4], has been applied in a number of tertiary studies. The criteria were initially based on four questions, with a fifth being added later. The five questions are:

---

[2]www.casp-uk.net
[3]Database of Abstracts of Reviews of Effects
[4]http://www.crd.york.ac.uk/CRDWeb/AboutPage.asp

TABLE 7.3: A Quality Checklist That Can Be Used across Multiple Study Types (Taken from Dybå, T. & Dingsøyr, T. (2008a)). Reproduced with permission.

	Criteria
1.	Is the paper based on research (or is it merely a 'lessons learned' report based on expert opinion)?
2.	Is there a clear statement of the aims of the research?
3.	Is there an adequate description of the context in which the research was carried out?
4.	Was the research design appropriate to address the aims of the research?
5.	Was the recruitment strategy appropriate to the aims of the research?
6.	Was there a control group with which to compare treatments?
7.	Was the data collected in a way that addressed the research issue?
8.	Was the data analysis sufficiently rigorous?
9.	Has the relationship between researcher and participants been adequately considered?
10.	Is there a clear statement of findings?
11.	Is the study of value for research or practice?

1. Are the review's inclusion and exclusion criteria described and appropriate?

2. Is the literature search likely to have covered all relevant studies?

3. Did the reviewers assess the quality/validity of the included studies?

4. Were basic data/studies adequately described?

5. Were the included studies synthesised?

Examples of the use of the DARE criteria include the broad tertiary studies reported in Kitchenham, Brereton, Budgen, Turner, Bailey & Linkman (2009), Kitchenham, Pretorius, Budgen, Brereton, Turner, Niazi & Linkman (2010) and da Silva et al. (2011) as well as a tertiary study by Cruzes & Dybå (2011b) which focused on research synthesis.

A number of other approaches to assessing the quality of systematic reviews are used within the medical domain, some of which are discussed in Dybå & Dingsøyr (2008b). In addition, we highlight two initiatives related to systematic reviews and meta-analyses within the clinical medicine field. One of these is the PRISMA[5] Statement which aims to help authors improve the reporting of systematic reviews and meta-analyses (Liberati, Altman, Tetzlaff, Mulrow, Gøtzsche, Ioannidis, Clarke, Devereaux, Kleijnen & Moher 2009). It is suggested that 'PRISMA may also be useful for critical appraisal of published systematic reviews'. However Moher, Liberati, Tetzlaff & Group (2009) do note that the PRISMA checklist is not a quality assessment instrument.

---

[5]Preferred Reporting Items for Systematic reviews and Meta-Analyses, http://www.prisma-statement.org/

A project undertaken by the Cochrane Editorial Unit (CEU) aims to specify methodological expectations for Cochrane protocols, reviews and review updates. As a result of this work, the CEU have produced a report describing methodological standards for the conduct of new Cochrane Intervention Reviews[6]. The report describes a checklist of 80 attributes relating to the conduct of reviews, indicating in each case whether they are considered mandatory or highly desirable.

7.2.2   Dealing with multiple study types

Many systematic reviews and mapping studies in software engineering include primary studies that utilise a range of different empirical methods. These typically include those methods described in Part II of this book. Where the primary studies are of a single type (for example, they are all case studies or all experiments) then a quality checklist can be selected or tailored for that specific study type. However, where a review includes multiple study types, researchers have to decide whether to use a single checklist or a set of type-specific checklists.

When a single quality checklist is used for a systematic review or mapping study, researchers have to consider which of the criteria (that is, which checklist items) are applicable for each study type. Of course this means that it is necessary to extract (and validate) the study type for each primary study before carrying out a quality assessment. When scores for a particular study are aggregated across the checklist items against which the study is assessed, the number of applicable items needs to be taken into account through a normalising process (see the third example in Section 7.4 which illustrates this approach).

Where multiple quality checklists are used, the same requirement to determine the study type arises. In this case, the study type is used to select the most appropriate checklist.

One problem that arises when there are multiple study types is that aggregated scores cannot be compared in a meaningful way across the different types. So, it becomes quite challenging to interpret these when considering the findings of a review. See Part III for further discussion about using quality assessment results from different types of study.

7.3   Procedures for assessing quality

Here we consider three aspects of the process of assessing the quality of empirical studies. These are:

---

[6]http://www.editorial-unit.cochrane.org/mecir

- *Scoring* studies against the checklist(s) used

- *Validating* the scores

- *Using* quality assessment results

Scoring studies

If a single checklist is being used, then each study will be scored against each criterion that is appropriate for the study type. If multiple checklists are used, then reviewers have to select the appropriate checklist and score the study against the items in that checklist. A number of approaches to scoring have been taken by reviewers. Some use a simple yes(1)/no(0) score (see, for example, Dybå & Dingsøyr (2008a) and Cruzes & Dybå (2011b)) whilst others recognise partial conformance to a criterion. For example da Silva et al. (2011) use a 3-point scale (yes(1)/partly(0.5)/no(0)). Whatever scale is used, it is important to ensure consistency by documenting the specific characteristics of a study that map to specific points on the scale.

Validating scores

As we have seen in Section 6.2, validation is an important element in maintaining confidence in the procedures and hence the outcomes of a review. The same options as are discussed for study selection are possible for validating quality scores. If quality assessment is being carried out by a team of researchers, then two or more members of the team can score each of the studies followed by a process of resolution. The process by which researchers obtain a consensus about the quality of a paper given a quality checklist has been investigated through a series of studies (Kitchenham, Sjøberg, Dybå, Brereton, Budgen, Höst & Runeson 2013). These studies found that using two researchers with a period of discussion did not necessarily deliver high reliability (that is, consistency in using a checklist) and simple aggregation of scores appeared to be more efficient (that is, involved less effort) than incorporating periods of discussion without seriously degrading reliability. The authors of the studies suggest using three or more researchers, where this is feasible, and taking an average of the total score using the numerical values of the scores. In contrast, a study by Dieste, Griman, Juristo & Saxena (2011) recommends against using aggregate scores and recommends only using validated checklist items.

Where quality assessment is being performed by a single researcher, such as a PhD student, then a test-retest approach to quality score validation can be used. This involves the researcher redoing the assessment of selected studies after a time delay. Alternatively, PhD students can ask members of their supervisory team to assess a random sample of the primary studies. Whether the assessment has been carried out by independent researchers, or where a lone researcher has taken a test-retest approach, the level of agreement between the scores can be checked (for example, using a Kappa analysis).

Using quality assessment results

As indicated in Section 7.1, results from the quality assessment process can be used within a systematic review in a number of ways. These include:

- specific quality criteria or the overall score can be used to exclude studies that are considered to be of low quality,

- analyses can be performed with and without low quality studies to determine the impact of such studies on the overall results,

- one of the research questions addressed by a review may focus on trends in the quality of primary studies relating to the topic of a review.

Whatever the role played by quality assessment, reviewers will need to consider the study type as well as the quality score for each of the primary studies that contribute to the findings of a review.

7.4   Examples of quality assessment criteria and procedures

Here we summarise three examples of quality assessment undertaken as part of software engineering systematic reviews. These cover each of the three types of systematic review: quantitative technology-focused reviews; qualitative technology-focused reviews and qualitative research-focused reviews (see Section 3.1 and Figure 3.1).

Quality assessment performed by researchers undertaking tertiary studies is also briefly highlighted.

The first example is a quantitative systematic review by Kitchenham et al. (2007) of studies which compare the use of cross-company and within-company cost estimation models. This review uses the checklist shown in Table 7.4 which is split into two parts (Part I and Part II). The criteria in Part I relate to the quality of the primary study and those in Part II are about the quality of reporting. The parts are weighted differently, with Part I having a weighting of 1.5 and Part II having a weighting of 1. The table indicates the possible scores for each of the criteria.

Quality assessment was carried out in parallel with data extraction in the following way:

1. For each paper, a reviewer was nominated randomly as data extractor/quality assessor, data checker or adjudicator,

TABLE 7.4: A Quality Checklist for a Quantitative Systematic Review (Taken from Kitchenham, B. A., Mendes E.& Travassos G. H. (2007)). Reproduced with permission.

	Criteria
Part I	
1.	Is the data analysis process appropriate?
1.1	Was the data investigated to identify outliers and to assess distributional properties before analysis? Yes(0.5)/No(0)
1.2	Was the result of the investigation used appropriately to transform the data and select appropriate data points? Yes(0.5)//No(0)
2.	Did studies carry out a sensitivity or residual analysis?
2.1	Were the resulting estimation models subject to sensitivity or residual analysis? Yes(0.5)/No(0)
2.2	Was the result of the sensitivity or residual analysis used to remove abnormal data points if necessary? Yes(0.5)/No(0)
3.	Were accurate statistics based on the raw data scale? Yes(1)/No(0)
4.	How good was the study comparison method?
4.1	Was the single company selected at random (not selected for convenience) from several different companies? Yes(0.5)/No(0)
4.2	Was the comparison based on an independent hold out sample (0.5), random subsets (0.33), leave-one-out (0.17) or no hold out (0)?
5.	Size of within-company dataset? fewer than 10 projects (score 0), 10-20 (0.33), 21-40 (0.67), more than 40 (1)
Part II	
1.	Is it clear what projects used to construct each model? Yes(1)/No(0)
2.	Is it clear how accuracy was measured? Yes(1)/No(0)
3.	Is a clear what cross-validation method was used? Yes(1)/No(0)
4.	Were all model construction methods fully-defined (tools and methods used)? Yes(1)/No(0)

2. The data extractor/quality assessor read the paper and completed a form,

3. The checker read the paper and checked the form,

4. If the extractor and checker could not resolve any differences that arose, the adjudicator read the paper and made the final decision after consulting the extractor and checker.

The assignment of roles was constrained so that no-one performed data extraction or quality assessment for a paper that they had authored and as far as possible the work load was shared equally.

In the second example, Dybå & Dingsøyr (2008a) report a qualitative systematic review of studies relating to Agile software development. The reviewers used the criteria shown in Figure 7.3 and formulated quite detailed descriptions of the issues to consider when scoring studies against each of the criteria. Studies were scored using a simple yes/no scale. The detailed descriptions of issues used to guide the scoring process can be found in Appendix B of Dybå & Dingsøyr (2008*a*).

Dybå & Dingsøyr and another researcher used the first criterion ('Is the paper based on research (or is it merely a 'lessons learned' report based on expert opinion)?') as the basis for inclusion/exclusion and they calculated their level of agreement for this criterion (94.4%). Disagreements were resolved by discussion among the three researchers.

The third example is a systematic review that addresses a research process, specifically the systematic review process (Kitchenham & Brereton 2013). This review included primary studies of many different types such as case studies, surveys and secondary studies. It also included discussion and 'lessons learned' papers. The reviewers chose to base quality assessment on the generic checklist developed by Dybå & Dingsøyr (2008*a*) (see Figure 7.3) with the additional question:

> "What research method was used: Experiment, Quasi-Experiment, Lessons learnt, Case Study, Opinion Survey, Tertiary Study, Other (Specify)?"

The determination of study type was based on the reviewers' own assessments rather than on the type claimed by the authors of a paper. Checklist items 5–8 were also adapted to address the different study types. The revised checklist items were:

> Item 5. "Was the recruitment strategy (for human-based experiments and quasi-experiments) or experimental material or context (for lessons learnt) appropriate to the aims of the research?"

> Item 6. "For empirical studies (apart for lessons learnt) was there a control group or baseline with which to evaluate systematic review procedures?"

> Item 7. "For empirical studies (apart for lessons learnt) was the data collected in a way that addressed the research issue?"

> Item 8. "For empirical studies (apart for lessons learnt) was the data analysis sufficiently rigorous?"

In addition, an allowable 'score' of 'not applicable' was included for questions 4-8. For most of the criteria, the allowable scores for applicable items were Yes (1), Partly (0.5), No (0) with interpolation permitted. The exception was the first criteria (Is the paper based on research?), for which only the scores of Yes(1) and No(0) were allowable.

The two reviewers undertook quality assessment (and data extraction) independently. Disagreements were discussed until agreement was reached. The reviewers noted some problems with their approach:

- Although they identified broadly which questions were relevant for particular types of study, they found that for some studies the context meant that further decisions about appropriateness had to be made during the quality assessment. This point is discussed in some detail in the paper and resulted in the reviewers assessing independently whether a question was relevant for a particular study as well as determining scores for each relevant criteria.

- Their assessments of study type frequently differed from those of the authors of a study. For example, if a case study was based on an opinion survey they classified it as an 'Opinion Survey' rather than a 'Case Study', and if a study was a post-hoc analysis of a systematic review they classified it as an 'Example' rather than a 'Case Study'.

- They found that using the checklist sometimes resulted in small studies obtaining good scores even though by their nature they could provide only very limited evidence of the value of the technique or method being studied. For example, if a study was a preliminary feasibility study it could score well on all checklist items even though it could provide very limited evidence of real value of the method being studied. Additionally, some lessons learned and experience papers scored well because relatively few checklist questions were relevant.

The level of agreement achieved for quality assessment, using values for the number of questions considered to be appropriate and the average quality score for each paper, was measured using the Pearson correlation coefficient.

As indicated in Section 7.2.1, a number of tertiary reviews have used the DARE criteria for assessing study quality. See for example, Kitchenham, Brereton, Budgen, Turner, Bailey & Linkman (2009), Kitchenham, Pretorius, Budgen, Brereton, Turner, Niazi & Linkman (2010), da Silva et al. (2011), Cruzes & Dybå (2011*b*) and Verner, Brereton, Kitchenham, Turner & Niazi (2014). With the exception of Cruzes & Dybå, reviewers scored each primary study (that is, each systematic review or mapping study) against each of the criteria with possible scores being Yes(1.0), Party (0.5) and No(0). Cruzes & Dybå scored studies as either meeting a criterion (Yes) or not (No). A range of different approaches was taken to allocating independent quality assessors and to resolving disagreements.

Chapter 8

Extracting Study Data

The objective at this stage of the review process, which is highlighted in Figure 8.1, is to extract, from the reports of the primary studies, the data needed to address the research questions. The strategy for data extraction, including the data extraction form (or forms) needs to be defined and justified. The use of data extraction forms can help to maintain consistency (across studies and between extractors), and, where these are held electronically, the extraction and data recording can be performed in one step. Although the forms will have been piloted during the planning phase, it is possible that they will have to be revised during the data extraction stage when a broader range of studies are processed. Some tools that provide support for data extraction and subsequent data management are indicated in Chapter 13.

The structure and balance for this chapter is somewhat different from the three previous chapters with a greater emphasis being placed on providing a set of examples to illustrate the strategies that have been followed across a range of quantitative systematic reviews, quantitative systematic reviews and mapping studies. The data to be extracted for a review of any kind is very closely related to the specific research questions for that review and also to the requirements of the synthesis/aggregation phase. In the examples, we show this connection as well as illustrating a range of procedures for extracting, recording and validating the data.

As noted in Chapter 7, although data extraction is quite distinct from quality assessment, these two stages can be performed sequentially or together.

8.1   Overview of data extraction

Different types of data are usually extracted for the different types of review although all usually include some 'standard data' that records, for example, publication details for each paper included in a review and information

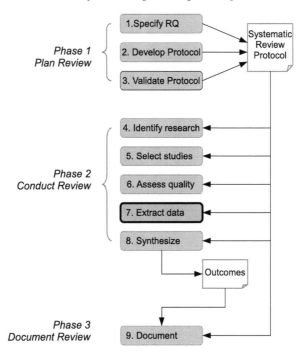

FIGURE 8.1: Data extraction stage of the systematic review process.

about the extractor and date of extraction. What other data is extracted depends very much on the research questions for a review. It should be noted also that sometimes the data needed will be spread across a number of papers.

For *quantitative systematic reviews*, data is most commonly in numerical form although it may include some qualitative data relating, for example, to the context of a primary study (such as the characteristics of participants in an experiment), or to opinions expressed by participants in a primary study, or to recommendations based on the findings of a primary study. This will particularly be the case where vote counting, as opposed to meta-analysis, is to be used as the method of synthesising the outcomes of the primary studies. Here qualitative data can be used to investigate possible explanations for differences in the outcomes of the primary studies (see Section 10.4.6). If the research questions relate, for example, to specific metrics such as defects found, time taken to perform a task or estimates of development costs, then this information is extracted for each study and recorded in a table, a spreadsheet or a special-purpose systematic review support tool.

For *qualitative systematic reviews* and *mapping studies*, data is often extracted in textual form or through the use of a set of classification schemes.

For *qualitative systematic reviews*, information about, for example, factors, barriers, motivators, recommendations or experiences is extracted and recorded. However, the extraction of this type of data can be more susceptible to bias and the data is less amenable to statistical analysis, making spreadsheets a less useful recording medium. For this type of study, data extraction and synthesis are very closely linked and are likely to be combined within a single process. See, for example, the meta-ethnography process described in Section 10.4.1 and thematic analysis, described in Section 10.4.4.

For *mapping studies*, data extraction and aggregation may be performed iteratively with the classification schemes (for example, relating to the technique used for a particular software engineering task, or the method used in an empirical study) being revised as more knowledge about the topic is gained through the extraction and aggregation process.

The commonly used procedures for data extraction and validation are mostly the same as those described for quality assessment (see Section 7.3), namely:

- Independent extraction by two (or more) reviewers followed by reconciliation through discussion or moderation

- For a lone researcher, taking a test-retest approach and comparing outcomes

- For a lone researcher such as a PhD student, engaging a member of the supervisor team to extract data for a sample of studies

Additionally, especially for qualitative data, extraction may be undertaken as a team, with two or more reviewers working together to reach agreement about the data to be extracted. Whatever approach is taken, if agreement between extractors (or extractions for the test-retest case) is poor then some review of data descriptions, and possibly of the research questions, may be needed.

8.2 Examples of extracted data and extraction procedures

We summarise the data extraction strategies for the following examples:

1. two quantitative systematic reviews which take rather different approaches to data validation,

2. two qualitative systematic reviews: a technology-oriented review and a research-oriented review,

3. a technology-oriented mapping study about the use of Open Source projects in teaching about software engineering.

Examples from quantitative systematic reviews

The first example in this category is a meta-analysis undertaken by Hannay et al. (2009) which aimed to determine the effectiveness of pair programming compared to solo/individual programming. The systematic review addressed the research question:

> How effective, in terms of quality, duration or effort is pair programming compared to solo programming?

Data was extracted about the type of treatment, the type of system, the type of tasks, duration of the study, number of groups, group assignment, type of subjects and their experience with pair programming, number of pairs, number of individuals, outcome variable, means, standard deviations, counts, percentages and p-values.

Data was extracted from all studies by three of the four reviewers and discrepancies were resolved through discussion amongst all four reviewers (that is, the four authors of the review paper). Reviewers performed separate meta-analyses for quality, duration and effort.

The second example of data extraction for a quantitative systematic review is taken from a study carried out by Hall et al. (2012) which reviewed the performance of fault prediction models. This systematic review addressed three research questions relating to context, independent variables included in the fault prediction models and the performance of specific modelling techniques. Three sets of data were extracted with a different procedure being followed for each set:

1. *Context data* - source of the data, maturity, size, application area and programming language of the systems studied. This data was collected by one of the reviewers.

2. *Qualitative data* - data relating to the research questions of the review that was reported in the findings and conclusions of the primary studies. Two reviewers extracted the data independently and discussed disagreements until these were resolved.

3. *Quantitative data* - predictive performance data for all models reported in a study. The form of the data depended on whether the study reported results via categorical or continuous dependent variables. For categorical studies, where possible, data about precision, recall and/or f-measure was recorded. For continuous studies, results in the primary studies were reported in terms of the number of faults predicted in a unit of code using measures based on errors (for example, Mean Standard Error) or differences between expected and observed results (such as Chi Square). Data was extracted and recorded using whatever measure was used in each study. For this data, two reviewers worked together to identify and extract the data from each study.

The reviewers intended to carry out meta-analyses of quantitative data; however, this subsequently proved problematic and so they chose to take a qualitative thematic approach.

Examples from qualitative systematic reviews

The first example in this category is taken from a review which aimed to 'plot the landscape' of reported knowledge about what motivates and demotivates software engineers (Beecham et al. 2008). The research questions for this review are shown in Section 4.3. Data was extracted about how each study answered each of the research questions, with the extractor recording information about:

- Software engineer characteristics,

- Software engineer motivators,

- Software engineering de-motivators,

- External signs or outcomes of motivated software engineers,

- External signs or outcomes of de-motivated software engineers,

- Software engineering as a motivator (e.g. what is motivating about the type of development used; task of coding, testing etc),

- Frameworks/models that reflect how software engineers are motivated.

Endnote[1] was used to record publication details for each paper and a results form was used to record how each study answered each of the research questions. An example of a populated form, which also shows the captured publication details and a potentially relevant study identified through backwards snowballing is described in Appendix B of the review protocol (Beecham, Baddoo, Hall, Robinson & Sharp 2006).

Data was validated by an independent expert on motivation in software engineering who recorded how each paper addressed each research question. Disagreements were discussed and for the small number of cases where they could not be resolved, a third independent researcher arbitrated. The approach taken to synthesising the data was to establish the frequency with which a characteristic or motivator was identified by the primary studies (most of which were surveys).

The second example in this category is a research-oriented qualitative systematic review focusing on the systematic review process (Kitchenham & Brereton 2013). The overall aim of the review was to identify, evaluate and

---

[1]http://endnote.com/

synthesise research about performing systematic reviews and mapping studies.The specific research questions address are listed in Section 4.3. Extracted data included:

1. Publication details

2. Review-specific data relating to:

   - Type of paper (problem identification and/or problem solution (PI) or lessons learned/opinion survey/discussion paper (E)),
   - Scope of study (mapping study, systematic review/both/update study/other)
   - Summary of aims
   - Main topics covered (multiple selections allowed from a list)
   - Method proposed - name or description
   - Validation performed? - yes or no
   - Actual validation method (as judged by the data extractors) - selected from a list or other (specified),
   - Claimed validation method
   - Summary of main results
   - Any process recommendations (determined by the data extractors)

Publication details were collected and recorded in a spreadsheet by the first author. For review-specific data, some discussion papers, lessons learned papers and opinions surveys (that is, E-type papers) were treated differently from other studies. If a paper covered a very specific topic and had a limited number of results then the data was collected, as for other studies, by both reviewers, and recorded in a spreadsheet. Disagreements were discussed until agreement was reached. If, however, the scope of an E-type paper was very broad (that is, if it covered many aspects of a review and/or included comments from a large variety of subjects), the spreadsheet was only partially completed and an additional data extraction form was used (see Table 8.1). For this third type of (textual) data, the first author extracted the data and the second author checked it.

Example from a mapping study

The example summarised here is a mapping study focusing on the use of Open Source projects in software engineering education (Nascimento, Cox, Almeida, Sampaio, Almeida Bittencourt, Souza & Chavez 2013). The study addressed three research questions:

1. "How are Open Source projects used in software engineering education?"

TABLE 8.1: Form for Recording Extra Textual Data (adapted from Kitchenham and Brereton (2013))

Issue Id	
Issue text	For each issue/problem raised/solution proposed specify this using the same text as the paper authors
Type	Advice, problem/challenge or value (benefit)
Suggestion for guidelines?	Yes or no
Novice issues?	Yes or no
Education issues?	Yes or no
Location in paper	Page number or table number/id
Stage in review process addressed	Research question/protocol/search/selection/data extraction/ quality assessment/data aggregation/synthesis/reporting
Importance	A ratio indicating number of votes out of the maximum possible number or a textual indication of relative importance
Related issue	Reference to any related issue
Comments	

2. "Are there any initiatives that combine open source projects with active learning in software engineering courses?"

3. "How is student learning assessed in such initiatives?"

In addition to publication details, data was extracted according to the following classification facets:

- *Software engineering topic* - based on the SWEBOK knowledge areas[2],

- *Research type* - using a set of approaches to research based on those of Petersen, K. Petersen, Feldt, Mujtaba & Mattsson (2008),

- *Learning approach* - using categories: active learning (general), case-based learning, game-based learning, peer/group/team learning, problem-/project-/inquiry-based learning, studio-based learning and other,

- *Assessment perspective* - where there is assessment, this can, for example, be from the perspective of the student (through peer or self assessment) or from the perspective of teaching staff,

- *Assessment type* - covering methods of assessing students (such as by

---

[2]www.swebok.org

examination, through developed software artifacts, interviews, exercises or surveys).

Nascimento et al. indicate that due to lack of time, data was extracted (that is the primary studies were classified) by only one reviewer. The authors recognise this as a limitation of their mapping study.

Chapter 9

Mapping Study Analysis

This chapter is the first of three consecutive chapters dedicated to data synthesis. We have split this topic into three chapters because the three different types of systematic review (quantitative, qualitative and mapping study) require very different procedures. Data synthesis is also one of the tasks that many software engineering researchers have identified as least well addressed by current guidelines, see Cruzes & Dybå (2011b) and Guzmán, Lampasona, Seaman & Rombach (2014).

Chapter 10 discusses qualitative synthesis which is suitable for systematic reviews of qualitative primary studies as well as systematic reviews of quantitative primary studies that are unsuitable for meta-analysis. Chapter 11 describes statistical methods used to synthesise primary studies that report quantitative comparisons of different software engineering techniques.

This chapter discusses the analysis methods used for mapping study reviews. We address mapping studies first because the analysis methods used to summarise results from mapping studies are generally quite straightforward but provide some insight into the problems experienced synthesising results from systematic reviews. The analysis of mapping study results is relatively simple because the data extracted from each primary study in a mapping study are much less detailed than the data extracted from primary studies in systematic reviews. However, more complex analyses based on text mining can help identify clusters of similar studies either to validate the study inclusion and exclusion process or identify subsets of studies for more detailed analysis.

The examples of data analysis presented in this chapter were analysed using the R statistical language. We strongly recommend using R since:

- It is free and open source.

- There are numerous ancillary packages developed by statistical experts.

- There are many textbooks describing R. We personally recommend *R in Action* (Kabacoff 2011).

- It provides flexible programming facilities.

9.1   Analysis of publication details

Many research questions can be answered by analysing publication details, such as:

- Author name and affiliation

- Publication date

- Publication type

- Publication source

Such data are usually analysed as simple tables of counts such as the number of publications per author, or per country of affiliation, or simple trend-based graphics such as the number of publications per year. For example, Mair, Shepperd & Jørgensen (2005) analysed empirical cost estimation studies to identify and investigate the data sets. They presented:

- A table showing the number of studies in each journal that was used in the search process.

- A line plot of the number of studies grouped in three-year periods.

When a mapping study is performed as part of a PhD thesis, a research student needs to know which papers are the most influential in their field, and to have read and understood them. Using a general indexing system such as Scopus, Web of Science, or Google Scholar, it is easy to discover how many papers have cited each primary study, which is a good way of identifying such papers.

Another type of analysis that can be useful when dealing with a large number of primary studies is to look for author networks, that is, groups of authors that have collaborated to produce a number of primary studies. This information can be used with classification data to identify whether groups of authors concentrate on specific topics, problems or methods. For example, in the context of a meta-analysis of machine learning defect prediction methods, Shepperd, Bowes & Hall (2014) used the author group as a moderator factor in their analysis. Surprisingly, it proved to be the most important moderator factor, accounting for 31% of the variation among studies. Although the primary studies all compared different prediction methods, prediction method as a factor accounted for only 1.3% of the variation.

It may also be useful to analyse the cross-references within a set of primary studies to look for clusters of studies and isolated studies. It is also worth checking whether the isolated studies should really have been excluded. Furthermore, analysis of the combined set of included and exclude studies can be used to check that inclusion and exclusion criteria have been used consistently. For example, if publication detail analysis shows that some excluded papers are among a cluster of included papers, they should be re-assessed for possible inclusion.

9.2 Classification analysis

The more interesting research questions are usually based on classifying the primary studies and may concern issues such as:

- Identifying the existing research approaches and/or concrete techniques used in a topic area and cross-referencing between the approach taken and the relevant primary studies.

- Identifying the experimental methods used in empirically-based studies.

- Mapping approaches and techniques to the overall software engineering process or to specific steps in a specific software engineering task.

For example, Mair et al. (2005), provide a large number of diagrams[1] relevant to describing and categorizing software engineering datasets. The diagrams show:

- The number and percentage of datasets that were available or unavailable to researchers.

- The number of datasets collected in three-year time periods.

- The size of the datasets in terms of number of projects.

- The dataset size, in terms of numbers of features (attributes).

- The frequency of dataset usage (that is, the number of studies that used each dataset).

Mair et al. did not specify *a priori* research questions for their mapping study, but, as experts in the topic area, they provided analyses of great interest to cost estimation researchers.

In another example, Elberzhager, Rosbach, Münch & Eschbach (2012) in a mapping study on methods to reduce test effort, asked the question:

"What are existing approaches for reducing effort when applying testing techniques, and how can they be classified?"

---

[1]The diagrams are labelled histograms but are actually frequency diagrams.

To address the question they extracted keywords from the abstracts of 144 primary studies and then looked for additional keywords by reading the introductions and conclusions of the primary studies. This allowed them to identify five groups of testing methods (specifically Test automation, Prediction, Test input reduction, Quality Assurance before testing, Test strategy). They then tabulated the number and percentage or papers in each category, but they also provided a narrative description of each approach.

To be helpful to readers, data displays should allow the reader to track papers to the categories that describe them. This is usually done by presenting all the extracted information in data matrices (which need to be either published in the review reported or included in ancillary information available to readers), but can also be incorporated into data analysis displays. For example, Elberzhager et al. looked at papers in each of the categories in more detail and identified more detailed subcategories. For test automation they identified different phases of code automation and displayed the information using a horizontal bar chart with the identifiers of each paper printed beside the relevant horizontal row. An artificial example of a horizontal bar chart is shown in Figure 9.1. This was obtained from the data and the code snippet shown in Figure 9.2.

Petersen et al. (2008) suggest the use of bubble plots to visualize relationships among categorical variables. An example showing the structure of a bubble plot can be seen in Figure 9.3.

A bubble plot assumes we have three categorical variables and want to plot two of the variables (X-Variable 1 and X-Variable 2) against the third (Y-Variable). The relationships are shown by the number of studies that share a specific X-variable category and a specific Y category variable. For example, in Figure 9.3, 15 studies exhibit both Y-Variable category 4 and X-Variable 1 Category 2. In this case, the value 15 means that 21.3% of the papers that have been categorized according to X-Variable 1. A bubble plot does not assume that every study is categorized against each variable (e.g., some studies may not exhibit any of the categories associated with a X-variable and other studies may exhibit several different categories of the same X-variable), nor does it display any direct relationship between the X-variables.

Bubble plots can be produced manually using a drawing package. Alternatively, R supports bubble plots of two-variables but to produce the bubble plot shown in Figure 9.3, the data must be organized as shown in Table 9.1, the X-Variable 1 has 5 categories which are mapped to the values xpos=-5, -4,....,-1 while X-Variable 2 has 6 categories mapped to the values xpos=1,2,...,6. The Y-variable has 5 categories mapped to the values yvar=1,2,..5. *val* identifies the number of primary studies that have the specific X-category and Y-category. The values of *xtpos* and *ytpos* identify the (x,y) co-ordinates on the bubble plot where the percentages associated with *val* should be printed so they are displaced from the bubble. The R code used should be based on the snippet shown in Figure 9.4.

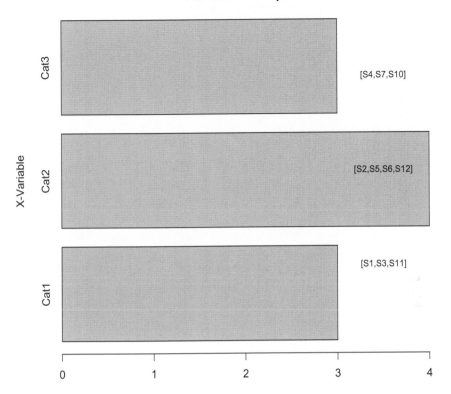

FIGURE 9.1: Example of a horizontal bar chart including study IDs.

A limitation of classification methods used in mapping studies is that although the set of categories relating to a specific feature or characteristic may appear to be mutually exclusive, primary studies are often more complex. For example, mapping study analysis often uses the classification of study types Wieringa, Maiden, Mead & Rolland (2006) proposed to classify requirements engineering papers. The categories include: Problem investigation, Solution design, Solution Validation, Solution selection, Solution implementation, Implementation evaluation. However, in practice, a paper discussing a "Solution design" will often include a section demonstrating the feasibility of the proposed solution which would be an example of "Solution Validation". This means that such a paper should be classified in both categories. It is often clear from bubble plots or tabular displays that the total number of classified papers is greater than the number of primary studies but it is not clear from bubble plots which papers exhibit multiple categories. Furthermore, if researchers categorise a paper in terms of their personal opinion of the 'main'

```
#Input the categorical data
XVar=c("Cat1","Cat2","Cat1","Cat3","Cat2","Cat2","Cat3","Cat3","Cat1","Cat2")
#Identify the number of studies in each category
counts=table(XVar)
#Creat a horizontal bar plot
barplot(counts, main="Horizontal bar plot", ylab="X-Variable",horiz=T)
#Add Text identifying the studies to the bar chart
text(3.5,1,"[S1,S3,S11]",cex=0.8)
text(3.5,2,"[S2,S5,S6,S12]",cex=0.8)
text(3.5,3,"[S4,S7,S10]",cex=0.8)
```

FIGURE 9.2: Bar chart code snippet.

goal of the paper, then some categories may be artificially underrepresented. Multiple classifications of primary studies or underrepresentation of certain categories make it more difficult to understand the implications of the reported frequency counts.

9.3   Automated content analysis

Recently, several researchers have suggested the use of text mining and associated visualization methods to analyse mapping study data, see Felizardo, Nakagawa, Feitosa, Minghim & Maldonado (2010) and Felizardo et al. (2012). These techniques can be used for analysing citations among papers as described in Section 9.1, however, in this section, we describe their use for content analysis. Content analysis and text mining can be used to:

- Check inclusion and exclusion decisions during primary study selection.

- Identify clusters of studies that might be suitable for more detailed analysis as a set of related studies.

Text mining and visualization require specialist tools (see Chapter 13). Felizardo et al. (2012) used the following process for content mapping using the *Revis* tool:

1. *Text preprocessing* is used to structure and clean the data. They used text from the title, abstract and keywords only. In addition, the text is analysed to create a vector of terms (words) present in the text which are weighted based on *term frequency-inverse document frequency measurement* which involves weighting words:

    - in direct proportion to its frequency in a specific primary study, but

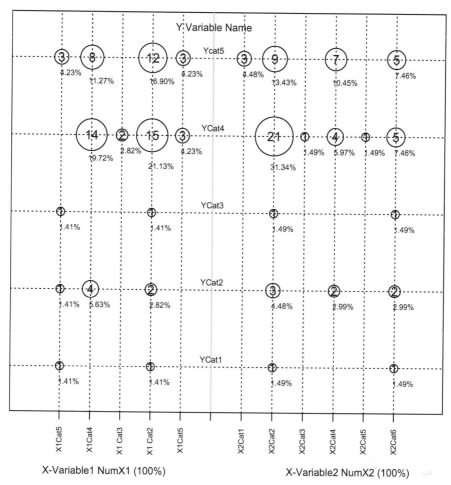

FIGURE 9.3: Example of a bubble plot showing the structure.

- in inverse proportion to its frequency in the other studies.

2. *Similar Calculation* which uses the vector of weighted words to calculate (dis)similarity among primary studies. Felizardo et al. used a method based on cosines: $distance(i,j) = 1 - cos(\bar{x}_i, \bar{x}_j)$ where $\bar{x}_i$ and $\bar{x}_j$ are vectors of weights for the ith and jth primary studies.

3. *Projection* which maps the m-dimensional vectors onto 2 or three dimensions that can be represented visually.

TABLE 9.1: Bubble Plot Data

xvar	yvar	val	xtpos	ytpos	percent
−5	5	3	−4.7	4.8	4.23%
−5	3	1	−4.7	2.8	1.41%
−5	2	1	−4.7	1.8	1.41%
−5	1	1	−4.7	0.8	1.41%
−3	4	2	−2.7	3.8	2.82%
−4	2	4	−3.7	1.8	5.63%
−4	4	14	−3.7	3.7	19.72%
−4	5	8	−3.7	4.7	11.27%
−2	1	1	−1.7	0.8	1.41%
−2	2	2	−1.7	1.8	2.82%
−2	3	1	−1.7	2.8	1.41%
−2	4	15	−1.7	3.6	21.13%
−2	5	12	−1.7	4.7	16.9%
−1	4	3	−0.7	3.8	4.23%
−1	5	3	−0.7	4.8	4.23%
2	1	1	2.3	0.8	1.49%
2	2	3	2.3	1.8	4.48%
2	3	1	2.3	2.8	1.49%
2	4	21	2.3	3.6	31.34%
2	5	9	2.3	4.7	13.43%
1	5	3	1.3	4.8	4.48%
3	4	1	3.3	3.8	1.49%
4	2	2	4.3	1.8	2.99%
4	4	4	4.3	3.8	5.97%
4	5	7	4.3	4.7	10.45%
5	4	1	5.3	3.8	1.49%
6	1	1	6.3	0.8	1.49%
6	2	2	6.3	1.8	2.99%
6	3	1	6.3	2.8	1.49%
6	4	5	6.3	3.8	7.46%
6	5	5	6.3	4.8	7.46%

The projection maps can be colour coded to show whether studies that appear similar based on content analysis have received the same inclusion and exclusions decision. They point out that clusters can be of two types:

1. Pure clusters where all primary studies received the same in(ex)clusion decision. Such clusters do not need to be reassessed.

2. Mixed clusters where some primary studies were included and some excluded. Felizardo et al. (2012) suggest reassessing any primary studies

```
bplot=read.table("filename.txt",head=T)
summary(bplot)
attach(bplot)

r=val/100
ny=c(1,2,3,4,5,6)

#Draws the circles - the spaces in the subtitle are intentional
symbols(xvar,yvar,circle=sqrt(r/pi), inches=.25,
xlab="",ylab="",xaxt="n",yaxt="n", sub="X-Variable1 NumX1 (100%)
X-Variable2 NumX2 (100%)",cex.sub=.8)

#Adds the numbers to the circles
text(xvar,yvar,val)
#Adds the grid lines
abline(h=c(1,2,3,4,5),lty=3)
abline(v=c(-5,-4,-3,-2,-1),lty=3)
abline(v=c(1,2,3,4,5,6),lty=3)

#Adds a central y-line
abline(v=c(0),lty=1,col="Yellow")

# Defines labels for the x-axis
labx=c("X1Cat5","X1Cat4","X1 Cat3","X1
Cat2","X1Cat5","","X2Cat1","X2Cat2","X2Cat3","X2Cat4","X2Cat5", "X2Cat6")
!
tckx=c(-5,-4,-3,-2,-1,0,1,2,3,4,5,6)

#Adds labels tick marks on the x-axis
axis(1,at=tckx,labels=labx,cex.axis=.6,las=2)

#Defines labels for the y-axis

laby=c("Ycat5", "YCat4","YCat3","YCat2", "YCat1")

#Specifies position of Y labels
nlab=c(5.1,4.1,3.1,2.1,1.1)

#Adds Y labels to plot
text(0,nlab,laby,cex=.65)
#Adds name to y-axis
text(0,5.4,"Y Variable Name",cex=.8)

#Adds the offset percentage information
text(xtpos,ytpos,percent,cex=0.6)
```

FIGURE 9.4: Bubble plot code snippet.

found in such clusters. If only one or two studies were in(ex)cluded, they needed to be reassessed in order to determine whether they should be reclassified to conform with majority decision.

In addition, isolated points need to be reassessed if they have been included.

9.4   Clusters, gaps, and models

We defined the main goals of mapping studies as finding clusters of studies suitable for more detailed studies and identifying gaps where more research is needed (see Chapter 3). In order to identify useful clusters and meaningful gaps, it is necessary to have some theoretical model of the mapping study topic against which the primary studies can be assessed. This may be a generic classification scheme such as that proposed by Wieringa et al. (2006), but it could be a classification scheme derived from an existing model of the software engineering processes addressed by the topic (for example the three layer model of cloud engineering), or of the way in which the existing software processes would be changed by the topic (for example the way test-before changes the overall testing process). We note however, that, although the identification of a large number of papers in a particular category is a strong indicator of a cluster, the absence of primary studies particularly in two-way tables or bubble plots does not necessarily imply a gap in the literature. It might mean that the specific combination of categories is either not meaningful or not important. To be identified as a topic suitable for further research, a gap needs a convincing explanation of why further primary research is likely to be important.

It is also possible that a mapping study might lead to the development of a model of the topic area, as an outcome of reading and classifying the literature. At the moment, this is an underutilised approach in mapping studies, but, as we point out in Section 10.4, Popay, Roberts, Sowden, Petticrew, Arai, Rodgers, Britten, Roen & Duffy (2006) suggest that the starting point of a narrative synthesis of a systematic review should be a model of the topic of interest. So if a mapping study is intended to be the starting point of a systematic review, it may be useful to consider whether its results can be represented as a model of the topic area, used to organise the primary studies.

Chapter 10

Qualitative Synthesis

This chapter discusses qualitative methods for synthesizing research studies. In most cases, qualitative synthesis methods are used when the individual primary studies used qualitative research methods, or used a variety of different experimental methods. In the context of software engineering, industrial case studies are a particularly important form of primary study because they provide more realistic information about the extent to which new methods and tools scale-up to the complexity of industrial scale software development than laboratory experiments. As discussed in Chapter 18 and Chapter 19, case studies often adopt qualitative methods. They, therefore, require qualitative approaches, such as the ones described in this chapter, to synthesise their results.

Qualitative synthesis methods are also useful for synthesising data from experiments, quasi-experiments, and data mining studies when the differences among outcome metrics, analysis methods, and experimental designs are too great to make statistical meta-analysis feasible. For this situation, we recommend *vote counting*. Vote counting is the practice of counting the number of primary studies that found a significant positive effect and the number that found an insignificant effect (or a significant negative effect) and assuming the effect is real if the majority of the studies are significant. Although vote counting is sometimes assumed to be a form of meta-analysis, many meta-analysts are strongly opposed to its use. The main argument is that although a significant finding provides evidence that an effect exists, a non-significant finding does not indicate that there is no effect, because lack of significance can be due to low statistical power. In addition, vote counting may give an idea of the direction of an effect but it does not give any indication of the magnitude of the effect, so it is not possible to decide whether an effect is practically important as well as statistically significant. However, in practice, many software engineering researchers, ourselves included, adopt vote counting when it is not possible to undertake a proper meta-analysis. We agree with Popay et al. (2006) that vote counting can be used constructively as part of a narrative synthesis, particularly if it can be associated with some form of qualitative moderator analysis.

10.1   Qualitative synthesis in software engineering research

Before discussing qualitative methods for synthesis, we discuss the extent to which qualitative synthesis is important for software engineering research. Cruzes & Dybå (2011*b*) reviewed the state of research synthesis in software engineering systematic reviews. They undertook a *tertiary study* that identified 49 systematic reviews published between the 1st of January 2005 and the 31st of July 2010. They found that the methods authors claimed to have used for synthesis were not always correct. They also reported that:

- 24 studies were mapping studies not systematic reviews

- 22 of the systematic reviews were not explicit about the synthesis method they used.

- Meta-analysis was used in only two systematic reviews (see Kampenes, Dybå, Hannay & Sjøberg (2007) and Dybå et al. (2006)), and, excluding mapping studies, all other systematic reviews used qualitative methods.

- Narrative synthesis was the most common form of synthesis (9 systematic reviews), followed by thematic analysis (8 systematic reviews) and comparative analysis(4 systematic reviews).

- Meta-ethnography and case survey were each used by one systematic review.

Cruzes & Dybå's study confirms the importance of qualitative synthesis for systematic reviews, but also, suggests that software engineering researchers are not good at describing the methods they use to aggregate and synthesise non-numerical findings.

Later studies indicate that the use of qualitative synthesis in software engineering systematic reviews continues to increase. Another tertiary study (da Silva et al. 2011) identified a second systematic review that used meta-ethnography (Gu & Lago 2009), while more recently Da Silva, F. Q. B.; Cruz, S. S. J. O.; Gouveia, T. B.; & Capretz, L. F (2013) reported a meta-ethnography of four primary studies presented as a worked example of the method. In addition, Cruzes, Dybå, Runeson & Höst (2014) present a study based on synthesising two case studies related to trust in outsourcing which used three different methods: thematic synthesis, cross-case analysis and narrative synthesis.

Also, the use of meta-analysis was underestimated with meta-analyses by Hannay et al. (2009), Ciolkowski (2009), and Salleh, Mendes & Grundy (2009) being missed by Cruzes & Dybå's tertiary study. In addition, at least, two more meta-analyses were published after 2010 (see Rafique & Misic (2013), and Kakarla, Momotaz & Namim (2011)).

Before discussing specific qualitative synthesis methods, we discuss some of the terminology used in the context of qualitative analysis. We then discuss the specific methods we believe are of most relevance to software engineering qualitative aggregation and synthesis. In this chapter, some of our methodological references come from the healthcare domain, in particular, nursing and healthcare policy. This is because this domain has a long history of qualitative research and has been grappling with the problems of synthesising qualitative research for many years. Furthermore, methodological studies using health care examples discuss topics that are familiar to most of us, for example, promoting healthy eating practices, or caring for sick children, which makes them easier to understand than examples from other domains.

10.2 Qualitative analysis terminology and concepts

Throughout this chapter, we will use the term *meta-synthesis* to apply to any method of qualitative aggregation or synthesis, that is, every form of aggregation or synthesis except quantitative meta-analysis. It is important to understand that most qualitative analysts view aggregation and synthesis as very different activities.

Aggregation is assumed to be similar to quantitative meta-analysis where information from different primary studies is combined together using counts and averages. For example, *quantitative content analysis* involves counting the number of times some specific words or phrases are mentioned in text. This is a rather quantitative approach to analysis and if it was used to obtain information from a set of primary studies would equate to an aggregation-based synthesis. Novice analysts usually find that aggregation is much easier than synthesis, but is only suitable for use with qualitative primary studies

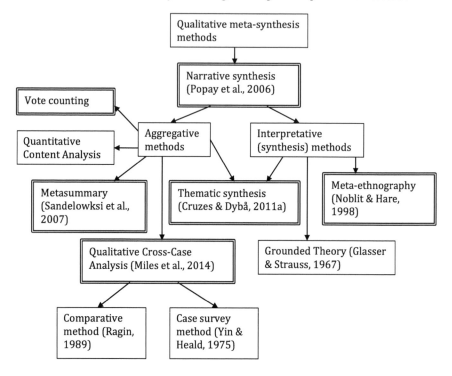

FIGURE 10.1: Methods for qualitative synthesis.

that have used fairly simple approaches to reporting their findings, such as content analysis or simply reporting the topics mentioned by participants.

In contrast, the goal of more purely qualitative studies is synthesis. *Synthesis* is referred to as an *interpretive* process which means using the concepts defined in specific primary studies to construct *higher order* models, that is, models that include concepts not found in any primary study. For example, in studies of globally-distributed software development, primary study authors may report problems observed by individuals working on distributed projects, whereas authors of secondary studies might use the information about reported problems to *infer* or *hypothesise* underlying causes of the problems which were *not mentioned specifically in any of the primary studies*. As a concrete example, Casey & Richardson (2008) re-analysed three case studies undertaken over a period of eight years. Although the only similarity between the cases was the shared aim of finding out what was actually going on and identifying what positive and negative factors influenced the software development strategy, the authors were able to identify the fear of job loss among staff employed by the client company as a factor that explained many of the problems observed between clients and vendors in each of the cases.

In general, the qualitative method used in the primary studies will influence the type of meta-synthesis that can be performed. Two types of qualitative method that are common in disciplines such as health care, psychology and social policy, are *ethnography* and *phenomenology*.

Ethnography is used to undertake longitudinal studies aimed at understanding the social and societal behaviour of human groups. Noblit & Hare (1988) developed *meta-ethnography* as a method for synthesizing different ethnography studies. In the context of software engineering, observational studies of agile teams might be based on an ethnography-based approach, see for example, Sharp & Robinson (2008) and Robinson, Segal & Sharp (2007).

Phenomenology is concerned with the way in which individuals perceive and interpret events. Phenomenology can underpin the use of *Grounded Theory*, which has the main aim of developing theory from the observed data. In the context of software engineering, Oza, Hall, Rainer & Grey (2006) present a Grounded Theory analysis of trust in outsourcing projects. Grounded Theory has had a major influence on qualitative research. Remenyi (2014) says that:

> "Grounded Theory not only offers a method by which social science research may be rigorously conducted but it also provides a more general explanation and understanding of how qualitative research works."

Terminology originating from grounded theory is used by many different qualitative synthesis and meta-synthesis methods and includes:

- *Coding* which involves applying descriptive labels to pieces of textual fragments, such as words, phrases or sentences. Miles, Huberman & Saldaña (2014) point out that words are the basic medium for all qualitative analysis, irrespective of the way in which the raw data was obtained. Initially, analysts look for codes that can be used to identify related textual fragments in different sources.

- *Axial coding* is an additional level of coding used to organise the basic codes derived directly from the text into more comprehensive concepts. This is also referred to as second order or second-level coding.

- *Theoretical sampling* or *purposeful sampling* aims to find data from a wide range of sources to increase understanding of the topic of interest. It assumes that information obtained from one source might raise novel issues leading the researcher to look for new types of data.

- *Theoretical Saturation* is the mechanism used to determine the completion of the theoretical sampling process. It occurs when obtaining additional data does not appear to be adding any new insight to the topic of interest. In secondary studies, the concepts of theoretical sampling and theoretical saturation are contrary to the concept of a search and selection process pre-defined in a study protocol. However, both approaches can be integrated if theoretical sampling and theoretical saturation are used as the final selection process applied to a set of related studies found by a pre-defined search process.

- *Continuous comparison* involves always comparing data from one situation with data found in another. In primary studies, this comparison is at the data level. In secondary studies, comparisons are based on the interpreted theories produced by primary studies (see the comments on *substantive theory* and *formal theory* below).

- *Memoing* refers to notes that a researcher makes to him/herself. They may be simple comments that one data item seems to resemble an item found in a previous text, or *analytic memos* that record initial ideas about higher level codes or themes.

- *Substantive theory* refers to the outcome of grounded theory. It is a theory that is derived from the data and is bounded by the context in which the data was obtained. It may not be generalisable to other situations.

- *Formal theory* is more generalised than substantive theory. Formal theories are sometimes referred to as *mid-level theories*. The originators of grounded theory suggested substantive theories produced in different studies could be synthesised into more general formal theories (Glaser & Strauss 1967). Kearney (1998) discusses the use of grounded theory to produce formal theories in the context of qualitative meta-synthesis.

A more detailed discussion of the philosophical basis of various qualitative meta-synthesis methods can be found in Barnett-Page & Thomas (2009).

10.3   Using qualitative synthesis methods in software engineering systematic reviews

If we attempt a qualitative meta-synthesis of interpretive qualitative studies, our qualitative meta-synthesis will involve interpreting the interpretations of the primary study authors. This, and the issue that synthesis may remove the contextual details necessary to fully understand qualitative findings, has led some qualitative researchers to suggest that the goal of qualitative meta-synthesis is inherently flawed.

However, many well-respected qualitative researchers believe that qualitative synthesis is essential to inform practice, see for example the discussion in Sandelowski, Docherty & Emden (1997). Nonetheless, we accept the warning of experienced qualitative researchers that undertaking qualitative meta-synthesis is difficult even for experienced researchers, let alone novices (Thorne, Jensen, Kearney, Noblit & Sandelowski 2004). However, we note pragmatically:

- Organisations such as the Cochrane Collaboration and the University of York Centre for Reviews and Dissemination both recommend incorporating qualitative synthesis with quantitative reviews in their systematic review handbooks, see Noyes & Lewin (2011) and CRD (2009). These reports make it clear that, in the context of health care, qualitative meta-synthesis can and should contribute to quantitative reviews of intervention effectiveness by helping to specify important research questions (that is, ones that matter to patients), providing evidence that explain variations in outcomes (for example, detailed investigations of variations in interventions, participants, and settings), and supplying complementary evidence related to aspects other than effectiveness (such as acceptability to patients). We believe it should be particularly useful in a domain such as software engineering, where relatively few primary studies are suitable for quantitative meta-analysis.

- In our experience, qualitative studies in software engineering report participants' viewpoints with relatively little interpretation being performed by the researchers. Such studies can be aggregated using the metasummary method (Sandelowski & Barroso 2003) and thematic synthesis (Cruzes & Dybå 2011a) which make less stringent demands on the expertise of the analysts. These techniques are discussed in more detail in Section 10.4.

10.4   Description of qualitative synthesis methods

This section briefly describes the qualitative synthesis methods we judge to be most relevant to software engineering researchers. These are methods that:

- Are currently being used by software engineering researchers.

- Or, are suitable for synthesising findings from software engineering primary studies.

- Or, are suitable for use by most researchers including relative novices. Note that we do not recommend any complete novice attempting qualitative meta-synthesis without having an expert mentor or supervisor.

Figure 10.1 shows the different qualitative methods discussed in this chapter identifying which are interpretive and which are aggregative. The double-lined boxes show the methods that are discussed in detail in this section. We have identified thematic analysis as both an aggregative and an interpretive method. This depends whether the synthesis stops at second-level coding or produces a higher-level synthesis.

10.4.1   Meta-ethnography

Importance to Software Engineers: This form of synthesis is well-suited to primary studies based on ethnological research, which are likely to occur when researchers study team behaviour over an extended time period. We have found three examples of software engineering systematic reviews have used meta-ethnography: Dybå & Dingsøyr (2008a), Gu & Lago (2009), and Da Silva et al. (2013).

Definition: Meta-ethnography is a method for synthesising ethnographic studies, which Noblit & Hare (1988) define to be "long-term, intensive studies involving observation, interviewing, and document review".

Process: Noblit & Hare (1988) define a seven stage process involving:

1. Getting started, that is, defining what is of interest.

2. Deciding what studies are relevant to the topic of interest.

3. Reading the studies. This means detailed reading and re-reading of the relevant primary studies

4. Determining how the studies are related. This involves listing the key *metaphors*, which may be phrases, ideas and/or concepts, in each study. Then looking at how they relate to one another.

5. Translating the studies into one another. Noblit and Hare describe this as comparing metaphors and concepts in one primary study with those in another. They emphasize that translation maintains the central metaphors and/or concepts in each primary study "in their relation to other key metaphors or concepts" in the same study.

6. Synthesizing the translations. Translations may result in agreement among studies, contradictions among studies, or may form parts of a coherent argument.

7. Expressing the synthesis which means reporting the results of the synthesis to interested parties.

Example: Da Silva et al. (2013) present a detailed report of their use of meta-ethnography to analyse four primary studies that investigated personality and team working. In Step 1, they defined their research question as:

> *How does individual personality of team members relate with team processes in software development teams?*

In Step 2, they used a previous systematic review and its unpublished extension as the basis to identify five relevant primary studies. They applied an initial screening to check that the primary studies formed a "coherent set". They then applied the quality criteria used by Dybå & Dingsøyr (2008a) and excluded one low quality study. In Step 3, all team members read the papers. They note that the papers were also read and reread during subsequent phases. During this phase they also extracted:

- *Contextual* data about each primary study. They suggest such information should be defined *a priori* and extraction should be performed by at least two researchers, and disagreements should be identified and addressed. They reported their contextual information in a cross-case matrix with rows identifying the concepts (specifically: Objective, Sample, Research methods, Design, Data collection, Setting, Country) and columns identifying each of the four primary studies. In most cases the cells included appropriate quotes from the primary studies.

- *Relevant concepts* associated with the research question identified in each study. The information was presented in a cross-case matrix with columns identifying the studies and rows identifying the concepts (specifically: Task Characteristics, Personality, Conflict, Cohesions, Team Composition, Performance, Satisfaction, Software Quality).

In Step 3, they considered relationships between the different studies. They first considered which of the six concepts were addressed by at least two primary studies (to make synthesis possible). They then investigated the definition of the six relevant concepts and extracted the operational definitions used in each primary study to check whether the terms were used consistently. Finally, they investigated the relationships between the concepts. They considered pairs of concepts and sought findings from the primary studies that discussed the interaction between a pair of concepts. They reported, for each primary study, the interaction between each pair of concepts reported in the primary study with a specific textual quote if available. This was the main input to Step 5.

In Step 5, they translated the concepts and relations from one study to another. Specifically, they compared each pair of concepts across all studies to produce their first-order synthesis as input to Step 6. In Step 6 they produced a second order synthesis which aimed to produce a synthesis that was more than the sum of its part. This involved creating a diagram that summarized the synthesis and narrative that described the "central story" (like grounded theory). Step 7 was realised by their journal paper.

They comment that in their view:

- Meta-ethnography is not straightforward to use. It requires experience with the methodology and "the philosophical stances that form the cornerstones of interpretative research".

- It is not practical to synthesize too many studies since "it would be easy to forget the meanings of previously synthesized studies as the synthesis proceeds".

They also note that, although two of their four primary studies were ethnographical ones, two were quasi-experiments, so they were able to use meta-ethnography in a mixed methods setting.

10.4.2   Narrative synthesis

Importance for Software Engineers: Cruzes & Dybå (2011*b*) identified narrative synthesis as the most frequently used qualitative synthesis method by software engineering researchers.

Definition: Narrative synthesis reports the results of a systematic review in terms of text and words. Popay et al. (2006) refer to it as "a form of story telling". They point out that any qualitative meta-synthesis involves some narrative synthesis even when more specialised synthesis methods are also used.

Process: Popay et al. propose a narrative synthesis methodology that is targeted at systematic reviews that are concerned either with the effectiveness of some intervention or with factors that influence the implementation of interventions.[1] Their approach involves four main elements:

1. Developing a theory of how, why and for whom the intervention works. This activity is usually done at an early stage in the review and is intended to help formulate review questions and identify the appropriate primary studies. The model is also intended to help both interpreting the review findings and also assessing the generality of the findings.

2. Developing a *preliminary* synthesis of the findings of the primary studies. In the case of effectiveness studies, this involves assessing the direction and size of effects. It may also involve identifying the results of any quality appraisal of the primary studies. For implementation reviews, this is aimed at identifying facilitators and barriers to adoption.

3. Exploring relationships in the data. This aspect of synthesis goes beyond the preliminary synthesis to explore the relationships among studies, both between the characteristics of individual studies and their findings, and between the findings of different studies.

4. Assessing the robustness of the synthesis. Robustness refers to the quality and quantity of the primary studies, the information reported in the primary studies, and the methods used in the synthesis.

The basic process model described above seems appropriate for software

---

[1] Their report is available on request from j.popay@lancaster.ac.uk.

engineering reviews. In particular, the idea of starting by constructing a model of the innovation is particularly interesting. In our view, mapping studies would be of much more value if they aimed to produce a model of the intervention they discuss, organising the literature to illuminate various aspects of the model.

Popay et al. propose a mix-and-match approach to undertaking the various process steps, some based on general approaches such as "grouping and clustering studies" and "tabulation", others based on ideas from a number of different methodologies. For example, they propose "transforming the data into a common rubric" as a technique for developing a primary synthesis, which in their example involved constructing effect sizes that could equally have been used for quantitative meta-analysis. They also recommend reciprocal and refutational translation based on Noblit & Hare (1988) as a technique for exploring relationships among the data.

10.4.3   Qualitative cross-case analysis

Importance for Software Engineers: Cruzes & Dybå (2011b) classify qualitative cross-case analysis as the qualitative analysis methods proposed by Miles et al. (2014). Although many of their analysis methods are aimed at individual primary studies, rather than synthesizing across multiple qualitative studies, the analysis methods documented by Miles et al. can be used for cross-case reporting, analysis, and synthesis. The methods are based on graphical and tabular displays of textual information. The displays are described in great detail and provide an operational description for the tables many researchers use in practice. For example, the table, that Cruzes & Dybå (2011b) used to compare the synthesis methods claimed by secondary study authors with the synthesis methods they actually used, could be described as a *two-variable cross-case matrix*.

Definition: Miles et al. (2014) define a variety of tables and graphics to summarise data and report findings from qualitative studies, many of which apply to cross-case analysis, and so can be used for qualitative meta-synthesis.

Process: Miles et al. (2014) propose an analysis method based on four elements:

1. *Data collection* which in this case means finding and reading relevant primary studies.

2. *Data condensation* which is the process of "selecting, focussing, simplifying, abstracting and transforming data". They consider data condensation part of the analysis process since it involves coding and summarising the data.

3. *Data display* which is an "organized, compressed assembly of information that allows conclusion drawing and action". Like data condensation, data display is part of analysis since it involves organising the rows and

columns of matrices in order to reveal patterns in the data, or drawing diagrams that show the relationships among named entities.

4. *Drawing and verifying conclusions.* Drawing conclusions involves identifying patterns, explanations, cause-event relationships and propositions. It starts as soon as data collection begins. Verification means testing conclusions with respect to 'their plausibility, their sturdiness, their confirmability— that is, their validity'.

The individual elements in the model are *flows of activity* and are not meant to be sequential.

The various displays they describe form the basis for organising the data, analysing the data and presenting the final results. Many of the displays are based on matrices which they define to be the intersection of two lists set up as rows and columns and as a 'tabular format that collects and arranges data for easy viewing in one place.' They define *Meta-Matrices* to be master charts for assembling descriptive data from different cases (which would correspond to different primary studies in the context of qualitative meta-synthesis).

They describe a great many different types of matrices and meta-matrices, including:

- Partially-ordered meta-matrices that stack up data from different cases in one table that can be reformatted and re-sorted to look for cross-case trends.

- Predictor-outcome matrices that identify the main variables believed to affect the observed outcome. Such matrices are qualitative versions of the effect size versus moderator tables that might be produced during a quantitative meta-analysis.

Miles et al. also identify numerous methods for graphically displaying qualitative data and findings. These involve named entities often within boxes (or circles) linked by lines indicating the direction of a relationship among entities, such as the order of events in time, or the influence of one entity on others. These are particularly useful, since software engineering researchers are often quite familiar with such graphics from process modelling and software design methods. One less common style of graphic that might be of relevance to software engineering researchers interested in categorizing objects such as faults, code changes, process types is a *folk taxonomy*. Miles et al. describe nine types of semantic relationships that can be used (for example, inclusion— where X is a kind of Y, spatial—where X is a place in Y, cause-effect—where X is a cause of Y) and provide an example of how a taxonomy can be constructed.

Example: As part of a comparison of thematic synthesis, narrative synthesis and cross-case analysis(Cruzes et al. 2014) report an example of synthesizing two primary studies related to trust in outsourcing. The overall goal of the synthesis was to:

"Understand factors of trust in outsourcing relationships."

Runeson and Höst performed the cross-case analysis. They point out that the major part of data reduction was already conducted by the primary studies. Furthermore, they were only synthesizing two relatively homogeneous and condensed papers, so they "tagged data directly in printouts of the papers". They extracted data of two types:

1. Characteristics of the cases studies (specifically, goal, target population and culture, number of companies and interviews, maturity of companies, methodological framework, data collection, data analysis, and the definition of trust).

2. Factors and subfactors reported as being associated with trust together with the frequency with which they were mentioned.

Moving to the data display step, this information was initially displayed in two separate unordered cross-case data tables.

For the trust related information, further data reduction was performed to analyse the semantics of the identified factors. Runeson and Höst identified synonyms and homonyms based on the definitions used in each primary study. Based on those definitions, they rearranged the factors table into an ordered meta-matrix showing the unique and common factors identified in each study for establishing and maintaining trust, ordered by the frequency with which factors were mentioned (with a caveat that this is a doubtful practice, if wrongly interpreted).

Data synthesis involved identifying the relations between factors reported in each of the primary studies and expressing them in a graph showing the primary study that identified the relation, and whether it was related to establishing or maintaining trust.

Conclusions and verification involved preparing condensed summaries of the views found in paper to highlight the main results. They comment that they found no contradictions between the studies, although they put different emphasis on the factors.

10.4.4 Thematic analysis

Importance for Software Engineers: After narrative synthesis, thematic analysis is the next most frequently used method of qualitative synthesis adopted by software engineering researchers. It fits well with analysing software engineering studies that are aimed at assessing the benefits, risks, motivators and barriers to adopting new software engineering methods.

Definition: Thematic analysis involves identification and coding of the major or recurrent themes in the primary studies and summarising the results under these thematic headings.

Process: Cruzes & Dybå (2011a) define a five-stage process for thematic analysis involving:

1. Reading all the text related to all the primary studies.

2. Identifying specific segments of text relevant to the research questions or topics that seem common to several studies.

3. Labelling and coding the segments of text.

4. Analysing the codes to reduce overlaps and define themes. Some themes are likely to be defined in advance as a result of the research questions, while others may arise as a result of reading the primary studies.

5. Analysing themes to create higher-order themes or models of the phenomenon being studied. The graphical displays discussed by Miles et al. (2014) can be used to represent such models.

Cruzes & Dybå provide a detailed explanation of the process including examples taken from thematic syntheses produced by software engineering researchers and a checklist identifying good practice for each step.

Examples: Staples & Niazi (2008) provide a reasonably detailed description of their thematic analysis methodology. Their systematic review investigated reasons individuals gave for adopting CMM.

With respect to reading the papers (Step 1) only one of the researchers read all the papers. The same researcher identified quotes (that is, text from each study) related to adopting CMM in each study (Step 2). Both researchers then reviewed every quote independently and identified a list of higher level categories that described a unique reason for adoption. The reason comprised a short name and description (Step 3). They note that agreement was initially poor, but they were able to come to agreement via joint discussion and "in some cases a third researcher". Next (Step 4), they reviewed the reasons and grouped them into five higher-level categories Customers, People, Performance, Process, and Product.

Subsequent analysis was based on analysing the frequency with which reasons were mentioned in the identified studies. Thus, they did not undertake Step 5.

Cruzes et al. (2014) present an example of thematic synthesis to synthesis two papers investigating trust in outsourcing. After initially reading the papers and copying textual extracts into the *NVivo* system (Step 1), they used the *NVivo* tool to help both to identify segments of text containing references to factors related to trust (Step 2) and to label (that is, code) the text segments (Step 3). They reduced overlap between codes and identified seven themes that grouped codes together (Step 4). They, finally created a higher-level model (Step 5) with three higher order themes. The higher-level model was presented as concept maps showing the relationships between higher order themes, second-level themes and the original codes.

10.4.5   Meta-summary

Importance for Software Engineers: Although Cruzes & Dybå (2011*a*) did not find any software engineering systematic review that used this method, it has properties that make it of relevance to software engineering problems. In particular:

- It is an aggregative method that may be easier for inexperienced researchers to understand than an interpretive method.

- It can be used to aggregate data from some types of qualitative and quantitative studies in the same meta-synthesis.

- It is appropriate for integrating findings from studies investigating barriers, motivators, risks and other factors associated with implementing a process innovation. In the context of software engineering research, there have been a large number of primary studies reporting the various problems found in globally distributed projects, and many secondary studies that have attempted to integrate the results of the primary studies (Verner et al. 2014). In our opinion, using this approach would have made the aggregation of primary studies much easier for the analysts to perform and for the readers to understand.

Definition: Metasummary is a quantitatively oriented aggregation method capable of integrating findings from *topical* surveys and *thematic* surveys (Sandelowski, Barroso & Voils 2007). *Topical surveys* are based on opinion-based questionnaires circulated to a relatively large number of participants. Analysis of topical survey data involves identifying the set of topics mentioned by the participants and counting how many participants mentioned each specific topic. This is usually done using content analysis and is essentially quantitative. *Thematic surveys* are typically based on researchers personally interviewing a relatively small number of participants. Analysis of thematic survey data involves looking for *latent patterns* in the interview data via first-order and second-order coding. Thematic analysis is more interpretative than content analysis but if it stops at identifying first-order codes, its findings still remain fairly closely related to the original data.

Furthermore, there is usually a disconnect between the methods that researchers claim to use and those they actually use. Sandelowski et al. (2007) suggest that the differences among methods are "typically honored more in the breach than in the observance". They point out that although qualitative surveys are meant to be "purposeful" and quantitative surveys are meant to be randomised, in the cases they investigated, most studies were actually convenience samples. Our experience with software engineering studies is consistent with their observations. Similarity between the findings and the actual methodology allows the metasummary method to aggregate results from both types of study.

Process: Metasummary is based on a five-step process:

1. Extract the findings from each study. Sandelowski et al. point out that findings in qualitative reports may be presented in other parts of the report rather than just in a separate results section. It is therefore necessary to separate relevant findings from other issues such as:

- Presentations of data, such as quotations or incidents.
- Reference to findings of other studies.
- Descriptions of analytic procedures, such as coding schemes.
- Discussion of the importance of findings.

2. Group topically similar findings together looking for equivalent findings.

3. Summarise and organise findings. Findings should be summarised using concise but comprehensive descriptions. They should be organised to show topical similarity (specifically, topics addressed by several studies) and thematic diversity (for example, favouring adherence or favouring non-adherence to a regime or process) and referenced to each primary study that mentioned the finding.

4. Calculate "effect sizes". Effect sizes are based on the number of *primary studies* that report a specific finding and *not* the number of participants mentioning the finding. This is consistent with the view that prevalence does not equate to importance. The *frequency effect size* for a specific finding is calculated as the proportion of *independent* studies that report specific finding compared with the total number of independent studies, that is:

$$FindingEffectSize = \frac{NumStudiesMentioningSpecificFinding}{TotalNumStudies}$$

The *intensity effect size* identifies which studies contributed most to findings. One intensity effect size metric is the proportion of findings with an effect size > 25% found in each study compared with the total number of findings with effect sizes > 25%, that is:

$$StudyIntensityA = \frac{NumStudyLargeEffectSizeFindings}{TotalNumLargeEffectSizeFinding}$$

where NumStudyLargeEffectSizeFindings is the number of findings in a particular study that had an effect size> 25%. A second intensity effect size metric is the proportion of findings found in a study compared with the total number of findings, that is:

$$StudyIntensityB = \frac{NumStudyFindings}{TotalNumFindings}$$

5. Report results. Findings can be displayed in summary matrices. An effects matrix would display each of the findings of each major type that is, favourable and unfavourable, indicating the effect size and the specific studies reporting the finding. A study influence matrix would identify the intensity effect sizes for each study, perhaps incorporating information about the nature of the study, for example, whether the study was

quantitative or qualitative, and summary information about participants such as nationality. An explanatory narrative is needed to describe the results and should discuss the impact of individual studies. In particular, studies that contribute little to the results, studies that contribute a great deal to the results, and studies that contribute many unique findings should be discussed to explain their relative contribution.

10.4.6 Vote counting

Importance for Software Engineers: Vote counting can be used in the context of quantitative systematic reviews when the variation among primary studies is too great for formal meta-analysis to be possible. A number of software engineering systemic reviews have reported results using variants of vote counting, see for example, Turner, Kitchenham, Brereton, Charters & Budgen (2010) and Kitchenham et al. (2007).

Definition: At its simplest, vote counting involves simply counting how many primary studies found a significant effect and how many did not. As discussed previously, simple vote counting has major methodological problems. However, it is more valuable when it is associated with a form of "qualitative" moderator analysis that investigates whether there are contextual or methodological factors that can help to explain differences in the outcomes of the primary studies using meta-matrix displays. We note that there is some difficulty in giving this form of analysis a name. Cruzes & Dybå classified several papers, that we would classify as "Vote counting", as examples of "Comparative Analysis" because they involved an investigation of possible moderating factors. If results are displayed in a tabular format, vote counting combined with moderator analysis is also a form of qualitative cross-case analysis (Miles et al. 2014).

Process: Like meta-analysis, vote counting assumes that a systematic review has identified a set of primary studies that each compare two software engineering interventions and it also requires that values of the outcome of the comparison, such as t-values, effect sizes or p-values, can be obtained from each primary study. Tabular displays are used to present the outcome values for each study which can be sorted or colour-coded according to which intervention was preferred. Popay et al. (2006) suggest a five-point scale to describe the outcome of the primary study:

1. Significantly favours intervention

2. Trends towards intervention

3. No difference

4. Trends towards control

5. Significantly favours control.

Additional moderating factors can be added to the displays to investigate whether there are any that appear to be associated with specific outcomes. Often, it is only possible to provide a narrative discussion of possible moderating factors. However, sometimes it may be possible to perform a more sophisticated synthesis. Cruzes & Dybå (2011*b*) suggest two such possibilities:

1. The *comparative method* (Ragin 1989) which uses Boolean truth tables to assess the combinations of moderators (modelled as boolean variables) that are associated with a successful or unsuccessful outcome of a case (for example, a primary study). The method assumes that there may be different combinations of factors that cause a particular outcome. It is able to cope with situations where some logical combinations of moderators do not exist among the set of cases. However, it appears to require that all important moderator variables are known. The technique is extremely complex but may resonate with researchers from computer science who are used to Boolean algebra and truth tables. The aim is to be able to say that a successful intervention occurs only when certain factors are present and other factors are not, using statements of the form "success occurs if and only if A OR (B AND NOT(C))=TRUE". Such statements imply that an underlying causal relationship is expected, rather than a statistical association exists among factors.

2. The *case survey method* (Yin & Heald 1975) uses standard statistical methods (for example, chi-squared tests, or logistic regression) to associate moderator values with binary or ordinal case outcome variables. The case survey method requires the availability of a large number of cases with the same moderator variables, which limits its applicability. The aim is to assess the frequency with which certain context factors are associated (or not) with a successful intervention and to provide a statistical assessment of whether the frequency is significantly different by chance.

Example: Kitchenham et al. (2007) reported a systematic review that compared the accuracy of cost estimation models built from data collection from a variety of different companies (cross-company models) with the accuracy of cost estimation models built from a specific company (within-company models). They grouped the primary studies into three groups: one for which the within-company models were significantly more accurate than the cross-company models, one for which there was no significant difference between the within-company and cross-company models, and one group of studies that were inconclusive (specifically, did not report any statistical analysis). They also produced a matrix display that identified the values of various study-related factors for each primary study, such as: the number of projects in the within-company dataset and the cross-company dataset, the size metric used, the type of model (linear or non-linear) derived from within and between company data, and the size of projects in each dataset. They also constructed a summary matrix display of the factors that seemed to be associated with within-company models outperforming cross-company models and those that seemed associated with cross-company models performing as well as within-company models identifying which studies contributed to each conclusion.

10.5 General problems with qualitative meta-synthesis

This section discusses two problems that need to be considered in most qualitative meta-syntheses:

1. What to do about primary study quality.

2. How to validate the final meta-synthesis.

10.5.1 Primary study quality assessment

There appears to be no consensus among qualitative meta-synthesists about how to assess the quality of primary studies or, even whether quality should be assessed at all. For example, see Thomas & Harden (2008) and Spencer, Ritchie, Lewis & Dillon (2003). Even researchers who use quality evaluation, on the basis that they wish to avoid drawing conclusions on unreliable data, are unwilling to use quality criteria to exclude studies. For example, see Thomas & Harden (2008) and Atkins, Lewin, Smith, Engel, Fretheim & Volmink (2008).

Empirical evidence casts some doubts on the value of quality assessment checklists for qualitative primary studies. Both Hannes, Lockwood & Pearson (2010) and Dixon-Woods, Sutton, Shaw, Miller, Smith, Young, Bonas, Booth & Jones (2007) compared different quality checklists. Hannes et al. (2010) compared three different structured methods:

1. The Critical Appraisal Skills Programme (CASP) qualitative checklist[2] which is a very widely-used checklist that was the basis of a checklist used by Dybå & Dingsøyr (2008a) for their systematic review of agile methods.

2. A checklist compiled by the Australian Joanna Briggs Institute (2014)[3].

3. The Evaluation Tool for Qualitative Studies (ETQR) which was developed by the Health Care Practice Research and Development Unit from

---

[2](www.casp-uk.net)

[3]www.joannabriggs.org

the University of Salford, in collaboration with the Nuffield Institute and the University of Leeds.

Based on an analytical evaluation, they concluded that CASP was least able to evaluate certain aspects of validity.

Dixon–Woods et al. (2007) undertook a comparison of two structured checklists and a subjective evaluation. They found only slight agreement among the three methods and that the structured methods used which were CASP and a UK Cabinet Office quality framework (Spencer et al. 2003), did not show better agreement than expert judgement. Qualitative analysis indicated that reviewers found it difficult to decide between the potential impact of findings and the quality of the research or reporting practice. They also reported that structured instruments appeared to make reviewers more explicit about the reasons for their judgements.

In a qualitative study of researchers making decisions about the quality of studies for inclusion in a meta-ethnography, Toye, Seers, Allcock, Briggs, Carr, Andrews & Barker (2013) identified two issues of importance to reviewers: firstly, *conceptual clarity*, which relates to how clearly the author articulated an insightful issue, and secondly *interpretive rigour*, which relates to the extent to which the interpretation could be trusted. These two issues are clearly related to the impact of findings and the quality of research practice mentioned by Dixon–Woods et al. (2007).

It is, however, encouraging that both Thomas & Harden (2008) and Atkins et al. (2008) have commented that poor quality studies contributed less to their synthesis than better quality studies. Overall it seems that evaluating quality is mainly useful for sensitivity analysis, where the contribution of the individual studies can be compared with their quality. This is also consistent with the suggestion, in the context of metasummary, that the analysts should discuss the impact of individual studies on the overall results.

10.5.2   Validation of meta-syntheses

There are two aspects to validation of a meta-synthesis. Firstly the systematic reviewers, themselves, should ensure that they have "done a good job" and secondly, readers of the final systematic review report should find it trustworthy and useful.

Systematic reviewers need to reflect on the process they have used and identify any limitations of the process itself, or the way they used the process. Some of these reflections will be reported in the "Limitations" section of the final report, others may lead to additional synthesis activities such re-reading some excluded papers, or obtaining a second opinion on the plausibility of some of the reported findings

Readers of the final report of a qualitative meta-synthesis also need to be able to understand and to trust the findings. Qualitative systematic reviews should have similar properties to reports of qualitative primary studies mentioned by Toye et al. (2013) and Dixon–Woods et al. (2007), such as:

- Clearly reporting of insightful and valuable findings.

- Using a rigorous synthesis method.

For thematic analysis, Cruzes & Dybå (2011a) discuss trustworthiness of qualitative meta-synthesis in general, from the viewpoint of credibility, confirmability, dependability and transferability. They also provide a useful checklist, that researchers can use to assess the validity of their process at each stage in the thematic synthesis including the final stage of assessing the trustworthiness of the synthesis. Their checklist includes four questions about the trustworthiness of a synthesis:

1. 'Have the assumptions about, and the specific approach to, the thematic analysis been clearly explicated?'

2. 'Is there a good fit between what is claimed and what the evidence shows?'

3. 'Are the language and concepts used in the synthesis consistent?'

4. 'Are the research questions answered by the evidence of the thematic synthesis?'

We note that these questions seem applicable to any qualitative meta-synthesis not just thematic synthesis.

Chapter 11

Meta-Analysis

with Lech Madeyski

Wroclaw University of Technology, Poland

This chapter explains the use of statistical methods to analyse data from primary studies that have measured the outcome of two different treatments (also referred to as interventions). These could be the results of randomised experiments, or more general correlation or data mining studies. In the context of software engineering, a treatment would be a method, algorithm, process or technique for performing a software engineering process, activity or task. One of the treatments would be referred to as a *control*, if it corresponded to the current standard software engineering technique. Outcomes are related to effectiveness, often in terms of time (duration), effort, or faults. The aim is to calculate an overall summary measure of the comparative impact of the treatments based on the data from all relevant primary studies together with an assessment of its precision.

Meta-analysis is used frequently in other disciplines but, as yet, it is rarely used in software engineering research. If evidence-based software engineering is to become more useful to software engineers in industry, it is important to undertake more primary studies of software engineering methods and techniques that are suitable for statistical aggregation. Thus, we hope meta-analysis will become more important in the future. We, also, hope that this chapter will make it clear that meta-analysis can be performed by any empirical researcher who has some experience of statistical methods. Although some readers might expect to find the topic here, we do not discuss *vote counting* in this chapter. Vote counting is discussed in Chapter 10.

11.1   Meta-analysis example

TABLE 11.1:   Example Data

$StudyID$	$Mean_T$	$Mean_C$	$Stdev$	$EffectSize$	$Variance$	N_T	N_C
Able	0.31	0.33	0.03	−0.66667	0.0009	30	45
Baker	0.30	0.35	0.21	−0.23810	0.0441	5	5
Carter	0.40	0.41	0.06	−0.16667	0.0036	25	30
Delta	0.45	0.43	0.10	0.0200	0.01	15	15

To introduce the concept of meta-analysis, we will begin with an artificial example. Table 11.1 shows the basic data for four studies. Each study is assumed to be a between-groups study with a treatment group comprising N_T experimental units which had a mean of $Mean_T$ for the study outcome measure and a control group with N_C experimental units which had a mean of $Mean_C$. The pooled within-group variance and its standard deviation (labelled StDev) are shown, together with the *effect size* which has been calculated for each study as:

$$EffectSize = \frac{Mean_T - Mean_C}{Stdev} \tag{11.1}$$

The effect size is a measure of the magnitude of the treatment effect. In this example, the effect size is called the *standardised mean difference*, but there are many other effect sizes that can be used.

The meta-analysis results for this dataset are shown in Figure 11.1 which uses a form of visualisation known as a *forest plot*. Forest plots were mentioned in Chapter 1 in the context of the Cochrane Collaboration logo and are the standard means of reporting the outcome of a meta-analysis. This figure reports the calculated effect size for each study and shows it graphically as a square contained within its 95% confidence interval. The square is inversely proportional to the variance, so the square of study "Able" is largest because it has the smallest variance. The small variance also means that the confidence interval about the effect size is the smallest of the studies (that is, the *precision* of the effect size is the largest among the studies). In contrast, study "Baker" has the largest variance, so it has the smallest precision, the largest confidence interval, and is given the least weight in the meta-analysis.

The overall effect size, shown by the diamond shape at the bottom of the figure, is the weighted mean of the individual study effect sizes. The weights are based on the inverse of the standard deviation, so this gives most weight to study "Able" which has the smallest variance. The centre of the diamond shows the estimate of the overall effect size, while its width shows the 95% confidence interval of the overall effect size.

All meta-analysis calculations are based on identifying an effect size together with its variance for each primary study included in the analysis. The most important point to remember is that although all primary studies in a specific meta-analysis must use the same effect size, the meta-analysis calculations are identical which ever effect size is used.

In this chapter, we will discuss the most common forms of effect size, how to calculate them and their variance, and the methods used to analyse them. A detailed understanding of statistical methods is not necessary to understanding the basic principles, but basic knowledge of statistical analysis and familiarity with the concepts of means, variances, standard deviations, and *t*-tests would be helpful for readers wanting to perform meta-analysis themselves.

Our example data were analysed using the `metafor` package for meta-analysis which can be used with the R statistical language (Viechtbauer 2010). If the data shown in Table 11.1 are copied into an R data table called `metadata` using the `read.table` command which copies data from a text file into R and the `metafor` package has been loaded, then the instructions in Figure 11.2 will recreate the forest plot shown in Figure 11.1. Note R treats anything on a line following the symbol "#" as a comment.

11.2 Effect sizes

In this section, we describe the most commonly used effect sizes. For effect sizes to be useful for meta-analysis, they need to:

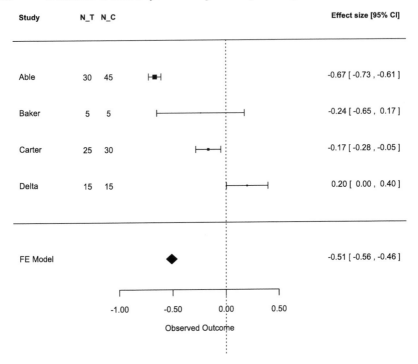

FIGURE 11.1: Forest plot example.

- Have well defined distributions, so the variance of the effect size can be calculated.

- Be capable of being calculated from statistics commonly reported in primary studies, such as *t*-values and sample sizes, without relying on the raw data being available.

11.2.1 Mean difference

The most straightforward effect size for between-studies (parallel) designs is the mean difference, which is simply the difference between the mean outcome variable for entities that used (or had applied) treatment T1 (that is, $Mean_{T1}$) and the mean outcome of the entities associated with treatment T2 (that is, $Mean_{T2}$):

$$EffectSize = Mean_{T1} - Mean_{T2} \qquad (11.2)$$

In this case, the entities observed in T1 are assumed to be independent of the entities observed in T2. In the case of a randomised experiment, a total of N entities will be allocated at random to either T1 or T2, with the

```
library(metafor)
#Copy example data into a data table
metadata=read.table("filename.txt",head=T)

#Do a fixed-effects meta-analysis
mod2=rma(yi=metadata$d, vi=metadata$v, method="FE")

#To do a random---effects analysis change the method parameter
#Use method="DL" for the DerSimonian & Laird method
#Use method="REML" for a restricted maximum likelihood estimator

#View the result
mod2

# To view documentation of the forest function use command help("forest")
#Create a forest plot of the result use

forest(mod2, xlim=c(-2,1.7), ilab=cbind(metadata$n1,
metada$n2), ilab.xpos=c(-1.3,-1.1), cex=.8, efac=1.8,
slab=c("Able","Baker","Carter","Delta"))

#Add extra text to the forest plot to labels extra data
text(c(-1.3,-1.1), 6 , c("N_T", "N_C"),  font=2, cex=.75)
text(-2, 6 , "Study", pos=4, font=2, cex=.75)
text(1.7, 6, "Effect size [95% CI]", pos=2, font=2, cex=.75)
```

FIGURE 11.2: Code snippet for a fixed-effects meta-analysis.

experimenters usually (but not always) ensuring that the number of entities is equal in each treatment group. In the case of software engineering studies, the entities being observed and measured can be software engineers, teams, processes, tasks, algorithms, or software components.

In order to perform a meta-analysis, we need to obtain the *variance* of the effect size. In this case, the variance of the effect size is the variance of the mean difference, which is the pooled within group variance:

$$V_{ES} = \frac{S_1^2}{N_1} + \frac{S_2^2}{N_2} \tag{11.3}$$

Where S_1 is the standard deviation calculated from the response variables from the N_1 participants using T1 and S_2 is the standard deviation calculated from the response variables from the N_2 participants using T2. If we can assume that the true variance of observations from each treatment group is the same, then S_1^2 and S_2^2 can be pooled to give a more reliable estimate of the true variance:

$$V_P = \frac{(N_1 - 1)S_1^2 + (N - 2)S_2^2}{N_1 + N_2 - 2} \tag{11.4}$$

and

$$V_{ES} = Var_P \frac{N_1 + N_2}{N_1 N_2} \qquad (11.5)$$

Equations 11.2 and 11.5 can be adapted to other form of experimental design. For example, suppose primary studies are based on a within subject before-after design i.e. all participants use method 1 to perform a task and then all participants use methods 2 to perform a similar task, there will be a total of N participants and analyses will be based on the average difference between each participant in time period 1 and time period 2. Assuming d_1 is the difference between the output values obtained from participant $d_i = x1_i - x2_i$ using method 1 and method 2, the mean difference is:

$$EffectSize = \frac{\sum d_i}{N} = \hat{d} = Mean_{T1} - Mean_{T2} \qquad (11.6)$$

In this case the variance of the effect size is based on the variance of the d_i values as follows:

$$V_d i = \frac{\sum \left(d_i - \hat{d} \right)^2}{N - 1} \qquad (11.7)$$

and

$$V_{ES} = \frac{V_d i}{N} \qquad (11.8)$$

The limitation of the mean difference is that it can only be used for meta-analysis if all primary studies used the same well-defined outcome measure. Since this is not always the case, we also need to consider effect sizes based on standardised mean differences.

11.2.2   Standardised mean difference

Most software engineering researchers have used the standardised mean difference as the basis for their meta-analysis when their output variable is numerical.

11.2.2.1   Standardised mean difference effect size

The standardised mean difference effect size for independent groups is the difference between the mean outcome variable for entities that used (or had applied) treatment T1 (that is, $Mean_{T1}$) and the mean outcome of the entities associated with treatment T2 (that is, $Mean_{T2}$) divided by the standard deviation of the outcome measure (S) which is referred to as the *standardizer*, that is:

$$EffectSize_{Standardised} = d = \frac{(Mean_{T1} - Mean_{T2}}{S} \qquad (11.9)$$

where if the variances of T1 and T2 estimate the same true variance, and $S = \sqrt{V_P} = S_P$ which is the square root of the pooled within-group variance (see Equation 11.4). This effect size is often referred to as Cohen's d, although some researches refer to this as Hedge's g.[1] If there is a control treatment, and it is believed that the treatment may change the variance not just the mean effect, it is preferable to use the variance of the control group, rather than the pooled within group variance. If the standardiser is based on the control group variance the resulting effect size is often referred to as Glass's δ.

Readers will notice that Equation 11.9 is quite similar to the equation used to calculate the value of t when preparing to do a t-test. There is, however, one extremely important difference, t is calculated using the *standard error* of the mean *not* the standard deviation:

$$t = \frac{(Mean_{T1} - Mean_{T2})}{S(\frac{1}{N_1} + \frac{1}{N_2})} \qquad (11.10)$$

which means that although the value of t can be increased by increasing the number of observations in each treatment group which decreases the standard error, effect size is not so affected. This makes sense because we would not expect the magnitude of the effect to be influenced by sample size (that is, if using a new method improves defect detection by 50%, the effect should not be changed by changing the sample size). The only difference would be that we would expect our estimates of the means and standard deviations to be more accurate with larger sample sizes, so our estimate of effect size would be more trustworthy (that is, would have greater precision). Many researchers point out that effect size is of more practical importance than the value of the t statistic, and recommend that it should be reported in preference to the t value or the p-value associated with the t-test. See the *Publication manual of the American Psychological Association* (2001), Cumming (2012) and the study by Kampenes et al. (2007).

However, we also note that:

$$d = t\left(\frac{1}{N_1} + \frac{1}{N_2}\right) \qquad (11.11)$$

This means that meta-analysts can still calculate the standardised mean difference, even if primary studies report only the t-values and the number of observations in each treatment group.

Another important point is that effect sizes based on the standardised mean difference are based on comparing *two* treatments. Thus, meta-analysis of primary studies that compare many different treatments require effect sizes to be calculated for each pair of treatments (or rather, for each pair of treatments the meta-analysts are interested in). This, again, makes sense since we would not (in general) be interested in effect sizes that depended on the outcomes of a third and irrelevant treatment. For example, if we have three

---

[1]There is no universally accepted terminology for effect sizes, so it is important always to specify the formula you have used when presenting a meta-analysis (Cumming 2012).

treatments, we might analyse the data using an analysis of variance technique and use an F-test to determine whether there are significant differences between the mean values. However, unless we inspect the means of the individual treatment groups, we would not know what the actual differences were. $F - test$ values calculated when there are more than two treatment groups cannot be directly converted into effect sizes because they are based on the effect of more than two treatments.

Other problems with designs that involve many different treatments is that investigating many different pairs of treatment effects:

- Risks finding spurious *significant* effect sizes by chance.

- Risks introducing dependencies between effect sizes if there is one control and multiple treatments.

Generally from a meta-analysis viewpoint, the simpler the statistical design the better.

11.2.2.2 Standardised difference effect size variance

The main complication with the standardised mean difference is that its variance is more difficult to calculate. According to (Borenstein, Hedges, Higgins & Rothstein 2009), a reasonable approximation to the variance is given by:

$$V_d \approx \frac{N_1 + N_2}{N_1 N_2} + \frac{d^2}{2(N_1 + N_2)} \tag{11.12}$$

This approximation relies heavily on the sample size. If the combined sample size in the two groups is less than 20, the approximation is likely to be poor (see Hedges & Olkin (1985), Morris (2000)).

Bearing in mind software engineering sample sizes are often small, it is preferable to use a different formula. The variance of d is based on the relationship between d and t (Morris (2000), Cumming (2012)). t is distributed according to the *non-central t distribution* and Equation 11.11 confirms that d is a multiple A of t, where

$$A = \sqrt{\tilde{n}} = \sqrt{\frac{N_1 + N - 2}{N_1 N - 2}} \tag{11.13}$$

Thus, the variance of d is

$$V_d = \tilde{n} V_t \tag{11.14}$$

and

$$V_t = \frac{df}{df - 2} \left(1 + t^2\right) - \frac{t^2}{[c(df)]^2} \tag{11.15}$$

where

$$c\left(m\right) \approx 1 - \frac{3}{4m - 1} \tag{11.16}$$

Equation 11.14is more accurate for small sample sizes, although it must be noted that the estimate of t is not likely to be very precise for small sample sizes.

Equations 11.11 and 11.15 imply that the variance of the effect size is a function of the effect size itself. The implication of this is that unless $d \approx 0$ the confidence limits on the standardised mean difference are not symmetric about the effect size (Cumming 2012).

The effect size and variance of standardised mean differences differ for different statistical designs Morris & DeShon (2002). Morris & DeShon discusses how to convert effect sizes from different designs into comparable effects sizes, and how effect sizes from different designs might be included in the same meta-analysis. However, such processes are rather complex and beyond the scope of this chapter.

11.2.2.3 Adjustment for small sample sizes

The standardised effect size d is known to be biased for small samples. It slightly overestimates the effect size. This can be corrected by applying the correction factor $c(m)$ (see 11.16)where m is the number of degrees of freedom used to estimate the standard deviation. The adjustment factor tends to 1 as the number of degrees of freedom increases, but for consistency most meta-analysis researchers recommend *always* applying the adjustment factor. Some meta-analysts refer to adjusted effect size as Hedge's g.

To use the small sample size bias adjustment, calculate:

$$d_{unb} = J \times d \tag{11.17}$$

Use d_{unb} as the best estimate of effect size. The variance of d_{unb} is calculated as:

$$V_{unb} = c(df)^2 \times V_d \tag{11.18}$$

11.2.3 The correlation coefficient effect size

Rosenthal & DiMatteo (2001) advocate the use of effect sizes based on the correlation coefficient r. The correlation based effect size for an experiment comparing two different treatments is called the *point-biserial correlation* and is calculated by associating with each outcome value a dummy variable taking the value 0 if the outcome was associated with one treatment (or the control) and 1 if the outcome value came from the alternative treatment. r has a number of advantages as an effect size:

- r is a statistic that is relatively well-understood.

- r is easy to interpret. There are heuristics to indicate the importance of an r value: $|r|$ is assessed as small if $|r| < 0.19$, medium if $0.19 < |r| < 0.46$, and large if $0.46 <= |r|$ (Kampenes et al. 2007).

- r can be used not only in the same situations as d or g, that is, in describing an association between a dichotomous variable and a continuous variable, but also in situations where it makes less sense to use d or g, for example, in describing an association between two continuous variables, or in situations where we want to generalize to more than two treatment groups.

- It is possible to calculate an estimate of r referred to as $r_{equivalent}$ from just the $p-value$ and the number of observations (Rosenthal & Rubin 2003).

r is not distributed normally. Therefore, it is customary to apply a variance stabilizing and normalising transformation:

$$Z_r = \frac{1}{2}ln\left(\frac{1+r}{1-r}\right) \tag{11.19}$$

The variance of Z_r is approximately $\frac{1}{N-3}$ where N is the number of observations from which r was calculated.

Using the transformed data, the meta-analysis calculations can be performed and the inverse transformation used to transform the results back to the correlation scale:

$$r = log\left(\frac{e^{2Z_r}-1}{e^{2Z_r}+1}\right) \tag{11.20}$$

11.2.4   Proportions and counts

If the outcome measures are proportions, for example, the proportion of software modules with one or more faults, or the proportion of failing projects in an organisation, there are several appropriate effect sizes, including the log risk ratio and the log odds ratio.

If modules were constructed using a treatment technique and a control technique and later tested for defects, data from such an experiment might be shown in a 2×2 table like Table 11.2, where failure represents a count of the number of modules exhibiting a fault in each treatment condition.

TABLE 11.2:   Binary Data

	Failure	*Success*	N
Treated	A	B	N_1
Control	C	D	N_2

Log risk ratio effect size

A risk ratio is the ratio of two risks (for example, the risk of a module having one or more faults). Using the labels shown in Table 11.2, the risk ratio is

$$RiskRatio = \frac{A/N_1}{C/N_2} \tag{11.21}$$

However, since the distribution of the risk ratio is extremely skewed, the effect size is based on the log of the risk ratio:

$$LnRiskRatio = ln(RiskRatio) \tag{11.22}$$

The approximate variance of the log risk ratio is:

$$V_{LnRiskRatio} = \frac{1}{A} - \frac{1}{N_1} + \frac{1}{C} - \frac{1}{N_2} \tag{11.23}$$

Taking the logarithms means that the effect size is the difference between the logarithms of the two risks. Meta-analysis calculations are based on the log risk ratio, and are then transformed back to raw data scale to present the results.

Log odds ratio effect size

Odds are defined as the proportion of entities that have a characteristic divided by the proportion that do not have the characteristic. Thus, if p is the proportion of modules that exhibited one or more faults, the odds of a module having a fault are:

$$Odds = \frac{p}{(1-p)} \tag{11.24}$$

If modules were constructed using a treatment technique and a control technique, data from such an experiment might be shown in a 2×2 table like Table 11.2. The respective proportions of modules with faults is $p_1 = \frac{A}{N_1}$ and $p_2 = \frac{C}{N_2}$, so the odds ratio is:

$$OddsRatio = \frac{\frac{p_1}{(1-p_1)}}{\frac{p_2}{(1-p_2)}} \tag{11.25}$$

Like the risk ratio, the odds ratio is extremely skewed, so the effect size is based on the natural logarithm of the odds ratio. Also, the effect of taking logs of the odds ratio is to construct an effect size equal to the difference between natural logarithm of the odds.

The approximate variance of the log odds ratio is:

$$Var_{LnOddsRatio} = \frac{1}{A} + \frac{1}{B} + \frac{1}{C} + \frac{1}{D} \tag{11.26}$$

11.3   Conversion between different effect sizes

It is useful to be able to convert between different types of effect sizes, if we want to include primary studies using different types of experimental design in the same meta-analysis.

11.3.1   Conversions between d and r

Conversion from a standardised mean difference d to the equivalent point bi-serial correlation coefficient r uses the formula:

$$r = \frac{d}{\sqrt{(d^2 + a)}} \tag{11.27}$$

where a is a correction for cases when $N_1 \neq N_2$,

$$a = \frac{(N_1 + N_2)^2}{N_1 N_2} \tag{11.28}$$

If only $N = N_1 + N_2$ is known, the only option is to assume $N_1 = N_2 = \frac{N}{2}$, in which case $a = 4$. The equivalent variance is:

$$V_r = \frac{a^2 V_d}{(d^2 + a)^3} \tag{11.29}$$

This conversion assumes that the treatment group identifier can be treated as a binary variable. Thus, any observation from the control group is treated as the binary bivariate variable $(0, O_{Ci})$ and any observation from the treatment group is treated as the binary bivariate variable $(1, O_{Tj})$.

Conversion from r to d is based on the equation:

$$d = \frac{2r}{\sqrt{1 - r^2}} \tag{11.30}$$

with variance:

$$V_d = \frac{4V_r}{(1 - r^2)^3} \tag{11.31}$$

This conversion assumes that r has a bivariate distribution and that one of the variables can be converted into a binary variable.

11.3.2   Conversion between log odds and d

Conversion from the log odds ratio to the standardised mean difference is based on the equation:

$$d = LnOddsRatio \times \frac{3}{\pi} \tag{11.32}$$

with variance

$$V_d = V_{LnOddsRatio} \times \frac{3}{\pi^2} \qquad (11.33)$$

The inverse conversion is:

$$LnOddsRatio = d \times \frac{\pi}{\sqrt{3}} \qquad (11.34)$$

with variance

$$V_{LnOddsRatio} = V_d \times \frac{\pi^2}{3} \qquad (11.35)$$

11.4 Meta-analysis methods

In this section we describe the procedures used to analyse the effect sizes. Firstly we introduce the models underlying meta-analysis, then we explain how to calculate the mean and standard error of an effect size.

11.4.1 Meta-analysis models

There are three basic analysis models: the fixed-effects model, the random-effects model and the mixed-effects model. All these models are supported by the **metafor** package in the **R** language (Viechtbauer 2010) which we have used in our meta-analysis examples.

The fixed-effects model is based on having $i = 1, ..., k$ independent effect sizes estimates, each corresponding to a true effect size, such that:

$$y_i = \mu + e_i \qquad (11.36)$$

where y_i is the observed effect in the i-th study, μ is the corresponding true effect, e_i is the sampling error which is assumed to be distributed normally with mean 0 and variance σ^2, so $e_i \sim N(0, \sigma^2)$. Thus, we assume that $y_i \sim N(\mu, \sigma^2)$.

However, meta-analyses are usually based on studies that used neither identical methods nor samples with exactly the same characteristics. Such differences can introduce additional variability, that is *heterogeneity*, among the true effects. The random-effects model treats the variability as being completely random, giving the model:

$$y_i = \theta_i + e_i \qquad (11.37)$$

and

$$\theta_i = \mu + u_i \qquad (11.38)$$

where u_i is the additional sampling variability caused by differences among studies and is assumed to be $\sim N(0, \tau^2)$. The extra parameter τ^2 represents the between-study variability around the underlying global effect μ. Using the random-effects model we can *estimate* μ which is the average true effect and τ^2 which is the total heterogeneity among the true effects. If $\tau^2 = 0$, there is no heterogeneity, so $\theta_1 = \ldots = \theta_k = \mu$ is the true effect.

Alternatively, if we find that the degree of heterogeneity is significant and we believe it is due to specific study level factors (for example, student subjects in some studies and practitioner subjects in other studies), we can include these factors as moderators in our analysis. This leads to a mixed-effects model of the form:

$$y_i = \beta_0 + \beta_1 x_{i1} + \cdots + \beta_p x_{ip} + u_i + e_i \tag{11.39}$$

where x_{ij} is the value of the j-th moderator for the i-th study. Again we assume that $u_i \sim N(0, \tau^2)$. However, in this case, τ^2 measures the residual heterogeneity among the true effects, that is, the variation among the effects that is not accounted for by the moderators.

11.4.2   Meta-analysis calculations

The calculations used for meta-analysis depend on the meta-analysis model used, *not* on the effect size metric used.

Fixed-effects mode

To calculate the average effect size using a fixed-effects approach use:

$$\overline{ES} = \frac{\Sigma_k W_i ES_i}{\Sigma_k W_i} \tag{11.40}$$

where $\overline{ES}$ is the estimate of the average effect size, $W_i = \frac{1}{V_i}$ is the weight assigned to the ith study which is the inverse of the variance V_i, ES_i is the effect size calculated for the ith study and k is the number of primary studies. The variance of $\overline{ES}$ is calculated as:

$$V_{\overline{ES}} = \frac{1}{\Sigma_k W_i} \tag{11.41}$$

The standard error of $\overline{ES}$ is the square root of $V_{\overline{ES}} = SE_{\overline{ES}}$. Then, the 95% upper and lower confidence limits are estimates as:

$$LCL_{\overline{ES}} = \overline{ES} - 1.96 \times SE_{\overline{ES}} \tag{11.42}$$

and

$$UCL_{\overline{ES}} = \overline{ES} + 1.96 \times SE_{\overline{ES}} \tag{11.43}$$

In addition, the test statistic for the null hypothesis that the common true effect μ is zero is:

$$Z = \frac{\overline{ES}}{SE_{\overline{ES}}} \tag{11.44}$$

Figure 11.1 presents a forest plot showing a fixed-effects analysis of the data presented in Table 11.1.

Random-effects model

In order to do a random-effects analysis, it is necessary to estimate the between-studies variance τ^2. The DerSimonian and Laird estimation method is based on calculating:

$$T^2 = \frac{Q - df}{C} \tag{11.45}$$

where T^2 is the estimate of τ^2, $df = k - 1$, k is the number of primary studies and

$$Q = \Sigma_{i=1}^k W_i ES_i^2 - \frac{(\Sigma_{i=1}^k W_i ES_i)^2}{\Sigma_{i=1}^k W_i} \tag{11.46}$$

and

$$C = \Sigma W_i - \frac{\Sigma W_i^2}{\Sigma W_i} \tag{11.47}$$

The data shown in Table 11.3 shows the working needed to calculate T^2 for our example dataset shown in Table 11.1:

$$Q = 506.83 - \frac{(-772.44)^2}{1511.6} = 112.10$$

$$C = 1511.6 - \frac{1322243}{1511.6} = 636.81$$

$$T^2 = 0.1713$$

TABLE 11.3: Calculating T^2

StudyID	ES_i	V_i	W_i	$W_i ES_i$	$W_i ES_i^2$	W_i^2	V_i^*	W_i^*
Able	−0.6667	0.0009	1111.11	−740.74	493.827	1234568	0.17222	5.8065
Baker	−0.2381	0.0441	22.676	−5.399	1.2855	514.189	0.21542	4.6425
Carter	−0.1667	0.0036	277.78	−46.296	77160.5	7.7160	0.17492	5.7168
Delta	0.0200	0.01	100	20	4	10000	0.18132	5.5150
Total			1511.6	−772.44	506.83	1322243		

Once T^2 has been calculated, the variance for each study is adjusted as follows:

$$V_i^* = V_i + T^2 \tag{11.48}$$

Then, the calculation of the weight per study, the mean effect size and the standard error of the mean are calculated in the same way as the fixed-effects parameters but using V_i^*:

$$W_i^* = \frac{1}{V_i^*} \tag{11.49}$$

$$\overline{ES^*} = \frac{\Sigma_k W_i^* ES_i}{\Sigma_k W_i^*} \tag{11.50}$$

$$V_{\overline{ES^*}} = \frac{1}{\Sigma_k W_i^*} \tag{11.51}$$

The `metafor` package will do a random-effects analysis by changing the parameter `method` in the call to `rma` to `method="DL"`.[2] Figure 11.3 shows a forest plot of a random-effects analysis of the artificial data in Table 11.1. R will display the results of the random-effects analysis by typing the name of the 'object' it was equated to by just typing `mod2` if using the code snippet in Figure 11.2, as shown in Figure 11.4.

The effect of using the random-effects model has to been to reduce the estimate of the mean effect size and to increase its standard error. Because, there is substantial heterogeneity among the studies, the value of T^2 is large compared with the within study variances. The last two columns in Table 11.3 show that the values of V_i^* and W_i^* are similar across the four primary studies. This has effectively equalized the weights given to the different studies, which means, compared with the fixed-effects analysis, the random effects analysis gives less weight to studies 'Able' and 'Carter' and more weight to studies 'Baker' and 'Delta'. The overall effect is to increase the variance of the mean effect size and the confidence limits about the mean, as shown by the elongated diamond in the RE model row of Figure 11.4.

11.5   Heterogeneity

It is not sufficient just to calculate a summary effect. It is important to understand the patterns of effect sizes. An intervention that consistently improves a software engineering process by about 15% is very different from one that has an average effect of 15% that sometimes improves the process by 25% and sometimes degrades the process by 5%. It is important to know whether effect sizes are consistent across studies or if they vary significantly among studies. Measures of heterogeneity tell us whether studies are consistent or not. Consistent studies are "homogeneous" and have low or zero heterogeneity.

Heterogeneity occurs when the variance among study effect sizes is greater than the variance within studies. One of the first methods proposed for measuring heterogeneity among priory studies was Cochran's heterogeneity statis-

---

[2]Using `metafor`, the REML method is preferable to the DerSimonian and Laird method for estimating τ^2, but it cannot be easily used for manual calculations.

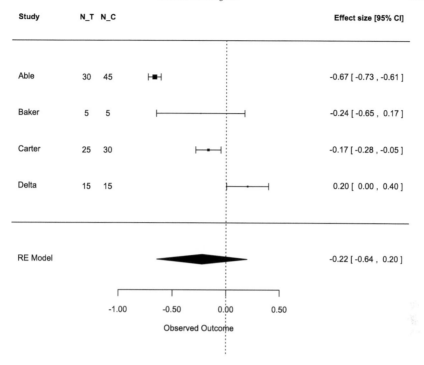

FIGURE 11.3: Forest plot example (random-effects model).

tic Q (Cochran 1954). The formula for Q based on a fixed-effects analysis is:

$$Q = \sum_{i=1,\cdots,k} W_i(ES_i - \overline{ES})^2 \qquad (11.52)$$

where ES_i is the observed effect size of the i-th study (irrespective of specific type of effect size), $\overline{ES}$ is the average effect size and W_i is the weight for the specific study, i.e the inverse of the effect size variance.

Under the null hypothesis that heterogeneity is zero, Q has a chi-squared distribution with $k-1$ degrees of freedom. Thus, theoretically Q can be used to assess whether the heterogeneity among a set of studies is greater than zero (this is called the Cochran Q-test). However, the Q-test is known to be of low power and some researchers prefer the I^2 statistic (Higgins & Thompson 2002). The formula for I^2 is:

$$I^2 = 100 \times \frac{(Q - df)}{Q} \qquad (11.53)$$

where Q is Cochran's heterogeneity statistic and $df = k - 1$. Negative values of I^2 are set to zero, so I^2 lies between 0% and 100%. Since under the null hypothesis that heterogeneity is zero, the expected value of Q is $k-1$, I^2 can be interpreted as the percentage of the total variation attributed to heterogeneity. Higgins, Thompson, Deeks & Altman (2003) provide some benchmarks for I^2

```
R Random-Effects Model (k = 4; tau^2 estimator: DL)

tau^2 (estimated amount of total heterogeneity): 0.1713 (SE = 0.1903)
tau (square root of estimated tau^2 value):    0.4139
I^2 (total heterogeneity / total variability):  97.32%
H^2 (total variability / sampling variability): 37.37

Test for Heterogeneity:
Q(df = 3) = 112.1003, p-val < .0001

Model Results:

estimate    se     zval     pval     ci.lb    ci.ub
-0.2226   0.2148  -1.0365   0.3000  -0.6435   0.1983
```

FIGURE 11.4: Random-effects analysis.

and assign adjectives of low, moderate, and high to I^2 values of 25%, 50%, and 75%, respectively.

Confidence intervals can be calculated for I^2 based on a transformation of I^2 to H^2:

$$H^2 = \frac{1}{(1 - I^2)} = \frac{Q}{(k - 1)} \qquad (11.54)$$

The details of how to construct the confidence intervals can be found in the additional material associated with Higgins & Thompson (2002). H^2 can be interpreted as the ratio of the observed variability to the expected variability.

In addition, T^2, which is the estimate of the between study variance, can be calculated using Equation 11.45.

T (which is the term R uses to refer to τ), H and I statistics are all reported by the `metafor` package as shown in Figure 11.4. To obtain confidence intervals on these statistics, use the `confint(mod2)` command on the calculated model (called `mod2`), the output of which is shown in Figure 11.5.

metric	estimate	ci.lb	ci.ub
tau^2	0.1713	0.0357	1.7514
tau	0.4139	0.1890	1.3234
I^2(%)	97.3238	88.3433	99.7317
H^2	37.3668	8.5788	372.7818

FIGURE 11.5: Confidence intervals for measures of heterogeneity.

Since it is necessary to choose between a fixed-effects or a random-effects model, it is tempting to perform a fixed-effects model, test for heterogeneity, then do a random-effects analysis if heterogeneity is statistically significant. This is **not** recommended by Borenstein et al. (2009), particularly since the tests of heterogeneity have low power. The decision to use a fixed-effects or a random-effects model should be based on whether or not the outcomes of all of the primary studies arise from the same underlying distribution. If we believe the studies are all functionally identical and we are not interested in generalising to other populations, we should use the fixed-effects model, otherwise the random-effects model is preferable.

Furthermore, if the random-effects model is used, but heterogeneity is relatively low or non-existent, the random-effects model results will be very similar to the fixed-effects model. Thus, it would seem better always to use a random-effects model. However, if the number of studies is very small the estimate of τ^2, is likely to have poor precision. In such a case, Borenstein et al. suggest that a fixed-effect model would at least provide a descriptive analysis of the included studies, and would be preferable to vote counting.

11.6   Moderator analysis

If there is extensive heterogeneity among the primary studies, it is useful to consider whether there are any systematic differences among the primary studies that might have caused the observed heterogeneity. These are potential *moderator factors*. Possible moderator factors include:

- Differences in participant type, for example, using students in some studies and practitioners in others.

- Differences in the software engineering materials or tasks used in terms of complexity or difficulty.

- Differences in experimental design, for example, within-subjects studies in some studies and {experimentexperiment!between-groupsbetween-groups in others.

- Differences in the control or treatment used in different studies.

There are two different types of moderator analysis: subgroup-based meta-analysis and meta-regression. Both types of analysis are supported by `metafor`.

Subgroup analysis is very much like an analysis of variance with the moderator factors as *blocking factors*. In most cases, subgroup effects are based on a fixed-effects model across subgroups and random-effects model within subgroups, this is called a *mixed-effects* model.

Meta-regression is equivalent to multiple regression with one or more study-level moderators acting as the independent variables and effect size as the dependent variable. However, in the meta-analysis case we use weighted multiple regression.

It should be noted that even if all the primary studies are randomised trials, subgroup analyses based on post-hoc identification of possible moderators are observational studies (like correlation and regression studies) and therefore *do not imply causation*. They indicate areas for further research.

11.7   Additional analyses

Other analysis that should be part of a meta-analysis include: assessing the impact of publication bias and general sensitivity analysis.

11.7.1   Publication bias

Publication bias refers to the fact that papers reporting positive effects are more frequently submitted to journals and conferences and more frequently published than papers that report negative or non-significant effects. There are several methods to address this issue:

- Funnel plots display the effect size (on the Y-axis) for each primary study against its standard error (on the X-axis). A funnel plot should exhibit a triangular pattern of points about the overall average effect size since studies with the smallest standard error should be close to the average effect size but studies with larger standard errors should have greater scatter about the average effect size. If there are more studies with effect sizes less than the average effect size than studies with effect sizes greater than the average effect size (so that the distribution is asymmetric), this can indicate that there are so-called "missing studies".

- Trim and fill plots assess the most probable number and magnitude of missing studies, and plot the missing studies on the funnel plot together with an adjusted estimate of the average effect size.

- The meta-analysis can be restricted to larger studies. One way of doing this is to do perform a cumulative meta-analysis with the studies sorted in sequence of smallest variance to largest variance. If the average effect size stabilizes when the high precision studies are included and does not change as low precision studies are included, then there is no evidence of any systematic bias. If the estimate of the average effect size changes when low precision studies are included, then there is some evidence of bias among the studies.

All these methods are supported by `metafor`. However, none of the methods are likely to be very accurate unless there are a reasonably large number of primary studies. Hannay et al. (2009) present a funnel plot and a trim and fill plot analysis of their meta-analysis results.

11.7.2 Sensitivity analysis

Sensitivity analysis investigates whether the results are robust to the assumptions and decisions that were made during the analysis. The sort of sensitivity analyses that can be undertaken are:

- Identifying the impact of different experimental designs (which can be done using a mixed-effects analysis).

- Assessing the impact of outliers using influential case diagnostics which assess the impact on the various statistics (such as standardised residuals, estimates of τ^2, heterogeneity measures) of omitting each primary study in turn from the meta-analysis. For example, the study by Hannay et al. (2009) report the impact of omitting one study at a time on Hedge's g and its confidence interval.

Both of these methods are supported by `metafor`.

Chapter 12

Reporting a Systematic Review

The final phase of the systematic review process is to document or report the study in ways that are suitable for the intended audiences. The context for this phase is illustrated in Figure 12.1.

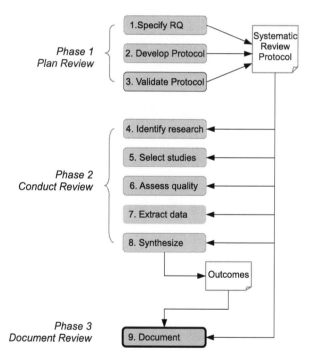

FIGURE 12.1: Reporting phase of the systematic review process.

Although, as we note below, the reporting of a systematic review may need to address a wider audience than would be expected for a conventional research paper, it still needs to answer the three key questions that Mary

Shaw identified in her tutorial on *Writing Good Software Engineering Research Papers* (Shaw 2003). These are:

- What, precisely, was your contribution [to your research field]?

- What is your new result?

- Why should the reader believe your result?

Indeed, one of her sub-questions expanding on the last of these does explicitly ask "what concrete evidence shows that your result satisfies your claims?", a question that is directly relevant to this chapter. As Shaw observes, getting one's work accepted for publication needs both interesting and sound results and also good clear communication of these to the reader. So far, we have been concentrating on how to obtain the results from a review (addressing the first two of the questions above), and now we need to consider how to communicate these results to others.

As we indicated in Section 4.4.9, a review can be reported in a number of ways. A journal paper, technical report or thesis chapter can usually provide full details of the process followed (including the reasons for the decisions taken and the approaches used for validation), together with the outcomes of each stage of a review plus the conclusions drawn from the results of a review. In contrast, conference papers (and sometimes journal papers too) are usually limited in length (often to 8 or 10 pages) and so can only cover some of the aspects of a review. And of course the use of these can be combined, with full details being given in a Technical Report, and the key aspects perhaps summarised as a conference or journal paper. Certainly, publication in some form in a refereed venue can contribute towards persuading others of the validity of your results.

So, in this chapter we consider three steps that need to be performed as part of the reporting phase.

1. *Planning* reports—which involves specifying the possible audiences and deciding what sort of document would suit their needs. Ideally, planning of the final reports will be initiated during the preparation of a review protocol.

2. *Writing* the actual reports.

3. *Validating* the reports—which involves asking internal, and possibly external, reviewers to assess the quality of reports.

The wider process of dissemination and publicising any important outcomes is outside the scope of this book. You can find a discussion of the issues involved in dissemination in Petticrew & Roberts (2006), and some aspects of this are also addressed in Chapter 14.

12.1 Planning reports

The nature of a systematic review means that the reporting of its outcomes is very likely to target a wider set of readers than would be the situation for a conventional research study, aimed simply at other researchers in a specific topic area, or in the case of a PhD thesis, the examiners. For example, Booth et al. (2012), identify six other groups of stakeholders that might be interested in medical and sociological research comprising: research funders, policy makers, practitioners, the research community, the media, and the general public.

While software engineering research should also be relevant to the same categories of stakeholder, the low levels of awareness usually exhibited by the media and the public means that, for the present, software engineering attracts little interest from these groups. Nevertheless, it is important that systematic reviewers make their results accessible to practitioners as well as the research community. In terms of policy making, the results may also be of relevance to standards organisations, and to authors of software engineering guidelines and text books. In contrast, mapping studies are usually of benefit only to the research community.

In planning therefore, it is important to identify the groups most likely to be interested in any outcomes (obviously, at this stage, you don't know what the actual results will be). Doing so may also provide an element of feedback to the design of the review. If changes to the research question(s) could provide outcomes likely to be of greater interest to one or more group, then this is the stage at which it should still be feasible to make such changes.

As already noted, to ensure that systematic review and mapping study reports are of benefit to the research community, they will need to be published in refereed journals and conference papers. All the details of the review process, data extraction and analysis/synthesis will need to be reported, including the rationale and criteria for excluding any 'marginal' papers (and hence, it is necessary to ensure that these details are all recorded while conducting the review—which in turn will also affect the planning for data extraction). In particular, the references for all the primary studies need to be provided, as well the data extracted from each paper, including quality data and data related to the specified research questions. In some cases, this can mean that the size of the resulting report exceeds journal and conference size guidelines. In such cases, the report should be supported by ancillary information such as a technical report and/or a database, if possible, held in an online repository. Planning should therefore take account of the likely reporting forms and, possibly, of the scale of these.

One way to make the results of a systematic review more accessible to other interested parties, particularly practitioners, is to produce shorter versions of the report concentrating on the practical implications of the results. These shorter versions can be directed to software engineering 'magazines' (as long as you do not violate originality requirements). Again, the readers of shorter versions of the report need be able to easily find a copy of the full report and be able to access any ancillary material.

12.2   Writing reports

The use of systematic reviews and, to a certain extent, of mapping studies, are advocated because they adhere to a rigorous methodology. So any resulting report should demonstrate clearly that you have used an appropriate systematic review process, and also that you have used it rigorously. In particular, a report should show:

- *Traceability*—providing the reader with a clear link from the research questions to the data needed to answer the questions; from the data to the data analysis; and finally from the data analysis to the answers to the questions and the study conclusions. This is relatively straightforward for quantitative systematic reviews and mapping studies, but may be much more difficult to demonstrate in the case of qualitative systematic reviews.

- *Repeatability*—ensuring that the methodology is defined clearly and in sufficient detail that other researchers could replicate it. This does not mean that other researchers would obtain exactly the same search outcomes and results. Time differences, at the very least, would make the results of searches different. Furthermore, differences are quite likely to occur for qualitative systematic reviews where researchers often use different synthesis methods which can result in different conclusions. Nonetheless, if using the same basic protocol, researchers should get broadly similar search results and be in a position to identify and investigate any divergences in conclusions.

  Booth et al. note that many people prefer the term 'replicability' to avoid implying claims for laboratory-like repetitions. (We discuss the issues relating to replication of primary studies in Chapter 21, where we also note that a 'differentiated' replication of a study may be useful for determining the boundaries or scope of an effect.)

The structure of a report/paper is usually fairly well-defined. While there may be variations, we would suggest that the following outline is one that addresses most of the above needs.

1. The *abstract*. The role of the abstract is to aid *selection*, by providing the reader (or search engine) with enough information to suggest relevance.

As noted earlier, we are advocates of structured abstracts (Budgen, Kitchenham, Charters, Turner, Brereton & Linkman 2008, Budgen, Burn & Kitchenham 2011), not least because their use encourages an author to include relevant information.

2. *Introduction.* A major role for this is to set the context, and following Mary Shaw's criteria, to make clear why this study is a useful contribution to a particular research field and why a systematic review is appropriate. So this is where we usually pose the research questions too.

3. *Background.* Usually this relates to the topic of the review, expanding on the description provided in the introduction, and where relevant, providing information about previous studies or reviews (whether expert reviews or systematic ones) and their contributions.

4. *Method.* This is where the core elements of the research protocol should be included in the paper, justifying the choice of the type of review, and the plan for its conduct, as well as the rationale for any other choices involved.

5. *Conduct.* This section is usually used to highlight any *divergences* from the plan, as well as to provide information about how well the team agreed about such issues as inclusion and exclusion (including providing kappa (κ) values to indicate the level of agreement where appropriate).

6. *Results.* This section usually describes the outcomes of searching and inclusion/exclusion, as well as of data extraction. As a section, this needs to be factual and thorough, leaving most of the interpretation for later sections.

7. *Analysis.* The outcomes from the synthesis process are described here. For a mapping study this may largely consist of tabulation and grouping. Some ideas about other forms of representation to use when reporting synthesis are covered in Chapters 9–11.

8. *Discussion.* This section is where the outcomes from analysis are considered within the wider context, and as such, has a large interpretive element. This is also where we assess the limitations of our study, through a discussion of the threats to validity.

9. *Conclusion.* This should seek to address how well the research questions have been answered, and what the answers are. This section provides the 'take home' message of a paper and so it is important for it to be concise and well focused, building upon the *Results, Analysis* and *Discussion* sections as appropriate.

Further ideas about how to report the outcomes from a systematic review may be obtained by consulting the PRISMA guidelines discussed in Section 7.2.1.

The way in which a report is written and presented is important too. There are many good textbooks about how to write technical reports and papers and so we have not tried to cover this aspect in any depth. However, there are some recurring issues that we have observed when refereeing or reviewing reports of systematic reviews, and so these are briefly discussed below.

Use diagrams and tables. Many authors illustrate the process involved in conducting the early stages of a review with a diagram that shows how this was conducted and the number of studies being retained at each stage of searching and inclusion/exclusion. Usually this can provide a good visual summary for the reader, and makes it easier to write the description of the process, since this can be 'written around' the figure. Figure 12.2 provides an example of such a figure, and is based upon one that we ourselves used when reporting a systematic review of empirical studies of the UML (Budgen, Burn, Brereton, Kitchenham & Pretorius 2011). We are not suggesting that this is the only way to structure such a figure, simply that they can play a useful role.

Tables can also provide useful summaries of complex processes. For example, when searching multiple sources for primary studies, tabulated results can illustrate the numbers found from each source, and also how many of these were unique (that is, not also found from other sources). Table 12.1,reproduced by permission of the Institution of Engineering & Technology, illustrates this using some of the values from a study we did into how reproducible systematic reviews were, published as (Kitchenham et al. 2012).

TABLE 12.1: Example of Tabulation: Papers Found at Different Stages

Papers papers	Digital Libraries	Scopus	Additional papers found from references	Additional papers from previous search	Duplicated reports
Search strings	1480	1275			
After initial screening	160	94	22		
Unit testing papers after 2nd screen	39	10	2	8	10
Regression papers after 2nd screen	25	6	2	2	9
Total testing primary papers	64	16	4	10	19

Take care with tenses. One of the arguments for producing a full and thorough research protocol is that it then helps when writing the final report.

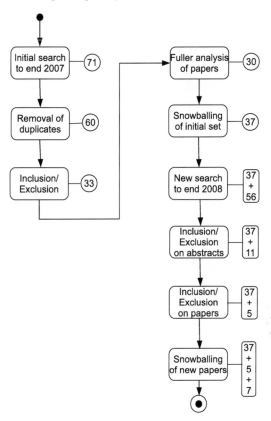

FIGURE 12.2: Example of a graphical model for the selection process. Reproduced with permission.

This is indeed very true. However, this is not just a matter of 'cut and paste', and one reason for this is that much of a protocol is likely to be written using future tense (what we are going to do), whereas a report is largely written in past tense (what we did). So yes, by all means reuse the protocol, but do take care to edit it so that the tense for the resulting report is consistent.

Avoid inventing new terms. Computing in general is a discipline that is rather apt to reinvent its wheels, and authors of computing (and hence software engineering) papers are often prone to try to introduce new terms to describe what they do. We strongly recommend that you avoid doing so when writing a report about a systematic review, largely because:

- the terminology used in systematic reviewing has been around for quite some time now (especially for the analysis elements), and should be quite adequate to describe what you are doing;

- using 'standard' terminology makes it easier for others to find your work (which after all, is usually what you want them to do of course)—it is not unheard of for systematic reviewers to be found complaining about failure by the authors of primary studies to use consistent terminology, while demonstrating just such a failure on their own part!

Conduct a review of the report. Because a systematic review can involve many people, different sections of the report may well have been written by different authors—and this sometimes shows in terms of differences of style, terminology and even grammar and tenses. Given that there are some marked differences between American and British English, using a mix of these can also be confusing[1]. So, once the report is drafted, an important task is that of an editorial review, carried out by one or two people, with the aim of making the paper more consistent and (one hopes) readable. The purpose of this type of review is editorial rather than technical, and we address the latter role in the next section. Also, where the paper is written in English, and none of the authors is a native-English speaker, we suggest that it may be useful to get a native speaker to go through the report, both to check grammar, and maybe also to revise any idioms that might not have translated well. (As we ourselves are essentially mono-lingual, we do admire the ability of others to write reports in another language, but as referees, we are also aware that even good papers can have some strange phrasing in places.)

12.3　Validating reports

Editorial aspects apart, all report authors have a responsibility to read and review a report, with the aim of ensuring that the following situation is true.

- The research questions are clearly specified and fully answered.

- The research methodology is fully and correctly reported.

- There is traceability from the research questions to data collection, data synthesis and conclusions.

---

[1]The phrase 'two nations separated by a common language' can be very true. As an illustration, fortunately unlikely to appear when reporting a systematic review, consider the use of the word 'momentarily'. To Americans, this means 'in a moment', while to the British it means 'for a moment'. So, consider the effect on British passengers of American air-crew announcing that "we will be landing momentarily"!

- All the tables and figures used to present the results are correct and internally consistent.

- In the case of systematic reviews, the conclusions are written clearly and are targeted both at researchers and also practitioners.

If possible, reports of systematic reviews should be independently reviewed. Within a research group (for example, a university department), colleagues may be willing to undertake independent reviews of reports on a quid-pro-quo basis.

Chapter 13

Tool Support for Systematic Reviews

with Christopher Marshall

Keele University

Software tools can provide valuable support for many aspects of systematic reviews and mapping studies. As we have seen in previous chapters, conducting a review involves the systematic storage, management, validation and analysis of what can be quite large quantities of data. Many of these review activities are error prone and time consuming when performed manually. The software systems that have been used to support systematic reviews and mapping studies in software engineering include basic productivity tools, such as word processors and spreadsheets, reference managers, statistics packages and purpose-built tools targeting either particular stages of a review or the review process as a whole.

In this chapter we focus mainly on special purpose tools that aim to support most review activities although noting that spreadsheets and reference managers are probably the most widely used tools to date. We note also that the tools landscape is changing quite rapidly with a growing number of tools, especially those targeting the software engineering domain, being developed, adapted and enhanced. Before looking at the tools for software engineering reviews, we summarise some of those used in other disciplines, specifically in the medical and social sciences fields.

13.1   Review tools in other disciplines

Tools used in the medical and social sciences domains range from the quite basic (such as 'paper and pencil', forms and spreadsheets) to complex special-purpose systems (such as EPPI-Reviewer[1] and Review Manager (RevMan)[2]).

Tools that provide a degree of automated support for specific activities within the systematic review process have been summarised and classified by Tsafnat, Glasziou, Choong, Dunn, Galgani & Coiera (2014). Some of these 'tools' are essentially approaches or methods that can be applied. For example, scoping studies (or mapping studies) can be used to identify research gaps and hence can direct reviewers towards research questions for which evidence is lacking. Software tools identified by Tsafnat et al. include:

- **Quick Clinical** - a federated meta-search engine,

- **Abstractr** - for initial screening of titles and abstracts (using a machine learning approach),

- **ParsCit** - supports snowballing through automated extraction of references,

- **ExaCT** - supporting data extraction using information highlighting,

- **Meta-analyze** - for meta-analysis of extracted data.

For data extraction in particular, a range of approaches have been compared by Elamin, Flynn, Bassler, Briel, Alonso-Coello, Karanicolas, Guyatt, Malaga, Furukawa, Kunz, Schnemann, Murad, Barbui, Cipriani & Montori (2009). The approaches included in the comparison are:

- Paper and pencil,

- Email based forms,

- Spreadsheet software,

- The Cochrane Collaboration's Review manager (RevMan),

- Database software,

- Web-based surveys,

- Web-based specialized applications.

---

[1]http://eppi.ioe.ac.uk/cms/er4
[2]http://tech.cochrane.org/Revman

Several experienced systematic reviewers from different countries assessed each candidate's setup cost, project setup difficulty, versatility, training requirement, portability/accessibility, ability to manage data, ability to track progress, ability to present data and ability to store and retrieve data.

Not surprisingly each tool type was found to have some benefits and some drawbacks. The authors of the study concluded that "specialized web-based software is well suited in most ways, but is associated with higher setup costs". They suggest that the selection of a data extraction tool should be informed by the availability of funding, the number and location of reviewers, data needs and the complexity of the review.

Below we summarise EPPI-Reviewer and RevMan which are two of the most widely used specialised applications.

EPPI-reviewer

The current version of EPPI-Reviewer, EPPI-Reviewer 4, is a comprehensive single or multi-user web-based system for managing systematic reviews across the social sciences and medical disciplines. It has been developed and is maintained by the EPPI-Centre at the Social Science Research Unit at the Institute of Education, University of London,UK. EPPI-Reviewer 4 is a "comprehensive online software tool for research synthesis" which supports all stages of the systematic review process. It can support narrative reviews and meta-ethnographies as well as quantitative reviews and meta-analyses. Functionality includes:

- *Reference management* - thousands of references can be managed and can be imported using a variety of formats. Web services enable direct access to PubMed[3]. Fuzzy logic is used to check for duplicate papers. The tool also manages linked documents so that multiple reports of a study can form a single 'units of analysis' (see Section 6.3).

- *Study selection and data analysis* - features include concurrent multi-user classification of studies, based on inclusion and exclusion criteria, disagreement resolution and the generation of summary reports. Common measures of effect using a variety of statistical methods can be calculated.

- *Synthesis* - a range of features are offered including running meta-analysis. Reports of categorical, numerical and textual data can be produced in a wide variety of formats. Also, text mining is used to support automatic document clustering (and can enhance the search process once there is a known set of relevant papers).

---

[3]http://www.ncbi.nlm.nih.gov/pubmed PubMed, which is supported by the US government, is a large-scale freely-available online database which indexes articles and books relating to medical and other life sciences.

- *Review management* - the system enables allocation of tasks to individual members of a review team, reporting of progress and export of review data to enable long-term storage.

Fees for use of the system are charged on a not–for–profit basis, and contribute to infrastructure, development and a range of support mechanisms.

Review manager (RevMan)

This is a substantial special-purpose system for supporting systematic reviews in the medical domain. It is mandatory (and free) for preparing and maintaining Cochrane reviews[4]. It can also be used in non-Cochrane mode. Information about using RevMan is widely available through a range of tutorials, webinars and user guides. The system offers support for:

- *Preparation of protocols* - the tool provides support for developing a protocol which adheres to the format required for a Cochrane review (as documented in the *Cochrane Handbook for Systematic Reviews of Interventions*[5]) or for developing protocols using other formats (in non Cochrane mode).

- *Text input* - information about a review can be copied and pasted from a word processor document or entered directly. Pre-defined headings (such as *Background* and *Description of the condition*) can be used or de-activated. The tool supports tracking of changes and spell checking.

- *Adding studies and references* - these can be added for included and (optionally) for excluded studies and for other papers that are cited in a review.

- *Tables* - two standard tables are available. These are the *Characteristics of included studies* table and the *Risk of bias* table. A *Characteristics of excluded studies* and other tables (generated using a *New Additional Table* wizard) can also be incorporated.

- *Data and analyses* - substantial support is offered for data input, meta-analyses and the generation of graphs (such as forest and funnel plots).

- *Finishing the review* - a *Summary of findings* table can be added when a review is complete. The table can be imported from a separate application designed to produce such tables (GRADEprofiler[6]) or can be created directly using RevMan's table editor.

RevMan is targeted at data management, meta-analysis and documentation of quantitative systematic reviews.

---

[4]http://www.cochrane.org/cochrane-reviews
[5]handbook.cochrane.org
[6]http://tech.cochrane.org/gradepro

13.2   Tools for software engineering reviews

Recent research[7] suggests that problems relating to systematic reviews (and mapping studies) faced in other disciplines are similar to those faced by software engineering researchers and so it may be that EPPI-Reviewer (and RevMan) could be used within software engineering too. We should bear in mind though that EPPI-Reviewer is not free to use (although users do suggest that payment provides a degree of confidence in the reliability and longevity of the tool). A mapping study has identified tools (other than 'standard' general-purpose tools such as spreadsheets and reference managers) that have been used to support software engineering reviews (Marshall & Brereton 2013). Information about these and a range of other tools (including those from other disciplines), is available through an on-line catalogue which classifies the tools and provides some useful links[8]. Tools identified by the mapping study are summarised in Table 13.1 which indicates for each tool the stage or stages of a review that are addressed, whether the tool was developed specifically to support software engineering reviews, the underlying approach used and a reference that can be followed for further information. As we can see, text mining is the most common underlying approach.

Four systems that target the software engineering systematic review process as a whole have been evaluated by Marshall, Brereton & Kitchenham (2014). These are:

- **SLuRp**[9] - supports many of the stages of the review process. One of the major strengths of this tool is that it supports review teams, enabling multiple reviewers to perform study selection and quality assessment independently, and providing a mechanism for managing the resolution of disagreements.

- **StArt**[10] - comes with a full installation wizard and is well supported by an introductory video providing an overview of the tool and its key features. The tool provides some support for protocol development and allows the results of searches to be imported using a range of formats. It also supports, to some extent, study selection, data extraction and synthesis. Major limitations of StArt are its lack of support for multiple reviewers working as a team and for quality assessment.

- **SLR-Tool**[11] - provides some support for protocol development, data ex-

---

[7]C. Marshall, P. Brereton and B. Kitchenham, *Tools to support systematic reviews in software engineering: a cross-domain survey using semi-structured interviews*, in preparation.

[8]http://systematicreviewtools.com/

[9]https://bugcatcher.stca.herts.ac.uk/SLuRp/

[10]http://lapes.dc.ufscar.br/tools/start_tool

[11]http://alarcosj.esi.uclm.es/SLRTool/

traction and quality assessment and data analysis. However, like StArt, it does not support multiple users working together on a review.

- **SLRTOOL**[12] - has a number of potentially useful features including some support for searching, however, at the time of writing, much of this potential seems to not yet be realised.

These tools are all free to use.

In summary...

Special-purpose tools to support software engineering reviewers are at present:

- few in number
- immature
- not widely used
- in some cases showing promise
- in need of further independent evaluation

General purpose tools such as spreadsheets and reference managers dominate.

In other disciplines, EPPI-Reviewer and RevMan support many systematic review activities including collaborative working.

---

[12]http://www.slrtool.org

TABLE 13.1: Tools to Support Systematic Reviews in Software Engineering

Tool name	Review stage	New tool?	Underlying approach	Reference
Project Explorer	Study selection Data extraction Reporting	No	Visualisation	Felizardo et al. (2010)
Revis	Study selection	Extension to an existing tool	Text mining	Felizardo et al. (2012)
SLR-Tool	All	Yes	(Incorporates text mining)	Fernández-Sáez, Bocco & Romero (2010)
Hierarchical Cluster Explorer	Synthesis	No	Visualisation	Cruzes, Mendonca, Basili, Shull & Jino (2007a)
Site Content Analyzer	Synthesis	No	Text mining	Cruzes, Mendonca, Basili, Shull & Jino (2007b)
UNITEX	Data extraction	No	Text mining	Torres, Cruzes & do Nascimento Salvador (2012)
SLuRp	All	Yes	-	Bowes, Hall & Beecham (2012)
SLONT/ COSONT	Data extraction	Yes	Ontology	Sun, Yang, Zhang, Zhang & Wang (2012)
StArt	All	Yes	-	Hernandes, Zamboni, Fabbri & Thommazo (2012)
Uses DBpedia, OpenCalais, Naïve Bayes tool	Study selection	No	Text mining	Tomassetti, Rizzo, Vetro, Ardito, Torchiano & Morisio (2011)

Chapter 14

Evidence to Practice: Knowledge Translation and Diffusion

The preceding chapters making up Part I of this book have addressed the various issues concerned with *adapting* the practices of the evidence-based paradigm to the needs of software engineering. In particular, they have described the role of a systematic review in amassing and synthesising evidence related to software engineering topics. So in this, the final chapter of Part I, we consider what should happen *after* the systematic review, and in particular, how the outcomes from a review (the *data*) can be interpreted to create *knowledge* that can then be used to guide practice, to help set standards, and to assist policy-making.

In other disciplines that make use of systematic reviews, this process of interpretation for practical use is often termed *Knowledge Translation* (KT), although as we will see, there are questions about how appropriate the "translation" metaphor is. Clearly, the way that KT is performed should itself be as systematic and repeatable as possible, and it should also reflect the needs and mores of practitioners, as well as of the different forms of organisational context within which they work.

In the interpretation of evidence-based practices for software engineering provided in Section 2.3, Step 4 was described as:

> *Integrate the critical appraisal with software engineering expertise and stakeholders' values.*

This essentially describes the role of KT, and while this does occur (we will examine some examples later in the chapter), the processes used tend to be rather *ad hoc* and to lack adequate documentation.

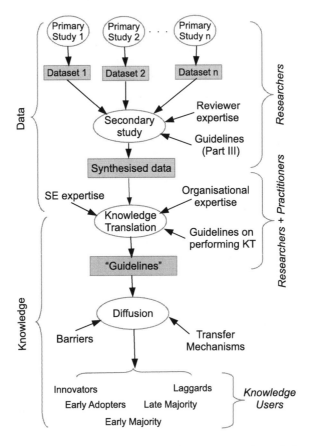

FIGURE 14.1: The pathway from data to knowledge.

Knowledge translation in itself is of course only part of the overall process of encouraging practitioners and others to make use of the evidence from a systematic review. Knowledge needs to be disseminated to be useful, and the processes through which new forms of knowledge become accepted and adopted by the relevant parts of society has been studied for many years, with the classic work on this being the book *Diffusion of Innovations* by Rogers (2003). The model shown in Figure 14.1 illustrates the sequence of quite complex processes involved in turning data into something that forms part of the professional's "knowledge toolkit".

Both KT and diffusion are large and complex topics, and we can only address them fairly briefly here. So, in this chapter we examine how KT is organised in other disciplines; discuss how it might be placed on a more systematic basis in software engineering; and review a number of examples of where KT has been performed to provide guidelines about software engineering practices. We also provide a short discussion of the nature of knowledge diffusion, and how this may occur for software engineering. Finally, we review the software engineering knowledge that has emerged from the first ten years of performing systematic reviews in software engineering and consider how this might help to inform and underpin better quality teaching, practice and research—which after all is the purpose of EBSE.

14.1   What is knowledge translation?

While there is quite an extensive literature exploring the concept of KT, we should observe that it is not the only term used to describe "post-systematic-review" activities. Other terms that are in use include "Knowledge To Action" (KTA) and "Knowledge Exchange" (KE), and such words as 'uptake' and 'transfer' are also used in this context. Part of the reason for the frequent use of "Knowledge Translation" appears to stem from it being a term used in the mandate of the Canadian Institutes of Health Research (Straus, Tetroe & Graham 2009), and because Canadian researchers have authored many papers on this topic.

A useful definition of KT is that produced by the *World Health Organisation* (WHO) in 2005, as:

> "the synthesis, exchange and application of knowledge by relevant stakeholders to accelerate the benefits of global and local innovation in strengthening health systems and advancing people's health" (WHO 2005)

(We might note that this actually refers to the 'exchange' of knowledge.) Other definitions are to be found, with a common thread being the emphasis on putting the knowledge into use.

The use of guidelines for performing KT so as to produce recommendations for practice has been investigated extensively for both clinical medicine and education. Rather confusingly, in the literature, both the procedures for producing recommendations and the recommendations themselves are apt to be referred to as *guidelines*. So in this chapter we will use *KT recommendations* for the guidance provided to the eventual users about how to interpret the outcomes from an individual systematic review. Similarly, wherever we refer to *KT guidelines*, these will refer to the set of *activities* used for deriving recommendations for practice or policy (KT recommendations), along with any guidance that might be provided about how to describe them.

An overview of how KT-related guidelines are used in clinical medicine is provided by the EU recommendations for drawing up KT guidelines on best medical practices (Mierzewski 2001). This also reviews the KT guidelines programmes used in different EU member states.

Within the UK, the National Institute for Clinical Excellence (NICE) has produced its own KT guidelines (NICE 2009), and these also provide a useful source of descriptions of translation models. International efforts towards evaluating KT guidelines produced by different organisations have included the AGREE II programme for *appraisal* of KT guidelines (Burgers, Grol, Klazinga, Mäkelä & Zaat 2003, AGREE 2009) and the assessment of these processes for the World Health Organisation (WHO) described in (Schünemann, Fretheim & Oxman 2006). Together these provide systematic approaches, for both producing KT recommendations, and also for evaluation of the procedures involved.

As noted earlier, the literature on this topic is extensive, which emphasises that, for healthcare, the process of translation is complicated by many factors. For example, Zwarenstein & Reeves (2006) observe that KT is often directed at producing KT recommendations for a single professional group, whereas the treatment of patients is likely to involve inter-professional collaboration. Similarly, Kothari & Armstrong (2011) observe that for KT research in more general health care, "developing processes to assist community-based organizations to adapt research findings to local circumstances may be the most helpful way to advance decision-making in this area".

We should also note that the appropriateness of this terminology has been challenged. Greenhalgh & Wieringa (2013) argue that the 'translation' metaphor is an unhelpful one and that its use "constrains thinking". Essentially, they argue that this term implicitly creates a model in which the only form of useful knowledge stems from "objective, impersonal research findings". In examining equivalent metaphors from other disciplines they emphasise the need to also involve such factors as "tacit knowledge of the wider clinical and social" situation when using such knowledge. In particular, they propose that a wider set of metaphors should be used, including ones such as "knowledge intermediation".

So, what their work highlights is that there are dangers implicit in simply adopting the 'translation' metaphor with its implication of researchers "handing down" scientifically distilled guidance. More realistically, the process of developing guidelines for use should be something that is shared between researchers and other stakeholders (which reiterates the earlier point about the use of 'exchange' in the definition from the WHO). And, taken together, what all of these studies also indicate is that systematising KT is a specific research activity in its own right, and that the process of KT involves much more than simply supplying the outcomes from systematic reviews to professionals.

14.2   Knowledge translation in the context of software engineering

In this section we examine how the activities of KT could be interpreted for software engineering. As a starting point, we have adapted the description of KT provided by the WHO, quoted in the previous section, as well as the variation used in Davis, Evans, Jadad, Perrier, Rath, Ryan, Sibbald, Straus, Rappolt, Wowk & Zwarenstein (2003), in order to define a process of KT for software engineering as being:

> The exchange, synthesis and ethically sound application of knowledge—within a complex system of interactions between researchers and users—to accelerate the capture of the benefits of research to help create better quality software and to improve software development processes.

The three key elements involved in achieving this are: the *outcomes* of a systematic review; the set of *interpretations* of what these outcomes mean in particular application contexts; and the forms appropriate for *exchanging* these interpretations with the intended audience.

In health care, a widely-cited paper by Graham, Logan, Harrison, Straus, Tetroe, Caswell & Robinson (2006) refers to this as "knowledge to action" (a term which we noted earlier). In this, the authors suggest that the process of KT (or KTA) can be described using a model of two nested and interlocked cycles that are respectively related to knowledge creation and knowledge application. Figure 14.2 shows how this concept can be interpreted for software engineering (Budgen, Kitchenham & Brereton 2013). The inner *knowledge cycle* is concerned with "knowledge creation" (which in the case of software engineering will be based upon primary studies and systematic reviews). The outer *action cycle* "represents the activities that may be needed for knowledge application", including the creation and evaluation of both KT guidelines and KT recommendations produced by using these. This element of the KT process is highly-interactive and holistic (Davis et al. 2003), being driven by the needs of the given topic, and hence evaluation will need to be a key element in maintaining consistency of practice.

Because of this, the positioning of the elements in the outer cycle should not be regarded as forming a sequence in the same way as occurs for the elements of the inner cycle. Rather, the outer cycle describes a set of activities that may well be interleaved and iterative.

Indeed, the value of Figure 14.2 lies less in its structure than in its identification of the various factors involved in performing KT. It also highlights the point that a transition to an evidence-informed approach to software engineering can only be achieved through a partnership of researchers, practitioners, and policy-makers.

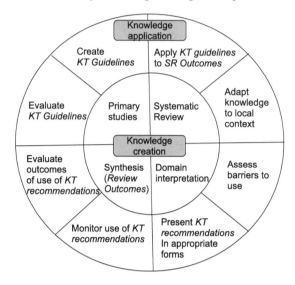

FIGURE 14.2: A knowledge translation model for SE. Reproduced with permission.

To conclude this discussion of KT, in the book by Khan, Kunz, Kleijnen & Antes (2011), the authors make some useful observations about the form that recommendations should have that are as relevant for software engineering as they are for clinical medicine. In particular, they highlight the following key points.

- Recommendations should convey a clear message and should be as simple as possible to follow in practice.

- What possible users "really want to know about recommendations is how credible (trustworthy) they are", noting too that 'credibility of a recommendation depends only in part on the strength of evidence collated from the review'.

In particular, they suggest classifying any recommendations as being either strong or weak. They define a *strong recommendation* as effectively forming a directive to adopt a practice or treatment, whereas a *weak recommendation* indicates that a decision about its adoption is something that needs to depend upon due consideration of other relevant factors.

Another approach to categorising the quality of evidence and strength of recommendations from a systematic review, again developed in a medical context, is the GRADE system (Grades of Recommendation, Assessment, Development and Evaluation), described in (GRADE Working Group 2004). A discussion and example of using GRADE in a software engineering context can be found in the paper by Dybå & Dingsøyr (2008*b*). To employ GRADE they used four factors: study design; study quality; consistency ("similarity of estimates of effect across studies"), and directness ("the extent to which the people, interventions and outcome measures are similar to those of interest").

The GRADE scheme specifies four levels for strength of evidence, and hence of recommendations. In Table 14.1 we provide the GRADE descriptions for each level, and suggest how these might be interpreted in a software engineering context. As expressed, this allows for the possibility that further studies may increase the classification assigned to any recommendations.

Evidence-based studies in software engineering are still far from reaching the level of maturity where strong recommendations can be generated with any confidence. Indeed, the nature of the effects that occur for the creative processes of software engineering is likely to make the generation of strong recommendations a relatively rare occurrence, indeed, it is worth noting that Dybå & Dingsøyr assessed the strength of evidence from their earlier study on agile methods as 'very low'.

TABLE 14.1: Strength of Evidence in the GRADE System.

Level	**GRADE definition**	**SE Interpretation**
High	Further research is very unlikely to change our confidence in the estimation of effect.	Supported by significant results from more than one good quality systematic review as well as by experiences from systematically conducted field studies.
Moderate	Further research is likely to have an important impact on our confidence in the estimate of effect and may change the estimate.	Supported by moderately significant results from at least one good quality systematic review, and by observational studies.
Low	Further research is likely to have an important impact on our confidence in the estimate of effect and is likely to change the estimate.	Supported by moderately significant results from at least one good quality systematic review.
Very Low	Any estimate of effect is very uncertain.	Only supported by results from one systematic review.

So, in the next section we examine some examples of where systematic reviews have led to the creation of some form of recommendations about how the outcomes might be used, and how the authors have qualified these.

14.3   Examples of knowledge translation in software engineering

The catalogue of 143 published systematic reviews described in (Budgen, Drummond, Brereton & Holland 2012) identified 43 reviews that contained material that could be used to inform teaching and practice. However, only three of these actually provided any form of *recommendations* about how the outcomes from the review could be interpreted in terms of practice. So, in this section we briefly describe each of the studies, the recommendations that they made, and how these were derived (where known).

14.3.1   Assessing software cost uncertainty

Cost and effort estimation have been studied quite extensively, which is perhaps not surprising given that they can have a major impact upon project success (and company profit). The study by Jørgensen (2005) particularly looked at uncertainty in forecasts, and some of the likely causes for this.

This particular study drew upon primary studies that came from both the software engineering domain and also a range of other domains that employ forecasting procedures. The paper presents a set of recommendations (termed 'guidelines' in the paper) derived from the analysis. For each recommendation, the paper identifies both the primary studies that provide supporting evidence and also identifies where this is discussed. The author also provides a rating using the same terms that we use above, together with explanations of what these mean in terms of the evidence provided.

Rather than present the whole set of seven recommendations here, we simply give two examples from them.

Recommendation 1: "Do not rely solely on unaided, intuition-based processes." This is rated as being *strong*, based upon the number of studies favouring it.

Recommendation 2: "Do not replace expert judgement with formal models." This is rated as *medium*.

Indeed, although the author describes the rating process as informal and subjective, it is actually quite systematic and provides an excellent model for use by other authors.

It is also interesting to note that the conclusions from a more recent summary of work in this area suggest that most of these recommendations are probably still valid (Jørgensen 2014*b*).

14.3.2   Effectiveness of pair programming

The meta-analysis on the effectiveness of pair programming described by Hannay et al. (2009) found wide variation in the form and organisation of the primary studies included, limiting the confidence with which any recommendations could be produced. There are also many other factors that influence whether or not such a technique might be considered effective, particularly the expertise of the programmers and the complexity of the task involved.

However, they did suggest that two recommendations were appropriate when pair programming was being used by "professional software developers", and in a context where "you do not know the seniority or skill levels of your programmers, but do have a feeling for task complexity". In this context they suggested that it was appropriate to employ pair programming for either of the following situations:

- When task complexity is low and time is of the essence

- When task complexity is high and correctness is important

No specific process for deriving these was described, although the analysis of the data implicitly supported them. In terms of the classifications suggested in Table 14.1 these should probably be considered as recommendations with *low* strength. However, in the future, they could be regraded as *moderate* if supporting outcomes from observational studies become available, given that this was a 'good' systematic review.

14.3.3   Requirements elicitation techniques

Our third example is a paper by Dieste & Juristo (2011) that examines elicitation techniques that are often used for determining system requirements. They present five recommendations (again, termed 'guidelines' in the paper), and for each one they identify the aggregated (synthesised) evidence that supports or refutes the recommendation. The authors have not attempted to assess the strength of these recommendations.

Again, we present two of these without attempting to include all of the supporting detail. We have also slightly reworded them, mainly to fit the role of a recommendation.

Recommendation 1: "The use of unstructured interviews is equally as, or more *effective* than, using introspective techniques (such as protocol analysis) and sorting techniques." The authors observe that it is reasonable to assume that this recommendation also applies to structured interviews.

Recommendation 3: "The use of unstructured interviews is less *efficient* than using sorting techniques and Laddering, but is as efficient as introspective techniques such as protocol analysis." Again, the authors observe that this should also apply to structured interviews.

One of the benefits (and complexities) of this paper was that it looked at studies that made comparisons between the different techniques, allowing the reviewers to provide an element of ranking in their recommendations.

14.3.4   Presenting recommendations

Looking at these three examples, we can see some common threads among them.

- The authors are experts at performing systematic reviews and have extensive expertise related to the topic of the review, assisting them with interpreting the outcomes of the review.

- Their reviews found quite substantial numbers of primary studies, so that the authors have been able to identify areas where these reinforced each other (or vice versa).

- They present supporting evidence for their recommendations, in two cases, directly listing the studies that agree/disagree with the recommendation. They also provide a discussion that explains how the primary studies support a recommendation (or otherwise).

All three examples also follow the advice of Khan et al. (2011) to keep the recommendations simple and to provide some indication of how trustworthy they are (in these cases, by discussing the underpinning evidence).

So, where a review team has the technical expertise to do so, there is scope to provide at least a basic element of knowledge translation of the review outcomes, in the form of recommendations. We provide a summary of key points for doing so in the box below.

Guidelines for Producing Recommendations

- Only do so if you have appropriate technical expertise.

- Only do so if your systematic review is a 'strong' one.

- Keep any recommendations simple and easy to follow.

- Identify the studies that support/refute each recommendation.

- Provide a separate derivation, related to the studies.

- Provide an indication of how strong a recommendation is (and what you mean by this).

- Identify the audience for the recommendation, and, where appropriate, whether the evidence comes from using practitioners or students as participants in the primary studies.

14.4   Diffusion of software engineering knowledge

Neither innovativeness nor quality will necessarily ensure that new devices or processes will be successful in being accepted by the communities most likely to benefit from them. So, even if we can produce strong recommendations that address topics of major importance to the software engineering community, their adoption still requires the community to be persuaded of their merits.

This situation is by no means unique to software engineering, and the terminology that is used at the base of Figure 14.1 is drawn from the ideas of *diffusion research* as set out by Rogers (2003). This is based upon the premise that the process of acceptance of innovative ideas and technologies tends to follow broadly similar processes, regardless of discipline. The topic overall is a large one and we can only touch lightly upon it here, where our main concern is to encourage *awareness*. The world does not automatically beat a path to the doorway of the person with a better mousetrap, agricultural practice or software development process. To gain recognition, that person needs to ensure that knowledge about their innovation gets to, and is accepted by, the people who will influence others.

The 'classical' diffusion model produced by Rogers recognises five major 'adopter categories' who are involved in the process of moving an innovation into the mainstream. Briefly, these are as follows.

- The *innovators* are people who like to try new ideas and are willing to take a high degree of risk in doing so. Communication between them is a strong element in sharing of new ideas, but it is likely that this will be across organisations rather than within them.

- The *early adopters* are opinion formers who have influence within organisations, and so they are the people whose opinion is sought by others who are considering change.

- The *early majority* are those who are more cautious than the preceding two groups, but still tend to be ahead of the average. They take longer to decide about changing their processes than the early adopters, and tend to follow rather than lead.

- The *late majority* are even more cautious, and only join in when they can see that their peers are taking up a change, and that they might even be disadvantaged by not doing so.

- Finally, the *laggards* are apt to be suspicious of change and may have only limited resources, which may also encourage caution about change.

So generally, the key to successful adoption lies in achieving buy-in from the people who fall into the first two categories.

An authoritative study in the field of health care by Greenhalgh, Robert, MacFarlane, Bate & Kyriakidou (2004), based upon a large-scale systematic review, suggests a wider and rather more proactive view of the way that knowledge can be transferred. In particular the authors suggest that it is useful to distinguish between the following three mechanisms.

1. *diffusion* where knowledge and awareness spread passively through a community, largely through natural means

2. *dissemination* through the active communication of ideas to a target audience

3. *implementation* through the use of communication strategies that are targeted at overcoming barriers, using administrative and educational techniques to make the transfer more effective

For recommendations produced from systematic reviews in software engineering, all three mechanisms can potentially play a useful role.

Pfleeger (1999) suggests that different adopter categories are motivated by distinct transfer mechanisms. She introduces the idea of the *gatekeeper* whose role is to "identify promising technologies for a particular organisation". Another part of their role is to assess the evidence presented for a new or changed technology. We should also note that identifying appropriate vehicles for communicating with the different gatekeeper roles is important, These vehicles might be social networks (particularly for innovators), trusted media sources (for early adopters) and some form of 'packaging' (which might be the incorporation of recommendations into standards) for the early and late majorities.

An associated issue here is that of *risk*. For the innovators and early adopters, taking up a new or changed technology may involve a higher level of risk. So any presentation of new knowledge has to help them make an assessment of how significant this risk might be in their particular context (and in exchange, what benefits they might derive).

While we cannot really delve deeper into these issues here, they are important ones for systematic reviewers to note. Publication of the outcomes of a review in respected refereed journals is only the first stage of getting knowledge out to users. In particular, the knowledge embodied in any recommendations may need to be spread through such means as social media, professional journals, and incorporation into standards. Another important vehicle is the use of educational channels (relating to 'implementation'), and we address this in a little more detail in the next, and final, section of this chapter.

14.5   Systematic reviews for software engineering education

One way in which software engineering does differ significantly from other disciplines that use systematic reviews is in the way that the topics of these are decided. For disciplines such as education, social science, and to some degree, healthcare, systematic reviews are often commissioned by policy-makers, who may well work within government agencies. To our best knowledge, there are so far no instances of systematic reviews in software engineering being commissioned by either government agencies or industry—this may be partly because secondary studies are still immature for software engineering, but also because IT-related decisions are rarely evidence-informed at any level. So, the available secondary studies do tend to have been motivated more by the interest of particular researchers than through any efforts to inform decision-making.

So, in presenting this 'roadmap' to available systematic reviews, the reader should remember that the distribution of topics is driven by 'bottom-up' interest from researchers, rather than 'top-down' needs of policy-makers. This may change in the future but for the present it is the situation that exists for software engineering.

And of course, even before we have finished compiling such a roadmap, it is inevitably out of date as new reviews become available.

14.5.1   Selecting the studies

The roadmap provided here is based upon a *tertiary study* that we performed in 2011, and published as (Budgen et al. 2012). The research question posed for this was:

> *What is available to enable evidence-informed teaching for software engineering?*

Although this was posed as an educational question, we did seek to identify any systematic reviews that could provide knowledge, advice or guidance relevant to either practice or teaching. We excluded any that were purely concerned with research issues (and of course, almost all mapping studies).

To identify candidate systematic reviews we used a two-part search procedure, organised as follows.

1. List all systematic reviews found in the three published 'broad' tertiary studies that were available to us (Kitchenham, Brereton, Budgen, Turner, Bailey & Linkman 2009, Kitchenham, Pretorius, Budgen, Brereton, Turner, Niazi & Linkman 2010, da Silva et al. 2011). Together, these covered the period up to the end of 2009. We also included one paper that was subsequently known to have been missed by these studies.

2. List the systematic reviews found in five major software engineering journals between the start of 2010 and mid-2011. While recognising that this would be incomplete, it was felt that it would identify the majority of published studies for this period.

Together, this produced a set of 143 secondary studies.

We then excluded any studies that addressed research trends, those with no analysis of the collected data and those that were not deemed to be relevant to teaching. Conversely, we included studies that covered a topic considered to be appropriate for a software engineering curriculum, using *Knowledge Areas* (KAs) and *Knowledge Units* (KUs) from the 2004 ACM/IEEE guidelines for undergraduate curricula in software engineering. This left us with 43 secondary studies.

Our data extraction procedure then sought to categorise these studies against the KAs and KUs, to extract any recommendations provided by the authors, and in the absence of these, any that we felt were implied by the outcomes, since few authors provide explicit recommendations. (As we have seen, *knowledge translation* is not a trivial task.) Data extraction was performed by pairs of analysts using different pairings of the four authors in order to reduce possible bias.

14.5.2   Topic coverage

Table 14.2 provides a count of the number of studies we categorised against each Knowledge Area.

TABLE 14.2: Number of Systematic Reviews for Each Knowledge Area

KA code	Topic	Count
QUA	Software Quality	6
PRF	Professional Practice	2
MGT	Software Management	13
MAA	Modeling & Analysis	7
DES	Software Design	1
VAV	Validation & Verification	7
EVO	Software Evolution	2
PRO	Software Process	5
	Total Studies	43

Within these numbers, there are some substantial groupings for particular Knowledge Units. In particular, nine of the studies classified as MGT were in the area of project planning, with a preponderance of cost estimation studies among these. (As this is an important topic, and one that teachers may not always be particularly expert in, this can of course be seen as a useful grouping.)

Fuller details of the studies are provided in Appendix A.

Further Reading for Part I

General Reading

There is a range of books addressing practice for systematic reviews, although usually written for use in disciplines other than software engineering. The following is a selection of books that we have found useful when learning our own way around this domain.

- Systematic Reviews in the Social Sciences: A Practical Guide. Mark Petticrew and Helen Roberts. This book is focused upon how to perform a systematic review, and so has been used by many software engineers, not least because it addresses many issues that are common to our discipline too. It was also available early on in the evolution of EBSE and hence helped to influence thinking about how this might be organised.

- *Systematic Approaches to a Successful Literature Review*, Andrew Booth, Diana Papaioannou and Anthea Sutton. This book approaches its subject matter from an "information science" viewpoint, rather than a discipline-specific one. And as might therefore be expected, it is particularly well provided with references to the research literature pertaining to systematic reviews. It provides a useful taxonomy of the different forms of systematic review, as well as offering practical guidance on the actual procedures.

- *Bad Science*, Ben Goldacre. Ben Goldacre enjoys a substantial (and well-deserved) reputation within the UK as something of an iconoclast, and as also being willing to take on a range of formidable opponents when it comes to mis-reporting of science (alternative therapies, pharmaceutical companies, newspapers, . . .). So you might ask why we see his book as useful, given that he is addressing *bad* practices. Well, one reason is that he sees an antidote to many of these bad practices as being to use a good one, with the use of systematic reviews being particularly important. Another is that he explains the benefits of a systematic approach in clear terms—as well as taking care to substantiate his arguments and to avoid excessive claims. So, while systematic reviews are not the main topic of the book, it gives some graphic ideas about the likely consequences of adopting some of the alternatives!

We also suggest that readers might find the following paper interesting.

Profiles in Medical Courage: Evidence-Based Medicine and Archie Cochrane, Richard A Robbins. Given that Archie Cochrane's work is considered to be the (substantial) pebble that set the evidence-based movement cascading down on to the world, it is quite interesting to know what motivated his views and how they emerged. Many articles have been written about him and about the factors that influenced his work, and this one is both representative and also very readable.

Undertaking a Systematic Review

The following two papers focus on the overall review process.

- *Lessons from Applying the Systematic Literature Review process within the Software Engineering domain*, Pearl Brereton, Barbara Kitchenham, David Budgen, Mark Turner, and Mohamed Khalil. This paper reports 19 lessons learned from early systematic reviews and mapping studies in software engineering. Most of the lessons are still relevant, although, with respect to lesson 13, we would now suggest that data extraction is performed independently by two or more reviewers rather than by one reviewer with one checker.

- *A Systematic Review of Systematic Review Process Research in Software Engineering*, Barbara Kitchenham and Pearl Brereton. This article also reports experiences of undertaking systematic reviews and mapping studies; however, it does so in a more formal way, reviewing studies that assess techniques that could be used to improve the review process. The paper discusses advice on performing reviews, proposes improvements to the guidelines and provides references to the 68 papers, reporting 63 studies, that address the review process.

A discussion of search strategies and of digital libraries and indexing services used in software engineering research can be found in the first of the following two papers. The second paper provides useful information for reviewers who are considering the use of a quasi-gold standard approach to validating their searches.

- *Developing Search Strategies for Detecting Relevant Experiments*, Oscar Dieste, Anna Grimán, and Natalia Juristo. This paper covers a broad range of topics relating to search strategies. Concepts such as optimal search, recall (sensitivity), precision and a gold standard are discussed as are some of the weaknesses of the commonly used digital libraries. Also, the precision and recall for a range of different search terms are compared using the set of papers found by Sjøberg et al. (2005) as the gold standard.

- *Identifying Relevant Studies in Software Engineering*, He Zhang, Muhammad Ali Babar and Paolo Tell. A detailed description of the proposed 5-step quasi-gold standard approach to searching and search validation is provided in this paper. The approach is clearly demonstrated and evaluated through two participant-observer case studies. Results from the case studies suggest that the approach is promising.

We are not aware of any software engineering research papers that focus specifically on study selection, other than those that propose the use of text analysis approaches. Although the approach looks promising most of the evaluation studies to date are rather limited.

Study selection is covered by a number of books on the use of systematic reviews in the social sciences and medical domain, including those by Booth et al. and Petticrew & Roberts listed above.

As indicated in Section 7.1, further discussion of the concepts of quality, bias and validity, can be found in *'Strength of evidence in systematic reviews in software engineering'* by Dybå & Dingsøyr.

Researchers undertaking reviews where the primary studies take the form of case studies can find further information about Runeson & Höst's checklists in their paper *'Guidelines for conducting and reporting case study research in software engineering'*.

The paper *Empirical Studies of Agile Software Development: A Systematic Review*, also by Dybå & Dingsøyr, provides details of their generic checklist and illustrates its use in a qualitative technology focused systematic review on Agile methods. This checklist can be used when a review includes primary studies that utilise a range of empirical methods.

Finally, we suggest looking at the SURE Critical Appraisal Checklists[1] and the work of the SURE team more generally, as they continue to develop and improve their checklists for quality assessment of a range of different types of study.

---

[1] http://www.cardiff.ac.uk/insrv/libraries/sure/checklists.html

Synthesis

In the paper *Systematic Mapping Studies in Software Engineering*, Petersen et al. (2008) provided a major incentive for the adoption of mapping studies in software engineering research. They proposed a review process for mapping studies and introduced the idea of representing classification information using bubble plots and the use of the classification system recommended by Wieringa et al. (2006) in their paper "Requirements engineering paper classification and evaluation criteria". This paper needs to be read before being adopted by mapping study analysts both to ensure that the proposed categories are appropriate to the mapping study topic area, and to understand the specific categories, in particular "Solution Validation" and "Implementation Evaluation". This is important because the differences between *validation* and *evaluation* is seldom well-understood and may be used inconsistently by primary study authors.

Qualitative meta-synthesis of qualitative primary studies is a complex and difficult task and there are a great many methods and techniques that can be used. When planning a qualitative meta-synthesis, it is useful to read a study that has used the same basic approach:

- For a meta-ethnography we recommend reading the paper *Using Meta-Ethnography to Synthesize Research: A Worked Example of the Relations between Personality and Software Team Process* written by Da Silva et al. (2013) which is organised as an example of the method.

- For a thematic analysis, we recommend the paper entitled "Recommended Steps for Thematic Synthesis in Software Engineering" written by Cruzes & Dybå (2011a) which provides a detailed explanation of the method. In addition, the paper "Using Qualitative Metasummary to Synthesize Qualitative and Quantitative Descriptive Findings" written by . (Sandelowski et al. 2007) presents an alternative to thematic analysis that can be used if the primary studies include little interpretive synthesis (that is, their outcomes are primarily lists of topics mentioned by participants, associated with frequency statistics).

If the primary studies include quantitative findings, but a meta-analysis is not possible, vote counting is an appropriate technique. We recommend reading the papers:

- *"Does the Technology Acceptance Model Predict Actual Use? A Systematic Literature Review"* written by Turner et al. (2010)

- *"Cross versus within-Company Cost Estimation Studies: A Systematic Review"* Kitchenham et al. (2007).

We would also recommend reading the technical report entitled "*Guidance on the Conduct of Narrative Synthesis in Systematic Reviews*" Popay et al. (2006) both for its examples of vote counting and qualitative moderator analysis and, more generally, for its broad overview of techniques that can be incorporated into a narrative synthesis and the worked examples of those techniques.

There are many good textbooks dealing with meta-analysis which are suitable for non-statisticians. Chapter 11 was based mainly on the book entitled "*Introduction to Meta-Analysis*" written by Borenstein et al. (2009). However, "*Understanding the New Statistics. Effect Sizes , Confidence Intervals and Meta-Analysis*" written by Cumming (2012) is particularity useful for statistical novices. Cumming is a strong advocate of using confidence intervals, effect sizes and meta-analysis to analyse statistical experiments rather than *null hypothesis significance testing*. Given the topic of this book, it is interesting to note that Cumming's textbook is *evidence-based* both in terms of its advice about statistical methods and in terms of its approach to learning and teaching.

For a good example of a software engineering meta-analysis, we recommend reading the paper *The Effectiveness of Pair Programming. A Meta Analysis* written by Hannay et al. (2009). This paper includes examples of forest plots based on both random-effects and fixed-effects models, sensitivity analysis based on assessing the impact on the meta-analysis results of removing one primary study at a time, and funnel plots to investigate publication bias including a trim-and-fill analysis.

Using Knowledge from Systematic Reviews

The paper *What Scope is There for Adopting Evidence-Informed Teaching in Software Engineering?* published as (Budgen et al. 2012) provides an illustration of a focused tertiary study. The aim was to identify where there was material from existing secondary studies that might be useful when teaching about software engineering. In this case the element of synthesis was provided by categorising the selected studies against a set of major topic areas. The material presented in Appendix A of this book is based upon an updated version of that study, that has been extended to include studies published up until the end of 2014.

Using EBSE in practice

In their paper *"Analyzing an automotive testing process with evidenced-based software engineering"* Kasoju et al. (2013) present the only direct application of evidence-based software engineering reported so far.

Kasoju et al. used a case study based on eight projects in a Swedish automotive company to investigate strengths and challenges/bottlenecks in the software test process. Data collection was based on interviews with project members.

Step 1 in EBSE is converting the need for information into an answerable question. From their analysis they identified the following:

- Strengths and weakness of the current process as viewed by teams using agile methods and teams using conventional waterfall processes. They note that some of the advantages identified by agile teams such as the developer and tester being the same person were regarded as problematic by larger teams.

- 10 "challenge areas" related to organization of testing, time and cost constraints, requirements, resource constraints, knowledge management, interactions and communication, testing methods, quality, defect detection and documentation. Each challenge area identified specific process problems. The specific process problems identified twenty-six issues for which evidence was needed.

Step 2 in EBSE involves tracking down best evidence. For this step, the authors used the top level research question "What improvements for the automotive testing process based on practical experiences were suggested by the literature?". They undertook a targeted search of testing studies with an emphasis on agile methods. In addition, they used the automotive software domain as one of their inclusion criterion. They looked for relevant literature in each of the 10 challenge areas and identified relevant information related to seven approaches to improving testing including requirements management, competence management, quality assurance and standards, test automation, test tool deployment, agile incorporation, and test management. They mapped the available evidence to each of the identified challenges.

Step 3 in EBSE involves critically appraising the evidence. For this activity they used *Value Stream Mapping*, which is a process analysis tool used for uncovering and eliminating waste. In the context of software engineering, waste relates to bottlenecks and delays. The authors constructed a "current state" map, identifying the process streams related to the five main test processes: test planning, test design, test build, test execution and reporting. They identified reasons for waste during the individual steps of each process and mapped them to the related challenges. Finally, they integrated the research evidence to produce a "future state" map. The future state map was based on the observation that the current testing approach did not suit the continuous flow of requirements, so there was a need for a new approach. They recommended an agile approach compatible with the best practices already used by small teams in the company, but including some more conventional aspects such as experienced-based testing, and documenting test plans. Although they were not able to validate the proposal solutions, the future state map was presented to the practitioners and includes their feedback.

Step 4 in EBSE is based on integrating the evidence with SE expertise and stakeholder values. Kajosu et al. do not mention this step explicitly but their use of automotive industry evidence, value stream mapping which was developed in the automotive domain, and current best practice from the projects used in their initial case study makes it clear that Step 4 was incorporated into Steps 2 and 3.

Finally, **Step 5** in EBSE involves evaluating the effectiveness and efficiency of steps 1-4 and seeking ways to improve the EBSE process. (Kajosu et al. refer to this as Step 4.) Reflecting on their study, they make the important point that problems were scattered across several different sub-areas of software engineering. This made a complete systematic review for all of the challenges impossible and was the main rationale for using a domain specific literature review. We entirely agree with their approach and believe it is completely consistent with the view that evidence-based practice should be tailored to SE expertise and stakeholder values. They also point out that it is difficult to use existing systematic reviews which are topic specific rather than problem specific. They stress the need for studies that address research questions that relate to practical issues such as " *Why testing windows get squeezed and what do we do about it?*" rather than academic concerns such as " *What do we know about software productivity?*".

Anyone seriously interested in evidence-based software engineering will find much of value in this paper. For software engineers working in the automotive industry, it should be essential reading.

Part II

The Systematic Reviewer's Perspective of Primary Studies

Chapter 15

Primary Studies and Their Role in EBSE

If evidence-based research is to be more than an academic exercise, and is to make a useful contribution to the discipline and practice of software engineering, then it needs to lead to evidence that can be used to:

- Deliver useful guidance about good practice, so that developers can determine what benefits they might get from using particular techniques, and be aware of the key factors influencing any effects;

- Enable organisations to make evidence-informed decisions about the adoption of policies related to software development and procurement;

- Underpin the work of standards bodies.

Because a systematic review is a *secondary* study, with its value coming from the objective synthesis of the outcomes of primary studies, these needs can only be met successfully if they are supported by:

- An objective and rigorous *process* for producing evidence; and

- A set of sound *empirical studies* that address the given topic.

Part I described how we can achieve the first of these through the procedures involved in conducting the different forms of secondary study normally employed for software engineering. In Part II, we now examine the second requirement, and identify what is needed if our primary studies are to provide useful input to a secondary study. To do so, we need to understand the forms and limitations of some major types of primary study, and how the reporting of their outcomes might best be organised to aid future synthesis.

Figure 1.5 presented a picture of the wider context within which systematic reviews are conducted. In Figure 15.1 we refine that model and illustrate the main relationships between primary and secondary studies that we will be examining in this chapter and the following chapters.

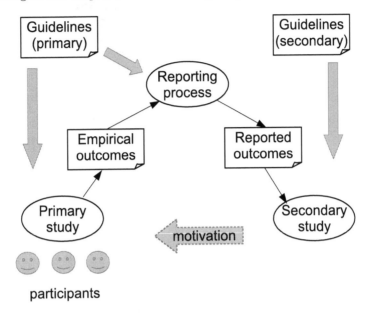

FIGURE 15.1: How primary and secondary studies are related.

We will not go into the fine detail about each form of primary study used in software engineering, not least because there are specialist textbooks that already provide that level of knowledge. Rather, our concern here is more with identifying and explaining what a systematic reviewer needs to know and understand about primary studies, the way that they are organised, and any constraints that affect them, particularly when assessing study quality. Like secondary studies, primary studies should be organised around a research protocol, and this in turn should take account of the likelihood of the outcomes being used in a secondary study. Indeed, as Figure 15.1 emphasises, the reporting process acts as a filter between the actual outcomes from a primary study, and what is eventually used for synthesis in a secondary study. Hence it is important to consider the organisation of reporting as a process in its own right.

How to report primary studies for the purpose of future synthesis is, quite understandably, rarely addressed in books about empirical practice. However, one of the frequent observations that systematic reviewers make when reporting on the conduct of their own study is the poor quality of reporting found for primary studies. There are some indications of a gradual improvement in the conduct of primary studies, as reported by Kitchenham et al. (2013), but clearly, from the comments of many frustrated reviewers, there is still room for improvement!

This chapter forms a brief introduction to the material in Part II. It provides an overview of some of the key forms of primary study used in software engineering and of their roles, together with an outline of what aspects of a primary study need to be reported, when considered from the viewpoint of its usefulness to a systematic review(er). We then elaborate further on both of these in the next few chapters when we look at some specific empirical forms. In addition, we briefly introduce the issue of *replication* from the viewpoint of the systematic reviewers, with this being followed up in greater detail in Chapter 21.

15.1   Some characteristics of primary studies

In Part I we focused upon the way that *secondary* studies are organised, and how this involves finding and aggregating the outcomes of relevant *primary* studies. So, the role of a secondary study can easily be defined in terms of its relationship to the constituent primary studies. Primary studies, however, exhibit a greater variation in form, and before examining some major forms, it is useful to briefly consider what it is that specifically characterises an empirical study as being a *primary* study, and two of the key concepts that are related to this.

The main distinguishing characteristic is that a primary study involves the researcher in making some forms of *measurement* that are related to the "attributes of interest" for the topic of the study. Hence we can expect that a primary study will usually involve using software engineering techniques or technology in some way, and collecting data about this use in a systematic manner. How exactly this is done will depend upon the form of study being performed.

Measurement is itself a large and complex topic in its own right, although we only require some very basic ideas about it here. In the context of software engineering, the book by Fenton & Pfleeger (1997) provides much more comprehensive coverage of these issues. For now though, we are mainly concerned with two particularly relevant concepts, namely those of the *attribute* and of the *measurement scale*.

An *attribute* is a measurable (or at least, identifiable) property of an entity, and hence measuring this in some way can be expected to tell us something about that entity. In empirical studies related to software engineering we encounter various forms of entity, including:

- A *process* or procedure that is performed by people, such as pair programming, writing test cases, or performing software inspection tasks;

- A *tool* such as a compiler, version control system, or web browser;

- A software *product*, or part of one, such as a method, an object, or a software service.

While each of these will have many properties and hence many attributes, an empirical study will usually only be concerned with those that are of relevance to the research question, although an awareness of the possible influence of the others may well be relevant. So part of the task of developing a research protocol for a primary study will be to identify the key entities, and hence the attributes of interest to the study. These attributes may in turn be ones that can be measured *directly* such as the number of lines of code in an object, or the number of dependencies that it has upon other objects; or that can only be measured *indirectly* by measuring other attributes and then combining these, as occurs for effort estimation.

Identifying and collecting suitable measures of attributes is an important element in designing a primary study. The choices made will also be important for the systematic reviewer, and will help determine how well one is able to aggregate the outcomes from different primary studies.

The second concept, that of a *measurement scale* is closely related to that of an attribute, since the choice of a suitable scale is related to the characteristics of the attribute. Measurement scales are summarised in the Glossary section, and here we just briefly review three forms of scale that are quite widely used in software engineering.

- *Nominal* scales are used for categorisation (gender, programming language, type of software component ...), and there is no sense of there being any ordering of the elements of the scale.

- *Ordinal* scales are widely used and often generated by the use of a Likert response format (discussed in Section 19.2), and these provide a sense of ordering, but do not have any specific 'distance' between each element in the scale. Such scales are often used when assessing more qualitative concepts such as *quality* (very poor, poor, adequate, good, very good) or when seeking to *rank* features of entities in some way.

- *Ratio* scales relate to 'countable' items, with a zero point and fixed intervals between the points on the scale. Examples of their use in software engineering include counting lines of code, measuring elapsed time, and counting the number of completed tasks. They are widely associated with *quantitative* attributes.

The scales employed for assessing an attribute determine the form of analysis that can be used in both the primary and secondary studies. And as indicated, these concepts are particularly important for primary studies, as these typically involve investigating the properties of software engineering entities.

15.2 Forms of primary study used in software engineering

Although the terms 'experimental' and 'empirical' are sometimes used as synonyms, especially by those unfamiliar with empirical studies, it is important to appreciate that 'formal' experimentation is only one way to conduct empirical investigations. As observed in Chapter 1, we may well begin our investigation into a possible phenomenon by conducting studies that are of a more informal *observational* nature, before undertaking more rigorous experimentation and collecting more systematic observations and measurements. Indeed, experimentation may not be the most appropriate way to address many of the issues we address in software engineering, such as the adoption of techniques and tools, and when performing assessments of specific software tools and systems. Fortunately, the 'empirical toolbox' is well provided with an assortment of empirical techniques that have evolved across a range of disciplines, and so the empirical software engineering researcher's main task is usually one of selecting from these to meet a particular need.

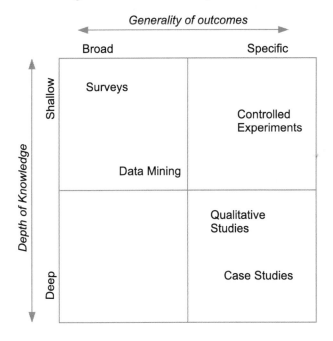

FIGURE 15.2: Primary study forms in the depth/generality spectrum.

When determining the form of study that is likely to be most appropriate for addressing a given research question, two key factors are:

- Determining the *depth* of knowledge required;

- Identifying the degree of *generality* needed for that knowledge.

Unfortunately, from a practical perspective, these are not independent choices. Given finite resources, it is unlikely that an empirical investigation can be both deep and also general in its scope. Figure 15.2 illustrates this concept for the forms of study that we address in the next few chapters, and that have been selected as being ones that are quite widely used in software engineering. We might add that although other empirical forms are in use, none of these overcome this constraint.

The positioning of the elements in Figure 15.2 should not be taken too literally, as these characteristics will clearly vary quite considerably for individual studies. What is more important is where they are positioned within the grid, and in particular, that none can simultaneously provide both depth of knowledge and generality. These characteristics are explained a little more below.

Controlled Experiments & Quasi-Experiments. These are well-suited to being used to answer quite focused research questions, often of a comparative nature (such as 'does the use of technique A lead to fewer software faults than occur when using technique B?'). However, for human-centric experiments, which are widely used in software engineering, the depth to which the issues around such a question can be addressed is constrained both by the willingness of participants to undertake lengthy tasks, and also by the need to recruit sufficient participants. We explore the use of these further in Chapter 16.

Surveys. The use of a survey makes it possible to answer certain types of research question by aggregating inputs from a large number of respondents. It may well be possible to obtain a sample from people who have a wide range of backgrounds, or who belong to very specific groupings, perhaps of those with particular types of skill or experience. Again though, in order to obtain a large enough sample of respondents who will complete the survey in full, the size of the survey task will need to be kept relatively small, making it difficult to explore issues in depth. In Chapter 17 we do examine ways in which selective "follow-up" activities might make it possible to add depth, at least for a reduced sample, but overall, surveys definitely fit into the 'shallow' category.

Case Studies. The use of case study research has increased considerably in software engineering in recent years, as we explain in Chapter 18. Indeed, for addressing some types of research question (such as those relating to adoption of techniques 'in the field'), this may well be the principal form of study used. Case studies do allow for deeper probing of an issue, and can be sustained over much longer periods of time than is usually practical for experiments. In exchange, the effort expended upon each case usually makes it impractical to conduct very many of these, so that the results usually lack generality.

Qualitative Studies. While experiments, surveys and case studies can, and do, collect qualitative data, they are largely regarded as being part of the 'quantitative toolbox'. Like case studies, qualitative forms can provide considerable depth of knowledge, but such studies are almost invariably very labour-intensive to conduct, making it difficult to use them to achieve other than very specific knowledge. We discuss what forms such studies may take in Chapter 19, where we also consider the types of question that may best be addressed through their use.

Data-Mining. The use of this is more appropriate to research questions that do not require human participants. Indeed, some surveys are effectively conducted by using a restricted form of data-mining. We discuss its use more fully in Chapter 20. In principle, with large amounts of data and large amounts of processing power, use of this form could be both general and deep, but achieving that is highly dependent on the available data and upon how detailed the questions that can be asked through its use can be. For that reason, the position given to this form in Figure 15.2 indicates that it can potentially probe more deeply than a survey, but will still not be very deep.

While other forms of empirical study, such as ethnography and action research, have been used quite successfully for software engineering tasks, their use is relatively uncommon, and so far, they do not appear to have been used to provide input to systematic reviews, at least, not to any significant extent.

Another important distinction between primary studies and the secondary forms that we have described so far, apart from making measurement, is that many of them involve direct contact with human participants. Because of this, we need to consider whether our studies are likely to involve any *ethical* issues, and if so, what forms these might take.

15.3 Ethical issues

When we involve people in our primary studies, we commonly refer to these as *participants*, rather than *subjects*. This is because involvement both in experimental forms of study, and also more observational forms, usually requires that the people actively perform tasks, in contrast to being recipients of some form of treatment.

Where our studies directly involve people in any way (and we include surveys in this too), we need to ensure that such a study is performed in an ethical manner—by which we mean that it will not disadvantage or harm those taking part (or conversely, give them an unfair advantage over their peers who do not participate, which is a situation that can arise where students form the source of participants).

Ethical issues have long been recognised as important by the major professional bodies, as witnessed by the ACM/IEEE-CS code of ethical conduct and professional practice (IEEE-CS/ACM 1999, Gotterbarn 1999). However, their emphasis is largely concerned with the responsibilities of the software engineer to the customer, employer and public, and when in the role of a producer of software systems. Our concern here is with a rather different aspect, which is the ethical responsibility of the researcher to those who are aiding them in their research.

Fortunately this is an issue that has long been recognised by other human-centric disciplines such as psychology, education and health. So most universities and similar organisations will have appropriate procedures for giving approval to studies that involve humans. Certainly any researcher planning a primary study of such a form will need to find out what their local procedures are, and will need to address these when writing the protocol, as well as seeking approval before actually beginning the study itself. The nature of our research means that this should usually be a fairly straightforward process, and any issues are more likely to arise from the way that the study is organised than from the study treatment or topic.

So, what sort of issues do we need to consider in our planning? We suggest a few examples of these below—most are related to experiments and also to the use of student participants, since this is a common situation, but they may also apply to surveys and observational forms of study too.

Informed participation. We need to ensure that any participants are fully aware of what is being asked of them in a study. So it is usually good practice to ask participants to begin their involvement by reading through a written explanation of what they are being asked to do, giving them the opportunity to ask any questions about this, and then get them to 'sign' consent to their participation (the act of signing may simply be one of pressing a button where the study is performed on-line). Indeed, the inclusion of such a consent form is often required when submitting an application for ethical approval.

Pressure to take part. Recruiting participants can sometimes be quite challenging, because in general both students and practitioners are unfamiliar with empirical studies. Since the researchers may also be teachers, their classes may feel that refusing to take part might count against their final grades. To avoid this, it is better if others (such as assistants) help to do the recruitment and avoid mention of who is actually leading the study.

Training benefits. Many software engineering experiments require some degree of training for the participants. Again, for students in particular, taking part could be perceived as offering the benefit of additional tuition and experience (creating further pressure to participate). This can largely be avoided if the training is made freely available to anyone, not just those taking part in the study.

Collecting demographic data. Our studies often require some element of 'context' about the participants, such as their level of technical knowledge, education etc. It is important to collect only the information that is relevant and needed for analysis, and also to ensure that any data recorded is managed in accordance with data protection legislation.

Reporting. This needs to avoid anything that might identify individuals in any way, particularly where we are analysing or comparing abilities or skills. This is probably a greater risk with case studies than with experiments or surveys, but the the authors of any forms of report do need to be aware of it.

As a final thought from the perspective of the subject matter of this book, the systematic reviewer also needs to be alert to any possible effects arising from some of these issues (such as participant recruitment). If participants felt pressurised to take part, could this have had an influence on the validity of the results from the study?

15.4 Reporting primary studies

While reporting about a study well has always been an important goal for researchers, the advent of secondary studies introduces a further perspective on this task. Since secondary studies form the focus of this book, we therefore address the issues related to their needs first, and then briefly consider more general needs of reporting.

15.4.1 Meeting the needs of a secondary study

As illustrated in Figure 15.1, the way that a primary study is reported can play an important role at almost every stage of any type of systematic review. Since we are concerned here with the use of data from primary studies in secondary studies, it may be helpful to begin by considering what the systematic reviewer will be doing while performing their review, and whereabouts in a report they might expect to find the information needed to support these activities.

1. To successfully *find* the report using a sensible set of keywords. (*Title, Keywords, Abstract.*)

2. To be able to confirm that the material in the report addresses the *research questions* of the systematic review easily and quickly, without needing to read the entire document. (*Title, Keywords, Abstract.*)

3. To be able to identify the *goals* and research question(s) of the study being reported. (*Abstract, Introduction.*)

4. For a systematic review (as opposed to a mapping study), to be able to confirm that the primary study is of *good quality*, and that its design and conduct conforms to appropriate "good practice guidelines". (*Methodology* and *Conduct* sections.)

5. For a systematic review, to be able to obtain the information needed for *synthesis*. (*Results* section.)

6. To be able to identify any *limitations, constraints* or *context factors*, such as any *divergences* from the original plan described in the experimental protocol that may relate to, or affect, the results. (*Discussion* section.)

7. To be able to use this report to help find *other* related papers. (*References, Discussion section, Related Work* section.)

The first three and the last of these are very general points that apply to all types of primary study, such as the way that title, abstract, keywords and references are used. The others, relating to data extraction and synthesis are more specific to individual types of study. So here we make some points about the parts of a paper used for this first group of activities, leaving the others to be covered more fully in the relevant chapters.

The Title. Clearly, getting all of the relevant information into a title, while still keeping a manageable length, can be rather a challenge. Our advice here is to use a sub-clause in the title such as ":An empirical study". So an example of this might be "Test-first versus test-last development: An experimental study". This provides the reader with information about the topic (testing), about its form (a comparison between two testing strategies) and an indication that the paper is an empirical one.

The Abstract. The abstract is a key element both for searching and also for making decisions about inclusion. It is important to recognise that an abstract is just that—a summary of the key elements of the paper, including the outcomes and conclusions. Unfortunately, many software engineering abstracts do seem to be rather poorly written, possibly because they have been produced in a hurry just before the paper is submitted. Empirical researchers need to be aware that their abstract is important and that writing it well does merit the expenditure of time and effort. We are strong advocates of the use of *structured abstracts* (Budgen et al. 2008, Budgen, Burn & Kitchenham 2011). A structured abstract is organised under a set of headings (such as: Context, Aim,

Method, Results, Conclusions) with one or two sentences per heading. We provide an example of such an abstract in Figure 15.3. Even if a paper is not an empirical one, this form is still useful, so we strongly advocate their use with all forms of software engineering papers.

Abstract

Context: In teaching about software engineering we currently make little use of any empirical knowledge.

Aim: To examine the outcomes available from the use of Evidence-Based Software Engineering (EBSE) practices, so as to identify where these can provide support for, and inform, teaching activities.

Method: We have examined all known secondary studies published up to the end of 2009, together with those published in major journals to mid-2011, and identified where these provide practical results that are relevant to student needs.

Results: Starting with 145 candidate systematic reviews (SRs), we were able to identify and classify potentially useful teaching material from 43 of them.

Conclusions: EBSE can potentially lend authority to our teaching, although the coverage of key topics is uneven. Additionally, mapping studies can provide support for research-led teaching.

FIGURE 15.3: Example of a structured abstract.

Keywords. How useful these are to a review team probably depends very much upon how constrained the choice is. Our experience is that where authors are constrained to use a 'standard' set of keywords, such as that produced by the ACM, these keywords are rarely very useful because the set is unlikely to contain any that are appropriate for use with empirical studies. However, when the choice of keywords is unconstrained then, as with the title and abstract, they can provide the means of positioning key terms so that they can be found by a search engine.

Introduction. In part, the introductory section of a paper should overlap with, and expand upon, the information provided in the abstract. Apart from explaining more about the rationale for undertaking a study, it should contain the research question for the paper, as well as any hypotheses or propositions (where relevant). It should also say something about the research method used, and indicate why the results are important.

Related Work. While this term is not always the one used (a section title such as *Background* is usually equivalent), the value of such a section is that it identifies the studies that the authors consider to be comparable with their own in some way, or that act as some form of baseline. For the reviewer who is seeking to snowball from a paper, this is usually a good section to consult (another is *Discussion*). As such therefore, this section acts as a useful index to the list of references.

References. These should be as comprehensive as possible, not just to aid systematic reviewers, but also to position the contributions of the work, and possibly to aid any future replication.

One last point related to finding primary studies is that wherever possible, authors should seek to use established terminology. Inventing eye-catching new terms or acronyms might seem like a good idea, but doing so increases the risk that your paper will be missed by a search.

15.4.2   What needs to be reported?

For this more general aspect of reporting, which is related more to the study itself rather than its use in secondary studies, we do have the benefits of more established guidelines. Usually, these have been based in some way upon analysis of published studies.

One of the first such sets of guidelines was that produced by Kitchenham, Pfleeger, Pickard, Jones, Hoaglin, Emam & J.Rosenberg (2002). These address the following six areas for a range of study types (we indicate the number of guidelines provided for each area).

- Experimental context (4 guidelines).

- Experimental design (11 guidelines).

- Conduct and data collection (6 guidelines).

- Analysis (5 guidelines).

- Presentation of results (6 guidelines).

- Interpretation of results (4 guidelines).

Later work by other empirical researchers has produced further guidelines, usually targeted at specific forms of study such as experiments and case studies. We will identify these in the relevant chapters, but will not go further into detail on this aspect.

15.5   Replicated studies

We will be examining this topic in much more depth in Chapter 21, so at this point, the main issue to raise is the potential value that replicated studies may provide. In any discipline, new knowledge is only widely accepted when the original study can be replicated by other researchers, preferably ones who are working independently in other institutions. For software engineering, the use of replicated studies has proved to be quite a challenge. Indeed, it has been observed that, in software engineering, replications conducted by the same experimenters usually obtain results consistent with the original studies, whereas those performed by other researchers do not always lead to similar results (Sjøberg et al. 2005, Juristo & Vegas 2011).

The reason why we mention this here, is that in software engineering we both need to conduct replications and also to report them. As referees for journals and conferences, we also need to recognise that the report of a replication makes a valid contribution to knowledge, and that researchers do not necessarily have to add new features to the issue under study in order to justify submitting a report on the study. This is something of a cultural problem for software engineering, for which it often seems that a paper needs to offer 'new' ideas if it is to be published, although this does occur for other disciplines too, further compounding the problems of *publication bias*.

On a practical note, this is one occasion where the "grey literature" may well be worth checking. There is reason to believe that many reports of replicated studies often do not get beyond "Technical Report" status, and so even when the inclusion criteria for a systematic review might normally require a paper to have been published in a refereed source, reports of replicated studies could well be considered as an exception.

Further reading

Each of the chapters of this second part suggests some further reading that is related to the specific topic of the chapter. For this chapter, we suggest that the book "*Researching Information Systems and Computing*" (Oates 2006) offers a very good overview of the wide spectrum of empirical research methods that can be, and largely have been, employed for software engineering studies.

Chapter 16

Controlled Experiments and Quasi-Experiments

Experimentation (in the formal sense), as conducted in software engineering, commonly involves a set of human participants being asked to perform software development-related tasks by using one or more specific procedures. For this chapter we will concentrate upon human-centric studies and their organisation, both because they are quite widely used (and hence likely to form inputs to a systematic review), and also because their use introduces many additional complications that then need to be considered when designing an experiment.

However, we should note that the methods used for some tasks (such as generating unit tests, or selecting unit tests from a 'pool' of tests, as good examples) are based on computer algorithms, and are, therefore, not human-centric. Such experiments are often organised as quasi-experiments where the different algorithms are each invoked under a series of different testing scenarios.

The organisation of this chapter is one that we will be using for the next few chapters where we examine some different forms of primary study. We begin by discussing the general nature of the particular form, and how it might be characterised, after which we look at how such studies are organised and conducted in software engineering. We then briefly examine the type of software engineering research question that might be answered through its use, followed by some examples of how these research questions are addressed, taken from the software engineering literature. We then return to the theme of this book and consider each form of study from the perspective of how it is used in secondary studies, and in particular, how data extraction and synthesis might best be organised in order to best reflect its characteristics. We also provide some suggestions for further reading.

16.1 Characteristics of controlled experiments and quasi-experiments

A controlled experiment is a way of organising an empirical study that (conceptually at least) is employed across a wide range of disciplines. The underlying philosophy is one of establishing a link between a *cause* and an *effect*, whereby we have some 'theoretical' model of how the two phenomena are linked, and want to determine how far our model is correct. Ideally, we make use of our model to create a *hypothesis* in the form of a statement about the way that particular changes in one quantity or phenomena (the 'cause'), will lead to changes in the other (the 'effect'). It is then the role of the experiment to *test* the hypothesis and enable us to decide whether it is true, or false.

16.1.1 Controlled experiments

In the context of software engineering experiments, we are usually concerned with comparing the efficiency or effectiveness of two different methods of performing a software engineering task, procedure, or process. In the context of an experiment, the methods may be referred to as the *treatments* or *interventions*. In order to make such a comparison, we need to identify some way of measuring efficiency or effectiveness. The specific measure or measures are called the *dependent* (or response) variable(s). For example, we may want to determine whether one method of reading a software specification will on average find more defects than another method. The simplest design for such an experiment involves comparing the average number of defects found in a software specification by a group of people using one reading method with the average number found by a group of people using the other method. If one of the treatments can be regarded as the normal or baseline reading method, it is referred to as the *control* method. Thus, a simple controlled experiment would have the framework shown in Figure 16.1.

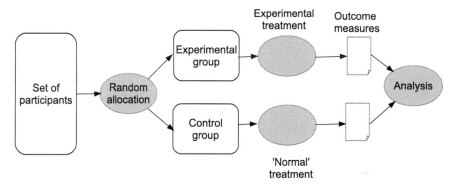

FIGURE 16.1: The framework for a controlled experiment.

In addition, relating to software engineering *methods*, treatments may refer to *participant characteristics*, such as software engineering experience, or *characteristics of the materials* used to perform the task, such as the complexity of design documents or code. If these characteristics can be mapped to ordinal-scale measures such as high, medium, or low, we can organise our experiments to investigate their impact on the effectiveness or efficiency of performing a task. Treatments that correspond to ordinal-scale characteristics are usually referred to as *factors*. We can also investigate the impact of different software engineering methods and ordinal characteristics in the same experiments by using two-way designs or factorial experiments.

In a controlled experiment *randomisation* is a major design element. There are two different types of randomisation:

Random sampling which involves obtaining a random sample of *participants* from a defined population of participants. We use the term *participants* rather than *subjects*, because in most software engineering experiments participants must use their personal skill and knowledge to perform a software engineering task. In contrast, human participants in some other disciplines, for example medicine, are often referred to as *experimental subjects* because they are usually passive recipients of some treatment (for example, a drug) prescribed by a doctor, and are required only to take the drugs as frequently, and for as long as, the doctor specifies. In a random sample, each member of the population has an equal probability of being selected for participation in the experiment. We note that, even in disciplines such as medicine or psychology, it is difficult to specify the *population* you want to sample, let alone identify all members of that population so that a random sample can be obtained. This means that samples are very seldom obtained by random sampling, and researchers rely mainly on convenience sampling. Convenience samples cause problems if we want to generalise the results of an experiment. This aspect of randomisation is related to the concept of *external validity* which will be discussed later in this chapter.

Random allocation to treatment which means assigning each participant to a treatment group at random. Random allocation to a treatment means that each participant has an equal probability of being in each treatment group. Even if we do not start our experiment with a random sample of participants, if we allocate them at random to the different treatment groups, we can still investigate cause-effect relationships. This aspect of randomisation is related to the concept of *internal validity* which will be discussed later in this chapter.

16.1.2   Quasi-experiments

The major difference between a controlled experiment and a *quasi-experiment* is the use of random allocation to treatment. Quasi-experiments are used when it is infeasible or impossible to perform random allocation to treatment. This situation can occur if the "treatment" we are interested in is an ordinal scale measure of the characteristics of the participants or the specific task, or specific materials.

One of the most common uses of quasi-experiments in software engineering is when we have relatively few potential participants and we want to minimise extraneous variability in response variables, so we ask participants to try out each treatment, and act as their own control. In this case we are deliberately deciding to use a quasi-experiment rather than a randomised experiment. It is important that researchers understand the difference between the two designs and the advantages and limitations of each design. A similar approach is taken in technology-centric testing experiments, when we have a number of programs with a known number of defects, and we evaluate the performance of a set of competing algorithms or tools undertaking the same automated testing task on each of the identified programs.

Quasi-experiments can also be used to track the impact of organisation-wide changes. For example, if working practices in a software development organisation are changed, so that projects before one date used one set of working practices and projects in a subsequent period used a different set of working practices, we can use a time-series based analysis to track the impact of the changes on project characteristics such as productivity and delivery time. This example illustrates how quasi-experiments allow us to perform experiments within an industrial setting.

Conducting formal experiments in the field is extremely difficult because of the complexity of large scale software engineering. Industrial software engineering requires a complex mix of skilled processes (for example, requirements engineering, design, coding, testing, maintenance). The outcomes of a specific process become the input materials for other processes. In large organisations, individuals work in teams that have specialist responsibilities either process-oriented ones (such as a systems architecture role) or product-oriented ones (such as having responsibility for specific applications or components). The allocation of individuals to specific tasks depends on their knowledge of existing components, or involvement in the preceding processes. Furthermore the impact of changes to early life-cycle activities may not be observable until much later in the development process. All these factors make field experiments difficult, so quasi-experiments, which were initially developed to assess the impact of legislative changes or social improvement programs, become a more practical option for evaluating the "real" performance of software engineering methods.

Although not always recognised as such, quasi-experiments are used quite widely in software engineering (Kampenes, Dybå, Hannay & K. Sjøberg 2009). Quasi-experiments can take many different forms (Shadish et al. 2002), although only a few of these are normally used in software engineering. In Section 16.2, where we look at how experiments and quasi-experiments are organised and conducted, we discuss some of the main forms of quasi-experiment that are of relevance to software engineering.

Regardless of the form of an experiment, the fundamental goal of its design is to enable adequate testing of the hypothesis. In turn, the task of testing the hypothesis remains a *statistical* one, since we need to differentiate between the effects that could happen at random, and those that might arise from the intervention. We discuss this further in the next section.

16.1.3 Problems with experiments in software engineering

In the context of performing software engineering tasks, human participants have much more impact on the outcome of software engineering experiments than do subjects in other disciplines such as medicine or psychology because they are usually asked to perform skilled tasks, the outcomes of which depend on their individual ability; they remember things they have done before and learn from them; they make decisions that might well be influenced by a range of factors, and may not even be conscious that some of these influences exist. In addition, when they take part in a study they are aware that the activities they are undertaking are being performed within some form of experimental context. These issues lead to specific problems with formal experiments:

- *Incompletely specified control treatments*, which arise when the control method is assumed to be well understood by everyone but may also be implemented slightly differently by each participant. For example, if the treatment being studied is the use of pair programming, and the control is the use of 'solo programming', then while we will probably have considerable methodological guidance as to how the pair programming tasks should be performed, solo programming tasks may not be well-defined. Lack of a defined control is likely to increase the variability of outcomes from participants using the control method and could make it more difficult to detect real differences between the methods. It might also make it difficult to undertake comparable replications of experiments and, so, diminish the opportunities for reliable meta-analysis.

- *Experimenter and subject expectation bias*, which occurs because people's expectation can influence the outcome of experiments (which is discussed in more detail below).

- *Difficulties performing controlled field experiments*, which make it hard to provide strong evidence about the behaviour of our methods and tools in practical industry situations. Commercial software development processes are often extremely complex, involving the development of systems and components over long periods of time, and they seldom give rise to small isolated tasks that can be used as the basis of formal experiments.

- *Software experiments are often small-scale laboratory experiments* involving small stand-alone tasks, so it is questionable, therefore, whether the results of such experiments "scale-up" to the complexity of software engineering tasks undertaken by practitioners[1]. This, of course, is related to the difficulty of performing field experiments. We note the some researchers strongly advocate the use of more realistic laboratory experiments such as using (and paying for) professional participants rather than student participants (Sjøberg, Dybå & Jørgensen 2007).

- *The need for special training to use new software engineering methods* means that the participants in a treatment group may be treated differently from those in a control group, although the basic experimental process is based on the premise that the *only* difference between the two groups is the use of the method. In particular, treating participants in the treatment groups differently can affect participant expectations. Furthermore, using the setting of a training course to investigate the impact of a new methods is also problematic.

---

[1]It is also possible that some methods that would be valuable in the context of a real industrial situation would appear to be an unnecessary overhead in the context of small isolated tasks.

When people take part in an empirical study, even as rather passive recipients of a treatment, psychologists have found that there are a number of human factors issues that can cause experimental bias, in particular subject and experimenter expectations, see for example, Rosnow & Rosenthal (1997).

Experimenter expectation bias occurs when the experimenters expect one treatment to be better than another and influence (often unintentionally) the experiment in order to "prove" that their expectation is correct. In such circumstances, they could assign the individuals most likely to benefit from a specific treatment to that treatment group. Random allocation to treatment aims to minimise the likelihood of this form of *selection bias*. However, if experimenters know to which treatment group each participant has been assigned, they can encourage subjects in the preferred treatment group to believe they are receiving the best treatment or using the better software engineering method which sets up *subject expectations* about how the experiment will progress. Unfortunately, even simple encouragement for, or against, a specific method can influence the outcomes of a human-intensive experiment.

In the context of medicine, in addition to random allocation, the main strategy to reduce experimenter and subject expectation is *blinding*. Blinding the participants means that they are unaware of whether they are receiving the treatment or a placebo, and also blinding the person administering the treatment, so that they cannot subconsciously provide clues to the recipient (*double-blinding*). However, in the context of software engineering, blinding is rarely a practical option and we need to consider other methods to reduce these effects. Rosnow & Rosenthal (1997) discuss strategies adopted in behavioural research to control experiment and subject effects including: increasing the number of experimenters, monitoring the behaviour of experiments, statistical analysis, maintaining blind contacts, minimizing experimenter-subject contacts and using *expectancy control groups*. These techniques are intended to reduce the extent to which individual experimenters can influence outcomes and to monitor and correct for any observed forms of bias.

16.2 Conducting experiments and quasi-experiments

A controlled experiment involves comparing what happens under one experimental condition with what happens under an alternative experimental condition. A simple example would be comparing the effectiveness of two unit testing methods (for example, functional testing versus structural testing).

The process of conducting an experiment involves:

1. *Formulating your research question as a testable hypothesis.* This requires defining *dependent* variable(s) and how it will be measured.

2. *Planning an experiment that is appropriate to test the hypothesis.* This involves:

- Defining the treatments at an operational level. For example, defining the coverage of the component required for the structural testing and the method to be used to identify functional tests from the parameters on the component interface. Also, if the treatment is one that can be applied with different degrees, levels, or amounts for example, a quality control process that can be either *stringent,* *standard,* or *lightweight,* quality control would be referred to as a *factor* with three *levels.*

- Specifying the design of the experiment. For example, how many software components will be tested (one or many), what software components will be used, how many participants will be required, how participants will be recruited, and how participants will be trained.

- Identifying any problems inherent in the plan (which is often referred to as the experimental *protocol*) and where possible, adopting procedures to minimise any problems.

- Defining how the data will be analysed. The analysis should be derived from the experimental design but should consider issues such as whether the dependent variable needs to be transformed or not and what sensitivity analyses will be performed.

3. *Conducting the experiment according to the plan.* It also involves recording any deviations from the planned process, such as participants dropping out of the study, or participants failing to adhere to the treatment to which they were assigned.

4. *Analysing the results of the experiment* in order to test our hypotheses.

5. *Reporting the study results,* including their implications for software researchers and practitioners, and a discussion of any limitations of the study which are usually discussed in terms of *threats to validity.*

We discuss the most important of these issues below.

16.2.1   Dependent variables, independent variables and confounding factors

Experiments are concerned with establishing whether some form of *cause* and *effect* relationship exists. Testing for this usually involves the experimenter in manipulating one or more *independent* variables associated with 'cause', and measuring one or more *dependent* variables associated with 'effect'.

The *independent variable(s)* are specified or controlled as a result of the activities of the investigator (for example, the number of errors 'seeded' in the case of a study on unit testing, the length of an item of software, the time allocated to a task, or the form of procedure to be followed).[2]

---

[2]We do not address controlled regression analyses, such as those used to establish the

The *dependent variable(s)* (also called *response* variables) are associated with 'effect' and are expected to change as a result of changes that the experimenter makes to the independent variable(s). If we can demonstrate that these changes are as predicted by the hypothesis, and are statistically significant, then the hypothesis can be considered to be supported by our results. Measuring any changes in the values of the dependent variables is therefore necessary in order to assess the outcomes of the experiment. In software engineering experiments, dependent variable measures usually relate to the performance of software engineering tasks such as staff effort, productivity, elapsed time or defect rates.

However, where experimentation involves people, there are likely to be other factors that might affect or influence the outcomes of a study, and which also need to be considered. We term these the *confounding factors*. A *confounding factor* is the presence of some (undesirable) element in an empirical study that makes it difficult to distinguish between two or more possible causes of an effect (as measured through the dependent variable). For software engineering two commonly-encountered examples of confounding factors are the skill levels of the individual participants and the extent of their prior experiences with the experimental subject matter.

Designing an experiment or quasi-experiment therefore involves careful choice of independent and dependent variables. These need to be appropriate measures in terms of the hypothesis, and the independent variables also need to be controllable, while the dependent variables should be readily measurable. The experimenter also needs to consider the presence of likely confounding factors and to seek to design the experiment so as to keep their likely effects to a minimum.

16.2.2   Hypothesis testing

The baseline for any statistical testing is the *null* hypothesis, which states that there is no effect from the treatment and therefore that any difference we see between the average value of the dependent variables for each of the treatment groups is a matter of chance. More formally, the role of a statistical test is then to distinguish between the *null* hypothesis and the *alternative* hypothesis, that states that an effect occurs because there is some form of cause-effect relationship present. This is illustrated in Figure 16.2.

Testing can be one-sided (one-tailed), where we are only interested in the effect of the treatment being to increase (or decrease) our measure. It can also be two-sided (two-tailed), where we are only interested in the question of whether the values produced by the treatment are different from those of the control. Hypothesis testing is treated as a decision process in which we have

---

optimum settings for industrial processes where the one or more independent variables vary over a pre-specified range of values and the dependent variable is obtained for each combination of independent values.

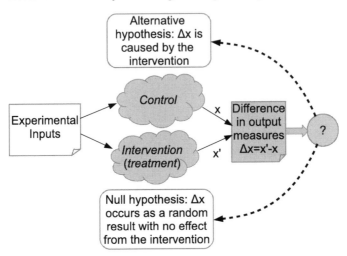

FIGURE 16.2: Hypothesis testing through use of an experiment.

two choices: reject the null hypothesis or do not reject the null hypothesis. We can also make two decision errors:

1. We can reject the null hypothesis when it is actually true. This is called a **Type I error**. By convention, if the chance of a Type I occurring by chance is 5% or less (a *probability* value of $p < 0.05$), we consider that the association does demonstrate the presence of a genuine effect. The probability of a Type I error is usually denoted by the use of the symbol α, and so if in a report of a study we see that the value of this is reported as (say) $\alpha = 0.01$ then we know that the statistical tests used by the experimenters have indicated a likelihood that there is no more than a 1% chance that the outcomes could have occurred by chance.

2. We can fail to reject the null hypothesis when it is really false. This is called a **Type II error** and the probability of a Type II error is referred to as β. The related concept of *statistical power* $(1 - \beta)$ is a measure of the probability of correctly rejecting the null hypothesis. While some software engineering experiments do report a value for statistical power, this is relatively uncommon (Dybå et al. 2006). However, it is important to note that α and β measure different probabilities, and if the value of β is low (for example < 0.5), the power of the experiment will be low, and we have high probability of making a Type II error.

Generally, we aim to set the value of the Type I error we will tolerate (which should be specified in advance in the protocol), while attempting to minimise the probability of a Type II error. However, the default choice of a value of 0.05 for α as a criterion for significance, although a widely used convention, is just that. Indeed, a significant result is not necessarily an important one. Given enough participants, it is usually possible to obtain a low value of α even when the effect of the treatment is not very large, and hence less likely to be of interest.

The arbitrary nature of significance levels, the ability to change significance levels by increasing the number of participants, and the disconnect between the significance of the results and their importance, are the reasons why many statisticians are strongly opposed to the standard null hypothesis testing framework described above. As an alternative, they recommend reporting effect sizes (which we discuss in the context of meta-analysis in Chapter 11) and confidence intervals. This approach is presented in detail by Cumming (2012).

16.2.3   The design of formal experiments

Standard design elements in formal experiments include:

- *Random allocation to treatment.*

- *Blocking* which is used both to allocate participants to homogeneous subgroups prior to allocation to treatment and to control confounding factors. Note, the term *matching* is used when participants are matched into similar pairs and one of each pair is assigned to each treatment group.

- *Treatments* which, in the context of software engineering, are the software engineering methods or procedures believed to affect the response variable and the characteristics of participation and software materials that can affect task outcome.

- *Covariates* which are measurements taken on the participants or the experimental materials prior to the experiment that are used to explain variations in the experimental response variable.

Although, by definition, all formal experiments include random allocation, simple random allocation does not guarantee equal numbers of participants in each experimental group. Since unequal numbers of participants can complicate statistical analyses, it is usual to restrict the randomisation procedure to ensure equal numbers of participants in each experimental group.

Another problem is that random allocation does not guarantee to deliver an unbiased sample, it only ensures that we ourselves have not introduced any so-called *selection bias*. It is possible that a random allocation could deliver biased treatment groups, for example by allocating better software engineers to one of the treatment groups. So, if we are aware that there are any characteristics of our participants or our experimental materials that could influence the results of the experiment, we should adopt other experimental design practices, such as blocking or the use of covariates, to control the impact of the characteristics. We use blocking to group participants into homogeneous sub-groups and we can then allocate participants in each subgroup at random to each treatment. If instead we can measure some characteristic, such as years of experience, we can use covariate analysis to adjust the outcomes of the experiment by allowing for any relationship between the covariate and the response variable

16.2.4   The design of quasi-experiments

Standard design elements for quasi-experiments include:

- *Time* since most quasi-experiments take place over a discernible time period.

- *Treatment*, often referred to as an *intervention*, which is a policy or method intended to cause some measurable factor to change.

- *Controls* which are units not receiving the treatment that are matched in some way to units receiving the treatment.

- *Pre-tests* which are measurements taken before the treatment condition is applied.

- *Post-tests* which are measurements taken after the treatment condition is applied.

The simplest form of quasi-experiment is to apply the treatment to a number of participants and then take a single post-test measure on each one. The obvious weaknesses of this design are that there is no way of knowing whether anything actually changed and, if anything did change, whether it was due to the passage of time rather than the treatment. Other design elements are used to address these basic weaknesses:

- One or more pre-test measures can be used to assess the situation prior to the treatment.

- Controls are added to see what happens if the treatment is not applied.

- Additional post-test measures are taken to see whether changes persist over time.

One form of quasi-experiment occasionally used in software engineering is based on measuring the results of each participant performing a task using the control method (which is equivalent to a pre-test) and then measuring the results of each participant performing a similar task using the new method (which is a post test). This is a very simple *within-subjects* design. It is called this because we measure both treatment conditions on the same subject, and each subject acts as their own control. However, this design still suffers from the problem that the effect of the treatment is confounded with the passage of time, and may be the result of other factors such as a learning effect.

A more sophisticated design is to use a *within-subjects cross-over* design, where half of the participants (selected at random) undertake a task using the control method and then undertake a similar task using the new method. The other half of the participants use the new method first and then the control method. Again, the outcome of the task(s) performed by each participant are measured after using both treatments, so they can act as their own control, but the design does not confuse the treatment effect with the passage of time. However, as we pointed out in Chapter 11, the standard error of the difference between the means of within-subject studies is not the same as the standard error of the difference between means of independent between-group studies, and statistical tests must be based on paired *t*-tests. This design is used quite often in software engineering experiments because of problems finding sufficient participants for between-groups experiments. It is a very reliable design as long as the order in which the participants used the methods did not affect the experimental results. For example, if having used the treatment method first, participants were likely to use it for the second task, even though instructed to use the control method (or vice versa), or using the control method first made using the treatment method easier (perhaps because the task was better understood), the results would be unreliable, see Kitchenham, Fry & Linkman (2003).

Quasi-experiments were introduced in sociology in order to assess the impact of large-scale interventions in complex social situations. In a software engineering context, we might want to assess impact on a specific organisation of undertaking a major process improvement. In such cases quasi-experiments can be based on taking a series of measurements on project outcomes (such as elapsed time, size, productivity, and defect rates) before the treatment is introduced and a series of measurements on project outcomes after the treatment is introduced. A *time-series analysis* of the average measures for projects per year should indicate whether the process change has been accompanied by an improvement in productivity, delivery times, or defect rates.

The basic time-series approach can be enhanced by including a control group as well as a treatment group—in a software engineering context, this might be appropriate if one department is trialling a major change to the software development process, while other departments are continuing to use the current software development process. In such a design, the control group keeps to the control method throughout the time period of experiment but the treatment group changes from the control method to the treatment method part way through the time period.

16.2.5 Threats to validity

To conclude this section, we consider how well we can determine whether the design and conduct of an experimental study can reliably detect an ex-

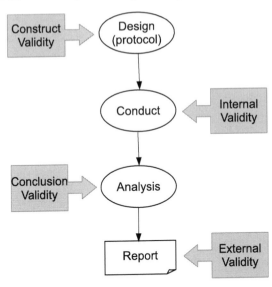

FIGURE 16.3: Threats to validity and where they arise.

perimental effect, and how confident we are that this is not also an artefact of the particular experimental context.

Given that there are likely to be many factors affecting a study, a key question for its design must be how trustworthy the outcomes from it are likely to be.? To be considered as valid, the results should be trustworthy for the 'population of interest' (for example, students, software developers, maintainers). A *threat to validity* is therefore a factor that may put the outcomes of a study in question. While these need to be considered for all forms of empirical study, threats to validity were initially identified and categorised to help researches to understand the limitations of quasi-experiments and, to a lesser extent controlled experiments, when any extrapolation of the outcomes beyond the immediate controlled environment can be more problematical. The anticipated threats to validity are usually identified in the experimental protocol, and their likely effects assessed as part of the design task. Further threats may then arise from the way that the experiment is conducted and analysed. So when reporting a study, it is conventional to report the experimenter's assessment of the threats to validity arising from its design, conduct and analysis in order to identify what confidence can be placed upon the results.

Four major forms of threat, associated with different phases of the conduct and reporting of an experiment, as shown in Figure 16.3, are as follows.

Construct validity is concerned with how well the outcomes of the study are linked to the concepts or theory behind the study—for software engineering there might be doubts about this if the model used in the study is inadequate (for example, performing a comparison between two procedures to determine which is 'better' without having a clear definition of what is meant by 'better'). It also relates to how well the concepts involved have agreed-upon operational definitions, for example if the term 'quality' is used without having a clear definition of its meaning for the study. Essentially this is an assessment of how well the concepts used in the study are defined.

Internal validity is where we seek to identify any factors that might have affected the outcomes (the dependent variable) without the researcher's knowledge and which might put any causal relationship that appears to exist between the treatment and the outcome in question—for software engineering, problems with this might arise because of the lack of a control group. This is therefore determined both by the design of a study, and also (to a greater degree) from the way that it is conducted.

Statistical conclusion validity reflects how well the experimenters were able to analyse the outcomes of a study, and also whether the way it was done is appropriate. The issue here is to assess how confident the experimenters are that any correlations between the variables that are found actually demonstrate that a cause-effect relationship does exist.

External validity is concerned with how well the conclusions from the study may be generalised to the intended population of interest. Here the threat is that the results may only be applicable to the particular context of the study—in software engineering these may often arise from the choice of participants: for example, using students rather than experienced developers may limit confidence that the results would apply to software development practice.

With experiments and quasi-experiments, as with other forms of empirical study, it is good practice for any report on a study to make an assessment of likely factors that could have influenced these, and how extensive this influence might have been. An extensive discussion of the issue of validity can be found in (Shadish et al. 2002).

16.3 Research questions that can be answered by using experiments and quasi-experiments

The research questions that can be addressed by experiments and quasi-experiments are, not surprisingly, very similar to the quantitative systematic review questions we discussed in Chapter 4. They are usually based on asking questions of the form "Is technique A better than technique B?". As discussed in Chapter 4, it is necessary to specify what is meant by the term 'better' and under what conditions any defined relationship might hold.

In this section, we present some examples of the research questions researchers have used in software engineering experiments and the detailed hypotheses relating to those questions. We have tried to select these based on the variety of the ways the research questions were tested and used, and also that they are studying topics that should be reasonably familiar to most software engineers and students.

16.3.1   Pair designing

There are many experimental studies that look at pair programming, often using student participants. However, the paper by Canfora, Cimitile, Garcia, Piattini & Visaggio (2007) looks at how well pairing works for design tasks, and in particular, describes a study performed using participants who were practitioners in industry. The authors stated two research questions:

1. 'Does pair designing require less effort than solo designing for a given task?'

2. 'Is pair designing better than solo designing in terms of the quality of the produced artefacts?'

Based on these they stated two null hypotheses, together with matching alternative hypotheses as follows.

- Null hypothesis H_{0a} states that "there is no difference in the effort employed between pair and solo designing", which they express as $\mu_{effort-pair} = \mu_{effort-solo}$.

- Alternative hypothesis H_{1a} states that 'there is a difference in the effort employed between pair and solo designing', expressed more formally as $\mu_{effort-pair} \neq \mu_{effort-solo}$.

- Null hypothesis H_{0b} states that 'there is no difference in the quality produced between pair and solo designing', expressed as $\mu_{quality-pair} = \mu_{quality-solo}$.

- Alternative hypothesis H_{1b} states that "there is a difference in the quality produced between pair and solo designing", or $\mu_{quality-pair} \neq \mu_{quality-solo}$.

In this example there are two practices to note. The first is that of stating the null hypothesis first (essentially a matter of style). The second is that the alternative hypothesis states only that there is a difference, and makes no assumptions about which will involve more effort or produce better quality software, which in turn will affect the form of statistical test used (as noted earlier, this is a two-tailed form).

16.3.2 Comparison of diagrammatical forms

The paper by Lucia, Gravino, Oliveto & Tortara (2010) examines how well using the *class diagram*, which is one of the notational forms used in the Unified Modeling Language (UML) compares with using the traditional 'database' notation of the *Entity Relationship Diagram* (ERD) during 'comprehension, maintenance, and verification activities on data models'. The paper does not state the research questions explicitly, and there are three hypotheses that relate to these three qualities, as listed below.

- H_{0_c} 'there is no difference between the support provided by ER and UML class diagrams when performing comprehension tasks on data models (Comprehension Support)'.

- H_{0_m} 'there is no difference between the support provided by ER and UML class diagrams when identifying the change to perform on data models to meet a maintenance request (Maintenance Support)'.

- H_{0_v} 'there is no difference between the support provided by ER and UML class diagrams when identifying the defects in data models during verification activities (Verification Support)'.

There are also three alternative hypotheses, and this time these are stated in a one-tailed form as it is expected that the UML notation will be more effective, as in for example:

Alternative Hypothesis H_{a_c}: $ComprehensionSupport(UML) > ComprehensionSupport(ERD)$

16.3.3 Effort estimation

The third example is based upon the paper by Grimstad & Jørgensen (2007). This paper states no formal hypotheses, but does state its research questions clearly, and bases the analysis upon these. So these are:

1. "How consistent are software professionals' expert judgment-based effort estimates?"

2. "Do more accurate estimators have more consistent expert judgment-based effort estimates than less accurate estimators?"

16.4 Examples from the software engineering literature

However lucid our descriptions, it is sometimes helpful to look at an example of the sort of features that have been discussed in this chapter. So in this section, we briefly review an example of the major form of randomised experiment together with three examples of quasi-experiment in order to illustrate the different forms of organisation discussed earlier.

16.4.1   Randomised experiment: Between subjects

An interesting example is that by Phalp, Vincent & Cox (2007). This study compared two different sets of guidelines for writing use cases, which are widely used to specify systems requirements. The experimental task consisted of reading the given set of guidelines, reading the descriptions of the system, and producing a set of use case descriptions.

A set of 60 student participants were divided into four groups of equal size. Two groups worked on one system description, and a second pair of groups worked on a different one. For each system description, one group used a smaller set of content guidelines (effectively, the control group) while the other used a more extensive set generated from a previous research project. Assessment was based upon assessing counts of usage of specific guidelines from each set.

16.4.2   Quasi-experiment: Within-subjects before–after study

This is the weakest form of quasi-experiment, where subjects use one technique to perform a software engineering task, then undergo training for a new technique, and then perform a task similar to the previous one, but using the new technique. An example study of this form is "The empirical investigation of perspective-based reading" (Basili et al. 1996).

The limitation of before-after studies is that the treatment effect is confounded with order. For software engineering this is a problem because if the task is skill-based, you would expect participants to improve their performance on the second task, irrespective of the treatment used. Also, if the initial task is not well-defined then the effect size observed between different studies will not be comparable. For example, if there is no training in the first technique then improvements in performance may arise simply because of the training and not because of the specific treatment. In other words, if the participants do not know how to undertake a task, then any systematic approach may improve their performance.

16.4.3   Quasi-experiment: Within-subjects cross-over study

As indicated earlier, this approach is used for comparing two (or more) techniques. Half of the participants, chosen at random, use technique A then technique B, while the other half use technique B first then technique A. An example of this is *An Internally Replicated Quasi-Experimental Comparison of Checklist and Perspective-Based Reading of Code Documents* (Laitenberger, Emam & Harbich 2001).

This is a stronger form of quasi-experiment since treatment and order are no longer confounded and there are well-defined approaches to analysing the data that allow both for period effects, that is, any systematic effect that occurs due to the fact that tasks are performed sequentially, and for having unequal numbers of participants in each group.

16.4.4 Quasi-experiment: Interrupted time series

This form is used when measures of some attribute are available (perhaps yearly or monthly) both before and after some intervention. This is regarded as the best way of assessing the impact of policy changes. An example of this is "Measuring the impacts individual process maturity attributes have on software products" (McGarry, Burke & Decker 1998). Although McGarry et al. did not provide any detailed statistical analysis, their time-based plots showed clearly that while two project outcomes appeared to have improved as the organisation achieved increased its CMM levels, other project outcomes were clearly the result of organisation changes pre-dating the CMM initiative.

16.5 Reporting experiments and quasi-experiments

Guidelines on how to report an experiment are provided in both Kitchenham, Pfleeger, Pickard, Jones, Hoaglin, Emam & J.Rosenberg (2002) and in Jedlitschka & Pfahl (2005).

Experiments and quasi-experiments usually produce a mix of quantitative and qualitative data, together with some form of statistical analysis. One of the advantages of these types of studies is that data collection can usually be quite tightly controlled.

In Chapter 8 we considered the data extraction process and looked at some examples of this. Obviously, one way that the researchers performing primary studies can help with this is with ensuring that their data is clearly presented. However, for the purposes of the secondary study, the data is not the only thing that needs to be provided, and experiments that are otherwise well reported sometimes fall short when it comes to some of the 'contextual' issues. Here, we provide a brief guide as to what needs to be presented in order to ensure that the experimental results can be adequately included in a Systematic Review.

1. The *research question* and any associated *hypotheses* (and null hypotheses). While this might seem obvious, this information is not always explicitly provided.

2. The *design* of the experiment, and in particular, the form used. Where a *dry run* was conducted beforehand using one or two participants in order

to test out the experimental procedures, the report should indicate the nature of any resulting changes. It is helpful to use terminology such as that provided in (Shadish et al. 2002), particularly when describing the design of a quasi-experiment.

3. The profiles of the *participants*, addressing such aspects as whether they were students or practitioners (or both), degree of experience with the topic, and any relevant information about allocation to groups.

4. The *conduct* of the study, providing information about any training provided, whether any participants dropped out (either between training and performing the study, or between sessions where a study took place over more than one session), or about any other divergences from the study plan.

5. The *results* themselves, organised as appropriate for the topic.

6. Details of the *analysis* methods used (and why they were chosen).

7. The *outcomes* from the analysis, together with any interpretation of these.

8. An assessment of any *threats to validity* or limitations.

Further reading

The authoritative text on the design of experiments and quasi-experiments, is widely viewed considered to be "*Experimental and Quasi-Experimental Designs for Generalized Causal Inference*" by Shadish, Cook and Campbell (Shadish et al. 2002). This book provides both philosophical underpinnings as well as good tactical guidance for the subject of experimentation in general. In particular, it explains the issue of design for quasi-experimental studies in depth, highlighting the strengths and weaknesses of each of the forms it identifies.

As a statistical textbook that emphasises alternatives to simple hypothesis testing, we recommend the book *Understanding the New Statistics. Effect sizes, Confidence Intervals, and Meta-Analysis* by Geoff Cumming (Cumming 2012).

Human participants form an important factor in most software engineering studies, and in their book *People Studying People. Artifacts and Ethics in Behavioural Research*, Rosnow and Rosenthal provide a summary of research bias in behavioural research. The range and extent of the possible problems is particularly sobering if we note that behavioural research does not have the requirement for skilled participants, nor the requirement to provide for special training for participants, that are needed in software engineering experiments.

For an approach that is more centred upon the needs of software engineering, although less extensive in coverage of the different experimental forms, the reader might turn to "*Experimentation in Software Engineering*", by Wohlin et al. (2012), which also gives some guidance on the issues involved with statistical analysis of such studies.

Finally, the limitations involved when experimental studies are conducted in industry are examined in the Systematic Review reported in (Dieste, Juristo & Martinez 2014). While the number of primary studies identified was small (10 experiments and five quasi-experiments), they provide important lessons about this type of study, which is one that will increasingly be needed if evidence-based studies are to be relevant for practitioners.

Chapter 17

Surveys

The basic idea of a survey is one that is familiar enough to most of us from everyday life. Surveys take many forms and play a number of different roles. At election time, surveys of a proportion of the electorate are used to forecast likely voting patterns. Businesses might use surveys that are targeted at particular groups of customers in order to plan their business and marketing strategies. And not all surveys require people to complete forms or answer questions—for example, a traffic survey might well involve counting the number and types of vehicles passing a particular point during different periods of the day. Similarly, a survey might study documents—for example, how many of the e-mails sent around an organisation have attachments, and what types of attachment are most commonly used. Indeed, the survey is a versatile tool and it is a form that is widely used in many 'human-centric' academic disciplines.

Surveys are also used in software engineering, although not particularly extensively. One reason is that where the purpose or topic of a survey requires the collection of information from people with specialist knowledge, which is likely to be the situation for software engineering, then this group may well be both quite small in number, and also quite difficult to identify reliably. Indeed, while our familiarity with surveys may make the option of conducting one look like a fairly easy choice, this is rarely the case, at least in software engineering.

In Figure 15.2 we positioned surveys as being both shallow and broad. We classified them as 'shallow' because respondents are rarely willing to spend very much time on responding to a survey (simply getting people to start responding to a survey is only part of the task; many will still not complete it). Similarly, if collecting data by observation, it is difficult to collect it in great detail. Hence surveys rarely enable us to probe very deeply into an issue. They were also classified as 'broad' because they are a useful means of exploring around a theme using multiple questions.

Surveys can, and sometimes are, used as input to secondary studies and hence we discuss them here. However, while there are many roles for surveys, they tend to be used in a rather limited way in software engineering. So we will constrain our discussion to those aspects that are of the most importance in this particular context.

17.1   Characteristics of surveys

The purpose of a survey is to collect information from a large group of people (or documents) in a systematic manner, and then to seek *patterns* in the data that can be generalised to a wider population than just the people making up the sample. The way a survey is organised will depend upon the question(s) it is seeking to answer, and the following represent two major purposes for undertaking a survey.

- *Experimentation:* where the role of the survey is to assess the impact of some intervention, such as the adoption of a new technique or programming language.

- *Description:* where the survey enables assertions to be made about some phenomenon of interest and its attributes—here the concern is not *why* the attributes have the forms they do, but *what* those forms are. For example, we might be interested in conducting a survey to determine which version control tools are used most widely in software development organisations.

For this chapter we will concentrate on surveys of people, since surveys involving documents are addressed in a later chapter (data mining).

Two important characteristics of a survey that matter to us when considering whether or how to use its outcomes in a secondary study are the way that data was collected, and how the sampling of respondents was organised. Between them, they determine what confidence we are able to place in the wider relevance of any findings.

Data collection for a survey can be performed in a number of ways, including observation. The mechanism used for data collection is usually termed the *instrument*, and in software engineering, surveys are commonly conducted by using the following two forms of instrument.

Questionnaires. The act of completing a questionnaire, either on paper or on-line is often thought of as almost synonymous with the idea of a survey. Many software engineering surveys use this approach, not least because it lends itself to the globally distributed nature of the software engineering community. If we need expertise on a particular topic, such as (say) use of modelling notations, then we may well need to access the international community in order to get an adequate level of response as well as to identify any regional variations.

Interviews. These have the advantage that they can be more tightly targeted at a particular group, and if using semi-structured interviews, it may also be possible to probe more deeply into the issues identified (usually related to "why"). However, interviews are time-consuming to perform, as well as requiring the researcher to possess some 'people skills', and it may also be more difficult to access the relevant group of respondents. Using phones or video links does offer scope for wider access, although these appear to have been little used in software engineering surveys.

Survey instruments usually make use of two forms of question. *Open* questions do not have any pre-determined set of possible answers, and are generally useful for collecting data for *descriptive* surveys. While they offer flexibility in proving issues, analysis of the responses usually needs a qualitative approach and any outcomes may be difficult to include in a systematic review. *Closed* questions may ask the respondent to select one or more values from a pre-set list (a *rating* question), a form that includes the use of *Likert* scales, or to order a set of pre-determined options (a *ranking* question). Closed questions can therefore be used in a quantitative analysis.

Collecting data is only part of the story of course, and an important element of this is to ensure that we collect data from the relevant group of interest. This is determined by the way that sampling is organised.

Sampling is important from two aspects. Firstly, in planning a survey it is necessary to decide what the population is that we are interested in, and from which we wish to draw our respondents. This is usually termed the *sampling frame*, and one of our problems in software engineering is that this may often be difficult to define clearly, and even when it is well defined, it might still be difficult to identify the individuals making up the sampling frame.

For example, we might be interested in conducting a survey of everyone who has had experience with using pair programming on a software development project (that is, not just experience from a teaching context). So these people would constitute our sampling frame. However, defining exactly what is required for potential respondents to qualify for membership of this (in terms of measures that can be used to define some basic level of experience) is quite difficult, as is what exactly we mean by such words as "using" too. And once we have a definition, there is no obvious place that we can go to in order to access this group of people. Some may belong to on-line special interest groups and can be accessed that way, others can perhaps be contacted through their organisation, but in general, we don't have any well-defined means of reliably identifying those people who make up the sampling frame—or even of estimating how many of them there are.

We will discuss sampling a bit more in the next section, but for the moment, assuming that we can define a sampling frame in some way, we need to define the size of sample that we need to obtain from this in order to be able to make sound inferences about the larger group. If we do know, or can reasonably estimate, the size of the sampling frame, then it is possible to determine the size of sample that we need (the number of respondents) in order to achieve the required *confidence interval* in our results. (That is, know how certain we are that the true population values will match those in our sample.) Table 17.1 shows some examples of the size of the sample we need in order to achieve a 95% confidence interval with an accuracy range of plus or minus 3 (that is, we are between 92% and 98% confidence of matching the true population with our sample). As a 'rule of thumb', a minimum of at least

TABLE 17.1: Sample Size Needed for 95% Confidence

Target population size	Sample size
50	47
5000	760
100,000	888
900,000	895

30 responses is really desirable if aiming to perform any form of statistical analysis.

So, given this information, we can start to make some sort of assessment of how the outcomes from a survey can be incorporated into a systematic review. In the next section, we look at one or two practical details that will also be significant when determining how we might use its outcomes.

17.2   Conducting surveys

Since a major concern for a survey is with determining how far we can have confidence in the outcomes, in this section we look a little more at how a survey might be conducted (including some aspects of its design).

We have already mentioned the issue of sampling and the basic challenge that this presents for most software engineering studies, where we often do not know the population size. A related issue is the *sampling technique* itself, used to obtain a sample from the overall sampling frame. There are two main categories of sampling technique, each of which can be performed by using a number of different strategies.

Probabilistic Sampling. This seeks to obtain a sample that is intended to be a representative cross-section of the population. Depending on the nature of the sampling frame, a *random* sampling of the population may suffice, or this might be *stratified*, to try and ensure that the sample has a similar profile to the overall population. (For example, if years of experience is a key factor, then this might be used as the basis for the stratification.) Another approach is to use *cluster* sampling, using a grouping of people with a like interest (for example, all the people who have ever attended a particularly relevant conference series).

Non-Probabilistic Sampling. While this form of sampling is generally considered as forming a much poorer basis for inference about the whole population, it may well be the only option available. One example of where this might occur is where we construct a website to collect the data and then post invitations to people to respond using this (termed *self-selection* sampling). Other strategies include *purposive* sampling (sending requests to people who have specific characteristics as well as being thought likely to respond); *snowball* sampling (asking respondents to identify others who might be willing to participate); and *convenience* sampling where people are recruited on the basis of being readily available or likely to be willing to take part.

Issuing a request to a set of likely respondents is only the first step. Surveys tend to have low *response rates*, with a figure of 10% usually being considered to be quite good. Ways to improve this include providing a good explanation about why participation is important when making the original request, and following up non-responders with a reminder after a suitable interval of time. With on-line surveys there may also be instances of partial completion, where people have started to enter their responses but have not completed the task. Criteria for deciding when such responses should be included need to be specified when writing the research protocol for the survey.

Design of the questionnaire, or the set of questions to be used in a semi-structured interview is an important design element. Questions need to be clear, and to check consistency of responses, it may be appropriate to use more than one question to address a topic. The set of questions should be assessed for the following qualities.

- Ensuring that the questions address a well-balanced sample of issues (*content validity*).

- Measuring the relevant attributes through the data collected by the questions (*construct validity*).

- Being confident that if the questions were given repeatedly to the same people that they would obtain the same answers (*reliability*).

So, when considering the use of the outcomes from a survey, these are all factors that should be reported and that can help a systematic reviewer to assess its usefulness.

17.3   Research questions that can be answered by using surveys

As indicated earlier, both experimental and descriptive surveys tend to be used to answer research questions that are concerned more with asking "what" rather than asking "how". It may be possible to dig deeper with a survey, particularly if using a small survey to probe the rationale behind responses to a larger survey, but this task is probably best performed by using interviews.

Surveys performed in software engineering tend to be descriptive and we provide two examples of such a form along with an example of the use of interviews.

Our first example is one that is used as an example by Kitchenham and Pfleeger in their review of the use of surveys in software engineering (Kitchenham & Pfleeger 2008). This is described in the paper "Components of Software Development Risk: How to Address Them? A Project Manager Survey" by Ropponen & Lyytinen (2000). This survey was conducted with over 80 project managers and asked the following two questions.

1. "What are the components of software development risk?"

2. "What risk management practices and environmental contingencies help to address these components?"

Both of these questions address *descriptive* issues (since they both begin with "what"). Kitchenham and Pfleeger describe this form of survey as *cross-sectional, case control* form, and observe that "most surveys in software engineering are of this kind of design". In a cross-sectional study, respondents are asked for information at a fixed point in time (which may be highly constrained, such as requesting information based upon a particular data and even time, or relatively relaxed, as when people are polled about some 'current' preference). A case-control design retrospectively asks about previous circumstances in order to seek patterns that might help explain a current phenomenon. (For details about the design options for both descriptive and experimental forms, see (Kitchenham & Pfleeger 2008) and (Kitchenham & Pfleeger 2002a).)

The survey itself used purposive sampling (mailing requests to people in Finnish software companies whose job title included either the word "manager" or some equivalent term).

Our second example addresses a more technical issue about the role of software design patterns and was motivated by the experiences of conducting a secondary study about their use (Zhang & Budgen 2012).

The survey, reported in Zhang & Budgen (2013), obtained responses from over 200 respondents, and sought to obtain opinions about how well used and effective were the well-known design patterns catalogued in the book from the

"Gang of Four" or *GoF* (Gamma, Helm, Johnson & Vlissides 1995). For this the research question was:

> "Which design patterns from the GoF do expert pattern users consider as useful or not useful for software development and maintenance, and why?"

Since this was based upon the experiences that users had had with these patterns, this can again be considered as being a "cross-sectional case-control" study. However, as we explain in Section 17.4, this can also be considered to have an experimental element too.

Our example of the use of interviews is the paper by Petre (2013), with the title "UML in Practice". Stimulated by a number of questions about how much the UML (Unified Modeling Language) is actually used by software developers, and by our own systematic review about the UML and its use (Budgen, Burn, Brereton, Kitchenham & Pretorius 2011), the author set out to conduct a series of interviews with software developers to establish whether they used the UML, and if so, how it was used. While the paper does not provide an explicit research question, the author states that the main questions she asked were:

1. "Do you use the UML?"

2. If the response was "yes" then the second question was "Can you tell me about how you use it?". If the response was "no", then the second question was "Why not?".

While the author does not claim that this study is a survey as such, it does provide an excellent example of the effective use of interviewing to probe reasoning that would not be easily extracted by the use of a more formal questionnaire.

17.4 Examples of surveys from the software engineering literature

We will continue to use as our examples the three surveys that were used to provide examples of research questions in Section 17.3. Only a brief description is provided for each one, identifying how it addressed some of the issues involved in designing and conducting surveys that were identified in Sections 17.1 and 17.2.

For each one, we provide an outline description of:

- The sampling frame
- The sampling technique(s)

- Data collection mechanisms

- Organisation

- Questionnaire design

based upon the information provided in each paper. Taken together, these do show some of the challenges that occur when conducting a study, especially in terms of obtaining 'suitable' samples of respondents, as well as in the design of the questionnaire.

17.4.1   Software development risk

The study described in the paper by Ropponen & Lyytinen (2000) aimed to sample a representative set of people with experience of project management within Finland. The sampling frame that was used was the membership list of the Finnish Information Processing Association, and the authors sent a copy of the questionnaire to those members of this organisation whose titles indicated that they had a managerial role, and also constrained the sampling so that they contacted no more than two people from any company. They do not indicate in the paper how they selected these when there were more than two, so their sampling method, although probably well defined, remains unknown.

The authors sent out 248 questionnaires (by post), and received 83 responses (a response rate of 33.5%, which can be considered to be very good). While the paper describes how they followed up a sample of non-responding people to find out why they did not respond (to check for possible bias), they do not indicate whether they sent out any form of follow-up prompt after making the initial request.

Their questionnaire was organised as a set of 20 questions about risk, based upon different scenarios and providing response options using a 5-point Likert Scale (ordinal scale), with values ranging from "hardly ever" to "almost always". In addition, they asked a set of "demographic" questions related to organisational characteristics as well as about the profile of experience possessed by the individual respondents. The paper also provides details of how the questions in the survey were validated by using a panel of experts, although as Kitchenham and Pfleeger observe, they "did not conduct an independent pilot study" (Kitchenham & Pfleeger 2002*b*).

17.4.2   Software design patterns

As mentioned earlier, the study described in this paper was motivated by the experiences from conducting a systematic review of studies of design patterns (Zhang & Budgen 2012). This review found very few experimental studies (11) and ended up augmenting the knowledge from these by using a small sample of observational papers (7). Since even well-known design patterns were hardly examined in any real depth, and some were not evaluated at all, the survey constituted an attempt to obtain further knowledge about how useful these widely-known design patterns were considered to be.

The desired sampling frame was that of software developers who had had experience of using some or all of the 23 design patterns catalogued by the "Gang of Four" in (Gamma et al. 1995). (These were the only design patterns that were examined in the set of studies included in the systematic review.) Lacking any obvious means of identifying or contacting all of these people, the surrogate sampling frame adopted was that of all of the people who had authored the papers on patterns that were found during the searching process used for the systematic review (whether or not those papers were empirical). So the initial sampling procedure could be described as indexsurvey!sampling!cluster sampling *cluster sampling* (people with like interests).

This initial sampling frame consisted of 877 unique authors, but with the passage of time, some contacts were no longer valid, reducing the size of the sampling frame to 681. Of these, 128 responded, giving a response rate of 19%. Respondents were also asked to pass the survey to others whom they knew were knowledgeable (*snowball sampling*). A further group were contacted through a special interest group mail-list (*cluster sampling* again), and the final total of usable responses was 206. All non-respondents from the original list of authors were followed up once (after an interval of two weeks). So, the survey employed a mix of sampling methods, and the size of the final sampling frame could only be estimated.

The questionnaire was administered by using an on-line site (*SurveyMonkey*), and a dry run was performed using two experienced assessors, leading to some changes in wording and in presentation. The survey began with some initial demographic questions about respondent roles and experience with patterns, and then used both *rating questions* and *ranking questions* to ask about the 23 patterns. For each pattern, respondents were asked to provide a rating of its usefulness using a 4-point Likert Scale (together with a 'no experience' option). In addition, they were also asked to provide a ranking of up to three patterns they considered to be the most useful, and of up to three that they considered to be of little use. For the ranking questions, respondents were also invited to provide comments to explain their choices (although few did).

While the survey was primarily descriptive, the authors argue that there was an experimental element, in that it was possible to make some comparison between the responses from different groups of respondents (classified either by experience or by role). This aspect was categorised as a *concurrent study in which participants are not randomly assigned to groups.*

Two limitations in the design of this study were the potential bias arising from the sampling (many of the respondents had higher degrees, which might be expected from the authors of papers, of course), raising the question of how representative they were of the target population, and the mixed forms of question used. The use of both rating and ranking questions made it impossible to check for consistency in the responses, and a design that simply used rating questions would probably have been better.

17.4.3   Use of the UML

Marian Petre's study (which won a "Best Paper" award) does carefully position itself against previous surveys conducted about the roles of the UML. The author interviewed some 50 practising software developers, taking no more than one per company. So, informally, the sampling frame was the set of all software developers, and the sample was obtained on a *convenience* basis (the author uses the term "opportunistic").

The key issue here is the word 'use', and this is where interviews allow for clarification of what is meant by this. Only 30% of respondents did use the UML, and the main three categories of use that they identified were: *retrofit* to meet a customer's requirement; *selective*, using the UML informally when it seemed appropriate; and for *automated code generation*, once the design has stabilised. This ability to categorise reasons for using (and also, for not using) something demonstrates a strength of using interviews, making it possible to follow up issues and obtain explanations.

So, although not formally presented as such, it can be argued that this was a descriptive survey, and as the author suggests, the results are indicative rather than definitive (the sample was relatively small). Responses were also verified by observation where practical.

While conducting such a survey on a larger scale would be quite challenging, one benefit of this type of study is to gain a deeper understanding of issues that can then be potentially explored more fully through a more extensive survey if desired.

17.5   Reporting surveys

This task involves reporting many of the same issues as for experiments (research question, design, conduct, results, analysis, outcomes, and threats to validity). Where appropriate, there may also be hypotheses to report as well. However, the report of a survey should also provide information about such issues as the population of interest, the sampling frame used, and the way that the sampling was performed.

Further reading

For those wishing to know more about survey design, there is an extensive literature on surveys. In most libraries, this is largely to be found catalogued under the headings of "social science" and "psychology", since for both of these disciplines the survey is an important tool. A widely-cited and very readable resource is the ten-volume *Survey Kit*, edited by Arlene Fink, although to obtain a basic understanding of key issues, the first volume alone should be sufficient. See (Fink 2003).

A detailed review of the use of surveys from a software engineering perspective can be found in (Kitchenham & Pfleeger 2008).

Chapter 18

Case Studies

From the perspective of the systematic reviewer, case studies represent an important category of primary study, largely because they provide a very relevant way to study many software engineering topics. Unfortunately though, the term *case study* has often been used rather loosely to describe what are really rather informal observational studies. And as a further complication, the term "case study" may also be used to refer to a large-scale teaching example, particularly when used in the context of business and management studies (a "teaching case study").

As indicated in Figure 15.2, for our purposes the case study can be considered to be a form of empirical study that allows deeper probing of issues, but at the price of reduced generality. It can be considered to be an observational study (as there is no controlled intervention), with the crucial distinctions being that it is *planned*, sets out to answer a specific research question, and collects data pertinent to that research question in a systematic manner. Case studies are widely used for research in the social and behavioural sciences, and in recent years, software engineers have looked to these disciplines in order to learn how to make the most effective use of this form of research. However, we should note that within the disciplines that make extensive use of case study research, there are two major schools of thought about its role.

- The *positivist* view enshrines a belief that there are general rules and patterns that govern human behaviour, and so any use of case studies can be geared towards identifying and exploring these. Software engineers, and scientists in general, are usually positivists.

- An *interpretivist* view takes the opposing position, whereby the outcomes of a case study can only be understood within the context of that study and there are multiple realities accessed through social constructs such as language. Hence the conclusions from a case study cannot be generalised or extended for wider use or application (including being used as an input to a systematic review). Some information science researchers tend to adopt this viewpoint.

So, case studies are used by both 'camps', but differently. Since software engineers have been predominantly influenced by the work of Robert K Yin, who is a major exponent of the positivist use of case studies (Yin 2014), and the positivist perspective does consider case studies as being suitable inputs for systematic reviews, not surprisingly, this chapter too will employ a positivist interpretation.

Yin's view of a case study is that it is an empirical enquiry that "is used to understand complex social phenomena" and hence that it:

- "Investigates a contemporary phenomenon (the "case") in depth and within its real-life context" (in contrast to an experiment, which seeks to isolate a phenomenon from its context);

- Is a particularly appropriate form to employ when "the boundaries between phenomenon and context may not be clearly evident" (in software engineering, the effects arising from software development practices and from their organisational context are often confounded).

This chapter is again written largely from the perspective of the needs of a systematic reviewer, and what knowledge they need about, and from, a case study. A much more extensive and detailed exposition of the use of case study research in software engineering (and especially about their design) is provided in the book *Case Study Research in Software Engineering: Guidelines and Examples* by Runeson et al. (2012). There is also a useful chapter on the conduct of case studies in Wohlin et al. (2012).

Drawing upon the definition provided by Yin, and also the definitions provided by Robson (2002) and Benbasat, Goldstein & Mead (1987), Runeson et al. suggest the following, software engineering-specific, definition for a case study.

> "An empirical enquiry that draws on multiple sources of evidence to investigate one instance (or a small number of instances) of a contemporary software engineering phenomenon within its real-life context, especially when the boundary between phenomenon and context cannot be clearly specified."

So, why are case studies so important to software engineering? In essence, it is because they are a form of study that can be used "in the field", to study how companies or individuals adopt and use new practices, as well as how they use well-established ones, often over a period of time (which we refer to as a *longitudinal* study). They can be used to address issues that do not readily lend themselves to the use of experiments or surveys, and to ask a range of different forms of question.

In exchange for this flexibility there are, of course, some associated trade-offs. The most obvious one is the lack of generality for any findings. Another, very important one, is the need to adopt a disciplined approach to conducting a case study, in order to ensure that any outcomes are both reliable, and also that they are as unbiased as possible. A case study does need to be designed in advance, and not fitted to a project retrospectively in order to confer greater respectability upon what is really an informal observational study. Recognising these issues is important for the systematic reviewer, who may need to be able to interpret and aggregate the outcomes from different case studies.

The rest of this chapter follows the same general format as the preceding ones by first examining the characteristics and forms that case studies take and how they are conducted, identifying the sort of research question that can be addressed through case studies, and providing an example for fuller illustration.

18.1 Characteristics of case studies

If we continue to focus our ideas around the positivist approach espoused by Yin (2014), then he suggests that some key characteristics of a case study inquiry are that it:

- "Copes with the technically distinctive situation in which there will be many more variables of interest than data points" (again this provides a contrast with an experiment, where the aim is to use as few variables as possible);

- As a consequence of having so many variables, it "relies on multiple sources of evidence, with data needing to converge in a triangulating fashion" (by 'triangulation', we mean that different sources of data are used to establish a conclusion, rather as a navigator uses 'fixes' on different points in order to determine his or her position);

- And as a further consequence, it "benefits from the prior development of theoretical propositions to guide data collection and analysis". (Here, we should view 'theory' as being a rather general concept, more akin to the concept of a 'model', rather than necessarily meaning something that is mathematical in nature. Indeed, the basis of such 'theory' could also be the outcomes from other primary or secondary studies.)

So, where an experiment assumes that relationships between different factors can be studied separately, a case study is particularly appropriate when the *interconnection* between the object of interest and its context is both complex and also likely to be an important factor. A case study is therefore well suited for use with a task such as studying the transition to the use of agile methods within a particular organisation. Here, important factors are likely to be the skills and attitudes of individual developers, the organisational context, the type of systems being developed, and the way that new practices are introduced. The effects of these are not readily separated, and even if they could be separated, studying them in isolation would be unlikely to provide an understanding of how they behave as a whole. A similar example, illustrating a similar software engineering context, is provided by the study of the adoption of object-oriented practices described in Fichman & Kemerer (1997).

These examples point to another significant characteristic of case studies, which is that the duration of a case study may well extend over a long period of time, particularly when studying issues such as the adoption of new technologies, or how a technology evolves. This ability to perform *longitudinal* studies of effects is both distinctive and valuable, but may also be quite demanding of effort on the part of the researcher.

So, having identified some key concepts about the nature of case study research, we now examine how this may be employed in the context of software engineering.

18.2   Conducting case study research

Case study research can be employed for a number of different purposes. Yin suggests that there are three primary roles that can be performed by a case study (Yin 2014), which are as follows.

- An **explanatory** study is concerned with determining *how* some process works, and with *why* it works successfully or otherwise. This role is one that cannot easily be performed using experiments or surveys, and so is a particularly important one for case studies.

- A **descriptive** study "provides a rich and detailed analysis of a phenomenon and its context", but without the element of interpretation and explanation of an explanatory study. So, here a case study can perform a role similar to that of a descriptive survey, being concerned more with *what* exists or occurs, although probing more deeply than a survey, and also being less general.

- An **exploratory** study is one that is mainly concerned with identifying the issues of interest for, and the scope for, a future, more extensive study (not necessarily a further case study, although possibly so). The benefits of using a case study for this role lie in its inclusive nature, drawing together many data sources, and so helping to identify possible links and relationships that can then be explored in more depth.

The first two forms may also benefit from the development of one or more *propositions*. These are derived from a theory or model, and a proposition "directs attention to something that should be examined within the scope of the study" (Yin 2014). For a case study, a proposition plays much the same role as a hypothesis does for an experiment, although it is unlikely to be tested statistically. Yin also makes the point that case study research should not be confused with qualitative research. A case study can use both quantitative and qualitative forms of data, as well as a mix of these.

Runeson et al. also suggest a fourth category as relevant to software engineering, derived from the classifications used by Robson (2002).

- An **improving** study is one that is conducted for the purposes of improving some aspect of the studied phenomenon.

In this form, the case study is performing a role similar to that of *action research*, although in a less iterative manner, aiming to learn from practice how to improve that practice.

Two further concepts that need to be examined before we look at the procedural aspects of conducting a case study are the distinction between *single-case* and *multiple-case* forms of study, and the choice of the *units of analysis* to be used in a case study. Yin categorises the four combinations of these as Type 1–Type 4 designs, as illustrated in Figure 18.1.

18.2.1 Single-case versus multiple-case

Yin suggests that single-case study, where only one case is employed (Types 1 and 2), can be considered as an appropriate design in situations such as those where:

- There is a *critical case* that needs to be examined, perhaps because it can help decide whether or not a theoretical model is likely to be 'correct';

- An *extreme* or unique case might exist, which is therefore worth studying for itself;

- Studying a *representative* case is sufficient to describe many other possible cases;

- Something new and previously inaccessible becomes available to an investigator (he terms this a *revelatory* case);

- A study may take place over a long period of time (a *longitudinal* case).

In software engineering research, neither extreme nor revelatory cases are likely to occur very often, and there may be few instances where a critical case exists. However, using single-case studies may well be appropriate for representative and longitudinal cases. The main risk in using a single-case study design is that the chosen case might turn out to be inappropriate (such

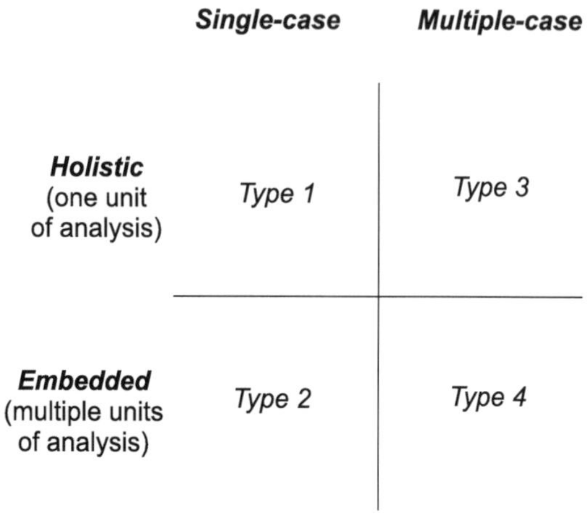

	Single-case	*Multiple-case*
Holistic (one unit of analysis)	*Type 1*	*Type 3*
Embedded (multiple units of analysis)	*Type 2*	*Type 4*

FIGURE 18.1: Characterising basic case study designs (adapted from (Yin, 2014)).

as a 'typical' case that emerges as being untypical). With so many possible factors to address, this may well be the situation in a software engineering context, so this is a design choice that needs to be viewed with caution when conducting a secondary study.

In contrast, multiple-case forms (Types 3 and 4) can provide more compelling evidence and also make it possible to use replication logic, whereby different cases produce the same results (or different ones, if there are good reasons for this, and if different results are predicted by theory). However, such a study is more limited in terms of the type of case that it can address (clearly its use is unsuited to extreme or revelatory cases), and it requires greater effort to manage multiple cases and their data.

The issue of *replication* is an important one and something that we return to in Chapter 21. For the moment, we might note that replication in case studies performs much the same role as replication in experiments, except that for a case study there is much less risk of having replications that are too close to the original study in terms of their parameters.

18.2.2 Choice of the units of analysis

This raises the question of what exactly constitutes a *case* in a case study? The case essentially forms the "unit of analysis" for a case study, and in software engineering this might be a particular company, or a project, or even an individual. As Yin observes, "the tentative definition of your case (or of the unit of analysis) is related to the way that you define your initial research questions". If these are concerned with how a company adopts a technology, or with how a particular agile process is used across different organisations, then the respective cases will be the company and the specific agile process.

The common distinction made here is between case studies that have a single *holistic* unit of analysis (whether used in a single-case or multiple-case design), and those that may have *embedded* units within the main unit. As an example of the latter, our unit of analysis might be a company, but we might then also study different projects within the company, with the latter being the embedded units.

In the same way that the research questions help to determine the appropriate units of analysis, so the choice of our research questions helps to determine what data needs to be collected. And that in turn leads us on to a brief description of how a case study can be organised.

18.2.3   Organising a case study

Yin identifies a five-step procedure for designing a case study. This structure is important, and needs to be clearly documented in the research protocol. By doing so, the researcher helps to ensure that a case study is conducted as a rigorous empirical study. It is also this *planned* element that particularly distinguishes case study research from observational studies and other less rigorous forms, and in turn, it is this additional rigour that should influence how we treat the outcomes from a case study when performing a systematic review. The five steps are as follows.

1. Determine the *study questions* by identifying the research question that represents the high level concern of the study. For example, a study question might be to "investigate changes to the development process that arise from adopting agile processes in an organisation".

2. Identify any *propositions* where these (if present) will be more detailed than the research question and will identify specific issues that need to be investigated by the case study. Continuing with the same example, in refining the general research question, the resulting propositions may be that the use of agile methods within the organisation should result in:

   - Reduced development time
   - Increased customer satisfaction
   - Improved developer motivation

   (forming three separate propositions that clearly address different stakeholder viewpoints).

3. Select the *unit(s) of analysis*. This step involves positioning the study within the set of four design types identified in Figure 18.1 and hence:

- Determining whether to employ single-case or multiple-case forms, and if deciding to use a single-case form, choosing between: a typical instance; an extreme instance; an instance that has elements that will help test out a theory; or a convenient instance

- Deciding upon the form(s) of unit to use: a company; a development project; a technology; a system etc.

4. Determine the *logic* that links the data to the propositions. Here the design task is one of determining what sort of data needs to be collected in order to evaluate the propositions. For example, addressing the proposition about "increased customer satisfaction" might involve conducting interviews with customers using questions about past and current satisfaction. And in order to do this, the different attributes comprising what might be considered to be 'satisfaction' for this particular stakeholder also need to be identified, along with ways of measuring them. There is a related issue of data collection and the role of the researcher that also arises here, since, depending upon how the study is organised, the researcher might be involved as either an observer or as a participant.

   Runeson et al. (2012) emphasise the importance of *triangulation* as a means to "increase the precision and strengthen the validity of empirical research" (and particularly of case study research of course). They note that this can take four forms.

   - *Data triangulation* involves making use of different data sources or of repeated measures.

   - *Observer triangulation* uses more than one person to collect the data.

   - *Methodological triangulation* makes use of a set of different data collection methods.

   - *Theory triangulation* employs different viewpoints or theoretical models for analysis.

   An interesting discussion of triangulation and an example of the use of data triangulation in a single-case holistic study (Type 1) is provided in (Bratthall & Jørgensen 2002).

5. Define the criteria to be used for *interpreting* the findings. This last step is concerned with analysing and evaluating the data we have collected in order to answer such questions as how well the findings support the proposition; to determine what 'level' of satisfaction is considered to provide support; and to establish whether the findings from the case study allow other explanations to be rejected. Here, the choice of analysis method is an important one, and Yin does warn that anyone with limited experience "may not easily identify the likely analytic technique(s) or anticipate the needed data to use the techniques to their full advantage".

This reinforces the point that a case study is a complex empirical study, and not simply something to be retro-fitted at the late stages of a project in order to provide a veneer of empirical evaluation!

So, when the systematic reviewer is considering whether or not a candidate primary study consists of a case study that is suitable for being included in a secondary study, these are some of the elements that we should expect to see reported. And of course, as with any empirical study, there is the need to assess and report on any threats to validity.

We should note here that *construct validity* is "especially challenging in case study research" (as a reminder, threats to validity were discussed in Section 16.2.5). Yin emphasises the need to address this by such mechanisms as *multiple sources of evidence*, with these being used in a convergent manner, so that the findings of the case study are supported by more than one source of evidence. *Internal validity* is also an important issue for explanatory case studies since, as always, establishing that a causal relationship exists does require confidence that the outcomes have not arisen because of some other factor.

Although ethical issues are not necessarily a major issue for the systematic reviewer, we should take note that they can form more of a challenge for case study research than most other forms of primary study. Where experiments and surveys usually deal with aggregated data, and participants are less readily identified, the same is not necessarily true for a case study. Because a case study addresses at most a few cases, there is greater likelihood of a reader being able to identify organisations or teams or individuals.

Where a systematic review is concerned with data that is in the public domain (usually through refereed conferences and journals) the use of case studies may not be an issue. However, greater care may well be needed when using material from the "grey literature".

18.3 Research questions that can be answered by using case studies

When considering the research questions that can be addressed by using a case study, it might be useful to note the observation by Runeson et al. (2012) about the distinctive nature of the most common objects of study for software engineering case studies. They identify four key properties.

- They are organisations *developing* software rather than *using* it.

- The organisation are *project*-oriented rather than *function*-oriented.

- The work studied is conducted by highly educated people performing *advanced engineering* tasks, rather than *routine* activities.

- Part of the reason for conducting a case study is to *improve* the practices, so there is an element of design research involved.

As one of our examples illustrates, we also consider research in software engineering is a valid object of study, although not a particularly common one.

The many different roles and forms that case studies can take means that there are many possible research questions that could be asked. So, rather than trying to be all-inclusive, we have selected a small number of examples.

An explanatory case study essentially addresses *how* and *why* questions in a context where the researcher cannot control the conditions for the study. Hence, explanatory case studies can be used to answer the type of question that other forms of empirical study can only address in a more limited way. They are also used quite widely in software engineering.

An example of this type of question is that addressed in the study by Moe, Dingsøyr & Dybå (2010) that examined the way that agile teams work. For this, the research question was:

> "How can we explain the teamwork challenges that arise when introducing a self-managing agile team?"

This study is an example of a single-case holistic case study (Type 1), for which the unit of analysis was the project team. It also used a 'theoretical model' in the form of a particular management science teamwork model, and involved some "theory building" related to this by suggesting two further elements that could be included in the model.

Our second example is a study of our own (Kitchenham, Budgen & Brereton 2011). Here the case study was used in a more 'methodological' role, providing a framework to help assess how mapping studies could be used to aid research. The specific research question was:

> "How do mapping studies contribute to further research?"

The study was a multiple-case study where a 'case' consisted of "a research activity following on directly from a preceding mapping study". In all there were five cases, with a mix of experienced and inexperienced researchers involved in each one. Since all of the cases were based on the same unit of analysis (a mapping study), this formed a Type 3 study. Data collection was mainly through the use of questionnaires. As a triangulation exercise, the findings were also compared with the experiences of two other researchers who had undertaken recent mapping studies.

While this was an investigation of the effectiveness of research practice rather than of software development practice, the use of a case study enabled a deeper understanding of the issues involved in using secondary studies in this role.

The case study by Bratthall & Jørgensen (2002) that was mentioned earlier in conjunction with the issue of data triangulation can be positioned as an exploratory case study. While the authors do not state an explicit research question, they do so indirectly as:

"there is a question regarding in which contexts different types of triangulation are beneficial"

after which they narrow this question down to data triangulation. (This too can be considered as having a 'methodological' element although it is firmly set in the context of a software development organisation.)

18.4 Example of a case study from the software engineering literature

Rather than attempting to review examples that cover the many variations of case study form, we expand here a little more on one of the examples that we discussed above, which is the study of a Scrum project discussed in (Moe et al. 2010). This allows us to highlight some of the key characteristics discussed in the preceding sections.

We first examine why this is a research challenge that is best suited to the use of a case study approach, and then examine some of the case study design choices it exhibits.

18.4.1 Why use a case study?

The emergence of the agile approach to software development has significantly altered the way that a software development team operates. In the 'traditional' plan-driven approaches as embodied in 'design methods', as the authors note, the role of the team "involves a command-and-control style of management in which there is a clear separation of roles". In contrast, an agile approach such as Scrum (the form studied here), is centred upon the use of self-managing teams for which there is no formulaic guidance—a team is expected to adapt its operation and forms of interaction to meet the needs of the development task as they evolve.

From an empirical perspective therefore, studying the teamwork challenges created by the need to adapt to a self-managing approach is one that is well-suited for a case study approach. For the researcher it is essentially an observational task, as opposed to a controlled one. There are very many variables involved in a complex set of interactions, again, as the authors observe, "the actual performance of a team depends not only on the competence of the team itself in managing and executing its work, but also on the organizational context provided by management". The study task is essentially a long-term one, and there is scope for triangulation of multiple data sources from the team and the researchers.

For this particular study, there was the added benefit that there is a range of theoretical material available, related to the study of teams. The researchers adopted a teamwork model based upon (Dickinson & McIntyre 1997), which identified seven core components of teamwork, structured as a learning loop. Combined with the real-life context, this all provided a very sound basis for adopting a case study approach.

18.4.2   Case study parameters

As we noted earlier, this was an explanatory single-case holistic study, in which the researchers studied one team within an organisation. Data collection took three forms: observations by the researchers (including attendance at project meetings); interviews with team members (and the Scrum master); and documents produced as part of the project records.

Analysis of this drew strongly on the management model, both as a framework as well as its role in providing a vocabulary for categorising data (an important benefit). One observation by the researchers was that the model was not really comprehensive enough, and in particular that it did not "describe certain important components, such as trust and shared mental models". (This might suggest that one problem for software engineering is to find management and organisational models that relate to a creative development process.)

Overall, the researchers were able to identify some challenges that arise when organisations move to the use of self-organising teams (and some limitations in terms of Scrum guidance). As such therefore, it can be argued that use of a case study did indeed allow in-depth field study of a complex phenomenon that could not readily have been performed in any other manner.

18.5   Reporting case studies

In Section 15.4 we discussed some generic issues concerning how primary studies needed to be reported in order to assist with the tasks of the systematic reviewer. These were more concerned with identifying whether or not a study was relevant than with its actual outcomes, and here we need to address the needs of the latter.

As for any empirical study, reporting its outcomes is only really useful if accompanied by a description of the methodological aspects, both as planned and as conducted. Within this, we would of course include the need to assess the threats to validity involved in the study.

For a case study, the reporting of both methodological elements and also of the outcomes can take many forms. Indeed, the very flexibility of the case study method makes it difficult to be overly prescriptive about how it should be reported, while at the same time making such reporting particularly important.

One aspect that can make life a little easier for the systematic reviewer is the careful use of terminology, particularly when reporting the methodological aspects. It is important to make clear how any terms are being used, perhaps by quoting a widely-known source such as Yin (2014) or a more domain-specific one such as Runeson et al. (2012). Both of these use much the same vocabulary and meanings, although drawing their examples from different domains.

Runeson et al. (2012) suggest the use of *checklists* when reporting a case study in order to ensure that all key material is included, and they also explain how the checklists have been systematically derived. (Actually, they identify useful checklists for many purposes, including design, data collection and analysis. However, our focus here is upon reporting of case studies.) Their checklist for reporting includes the following (Reproduced with permission.):

1. Are the case and its unit of analysis adequately presented?

2. Are the objective, the research questions and corresponding answers reported?

3. Are related theory and hypotheses clearly reported?

4. Are the data collection procedures presented, with relevant motivation?

5. Is sufficient raw data presented (for example, real-life examples, quotations)?

6. Are the analysis procedures clearly reported?

7. Are threats to validity analyses reported, along with countermeasures taken to reduce threats?

8. Are ethical issues reported openly (personal intentions; integrity issues; confidentiality)?

9. Does the report contain conclusions, implications for practice, and future research?

10. Does the report give a realistic and credible impression?

11. Is the report suitable for its audience, easy to read, and well structured?

Together, these should provide a clear report about the motivation, conduct and analysis involved in the case study.

If provided with this information, our remaining question is what further information is needed for the systematic reviewer? As always, this is partly dependent upon the type of review: for a mapping study where the main synthesis activity involves categorisation, the preceding list may well be sufficient. However, for a fuller degree of synthesis, the reviewer may also need both contextual information about the case and also some additional topic-specific material in order to be able to make best use of the case study in their review.

Further reading

The main source of knowledge about case study research comes from the social sciences, and this chapter has drawn heavily upon the positivist research model that has been developed by Robert K Yin. His book *Case Study Research: Design and Methods* (Yin 2014) has formed a major influence upon the adoption of this form by software engineers. It has also probably been a major influence in ensuring that case study research is viewed as a 'respectable' form of empirical study for many disciplines, being very solidly grounded in the research literature, as well as very readable. For a software engineer, the main limitations are probably the examples, which inevitably are drawn from more 'social' disciplines, although the number of citations to this work in the software engineering literature suggests that this has obviously not proved a major barrier.

From a software engineering perspective, the book *Case Study Research in Software Engineering: Guidelines and Examples* by Runeson et al. (2012) forms a major addition to the empirical software engineering literature. This book is organised as two parts: a first part that discusses the methodological issues of case study research in general, and how it can be and has been adapted for use in software engineering; and a second part that provides some substantial discussions about examples of its use. Like this chapter, it is heavily influenced by Yin's work, although balancing this with other influences from case study research in social science by Robson (2002), and in information systems by Benbasat et al. (1987). For a software engineer planning on conducting or reviewing a case study, this book provides an excellent resource, including checklists, examples and analysis models.

Chapter 19

Qualitative Studies

Software engineering research has been fairly late to recognize the value of qualitative studies. Carolyn Seaman was one of the first strong advocates of qualitative studies (Seaman 1999). She suggested that the advantage of qualitative approaches is that "they force the researcher to delve into the complexity of the problem rather than abstract away from it". She also pointed out that, in the context of software engineering, "the blend of technical and human behavioural aspects lends itself to combining qualitative and quantitative methods, in order to take advantage of the strengths of both".

This chapter discusses the role of qualitative studies in software engineering from the viewpoint of their use as primary studies in systematic reviews. We define what we mean by a qualitative study, and discuss how qualitative research is conducted, what sorts of software engineering research questions can be addressed using qualitative methods and what a meta-analyst requires from a qualitative primary study.

19.1   Characteristics of a qualitative study

A qualitative study addresses research questions that are related to the *beliefs, experiences, attitudes and opinions* of human beings either as individuals or in groups. Many qualitative studies are organized as case studies (see Chapter 18) or surveys (see Chapter 17). Qualitative researchers aim to gather in-depth understanding of human experiences, beliefs and behaviour using relatively small focused samples, for example *theoretical samples* rather than one-off random samples. Theoretical samples are based on soliciting information from participants likely to have insight into the research topic. The aim is to select participants that, taken together as a group, are able to shed light on all aspects of the research question. Theoretical samples may be extended to include other participants as the researcher develops his/her theories about the research topic. Theoretical sampling terminates when *theoretical saturation* is achieved, that is, when the researcher believes that no new information will be obtained by interviewing or observing any more participants. Thus qualitative research aims to identify as many different viewpoints as possible rather than the most-commonly expressed viewpoint.

19.2   Conducting qualitative research

In software engineering research, qualitative studies often involve either asking participants about their experiences and opinions as software engineers (or managers), or observing their behaviour when they are in their usual working environment.

There are various approaches used to ask questions including:

- *Self-administered questionnaires.* These usually have a semi-quantitative form where participants are asked to assess the extent of their (dis)agreement with a question based on a *Likert response format.* There is some confusion between a Likert scale and Likert response format (Carifio & Perla 2007). A Likert scale refers to a set of positive and negative questions or statements (referred to as *items*) about some underlying concept. A Likert response format is an ordinal scale used to assess each item in the Likert scale. A Likert response format is often based on a five point response of the form:

  1. Strongly agree

  2. Agree

  3. Neither agree nor disagree

  4. Disagree

  5. Strongly disagree.

Some researchers prefer a four point (even) set of responses removing the middle option. This is to make respondents think seriously about their preferences rather than simply take a "neutral" middle option. Other researchers prefer more response categories, up to a maximum of seven, to allow respondents to give more precise answers, since if increased precision proves unnecessary, little used response format categories can be collapsed.

Questions based on Likert response formats are called *closed questions* because they have a restricted number of possible answers. However, self-administered questionnaires may also include *open questions*, where the participant is asked to comment about some question or issue in their own words. Many surveys are based on self-administered questionnaires (see Chapter 17).

- *Interviews.* For these the researcher meets participants, usually in a one-to-one meeting (although it is possible to have more than one researcher or participant in some situations) and discusses the topic of interest. In *unstructured* interviews the researcher discusses the general topic with the participant and allows the discussion to have its own momentum, similar to a personal conversation. In *semi-structured* interviews the researcher has a list of questions related to the topic (s)he will ask, which provides an overall structure to the interview. In many circumstances, with permission of the participant, the researcher will take an audio-recording of the interview for later transcription; otherwise the researcher will have to take notes during the interview.

- *Focus groups.* Here a researcher meets with groups of participants and obtains group answers to his/her research questions. Focus groups are often used in marketing to get opinions about products or services.

Observational studies are based on researchers watching software engineers in their working environment, including team meetings. They may also involve reviewing documents used/produced by software engineers. They are often *longitudinal studies*, that is, studies that take place over a relatively long time period, monitoring the same participants or workplace throughout. Sometimes, working processes and meetings may be video-recorded (as may focus group sessions). If the researcher is part of the software development team, the study is called a *participant-observer study*. Participant-observer studies are often used for *action research*. In software engineering, action research occurs when researchers introduce a change in working practices, usually in order to address a perceived process problem, and then monitor the resulting practice to assess the impact of introducing that change.

Qualitative researchers acknowledge that what participants do, and what they say they do, may differ. To address this issue qualitative researchers recommend *triangulation*. Triangulation, in this context, means using information from different sources to cross-check results and validate findings, see for example, the paper by Bratthall & Jørgensen (2002) which provides empirical evidence that a multiple data source case study is more trustworthy than a single source case study. Triangulation also refers to a variety of methods to be used to validate qualitative findings. The methods are based on:

- Theories, that is, theories derived in a specific qualitative study can be compared with theories derived in other comparable studies.

- Methods, that is, using multiple methods to assess the same data.

- Researchers, that is, different researchers analyse the same data independently and compare their findings.

Regardless of how a qualitative study is organized, it must be converted into *text*. Videos or recordings need to be transcribed and initial notes need to be converted into detailed field reports.

19.3   Research questions that can be answered using qualitative studies

Qualitative studies in software engineering address questions related to the way people work, both in teams and as individuals, and how the environment in which they work helps or hinders their software engineering tasks. Research questions may typically address:

- The organisation of teams and their working practices, for example, how do agile teams manage to work successfully?

- Interactions among teams with different software engineering roles, for example, what is the best way to encourage effective working between testing and development teams?

- Opinions of people about their working experiences, for example, what do individuals find most useful about test-before programming?

- Opinions of people about barriers and risks associated with a new method or process, for example, what are the major problems facing companies providing global outsourcing solutions?

The main assumption behind qualitative methods is that such questions are best answered either by asking the software engineers and software managers involved in the tasks for their opinions concerning the topic of interest, or by observing their behaviour when performing their software engineering tasks.

19.4   Examples of qualitative studies in software engineering

In this section, we describe some qualitative software engineering studies. Firstly, we discuss three studies that used a mixed methods approach including both quantitative and qualitative data. Then we describe two papers that used completely qualitative approaches.

19.4.1 Mixed qualitative and quantitative studies

Seaman & Basili (1998) report a mixed qualitative and quantitative study looking at how companies organize document inspection meetings aimed at finding defects in software programs. They used a participant-observer method which included observing 23 inspection meetings, each of which covered 2 or 3 C++ classes. The intention was to "capture first hand behaviours and interactions that might not be noticed otherwise". At each inspection the observer recorded:

- Names and roles of each inspector, the amount of code inspected (in lines of code).

- Details of each discussion during an inspection meeting: its length, participants, the type of discussion, and notes describing the discussion.

After each inspection, the observer wrote extensive field notes.

The observer also undertook a series of semi-structured interviews with inspection participants which obtained information about organizational relationships, data on non-meeting-based inspection activities and a subjective assessment of the inspected code's complexity. Most of the interviews were recorded (but not transcribed) and used to assist with producing detailed field notes.

Data analysis was based on the constant comparative method advocated by Glaser & Strauss (1967) and the comparison method suggested by Eisenhardt (1989). Both of these approaches aim to develop *theory* from qualitative data rather than test hypotheses. The actual data analysis involved performing case-by-case comparisons aimed at finding patterns among the characteristics of the inspection meetings. This led to the construction of a network of relationships among variables identified during the analysis. Variables were categorized into three types: outcome variables (such as the length of meetings, time spent on specific discussions), organizational variables (such as the organizational distance among participants and the physical distance among participants) and two moderating variables: the length of code and the complexity of the code. Quantitative information (that is, the average time for long and short meetings) was incorporated into the network to further define the relationships among elements. The end result of the study was a set of "well supported hypotheses for further investigation", for example:

- *Hypothesis: The more complex the material being inspected, the less time will be spent discussing global issues.*

- *Hypothesis: The more experienced or skilled the author is perceived to be, the less preparation time will be reported.*

Rovegard, Angelis & Wohlin (2008) also used a mixed qualitative and quantitative approach to investigate change impact analysis (IA). They had three research questions:

1. *RQ1. How does the organizational level affect one's assessment of importance of IA issues?*

2. *RQ2. What difference does the perspective make in determining the relative importance of IA issues?*

3. *RQ3. Which are important IA issues and how are these addressed in software process improvement (SPI)?*

They used a three-step research process, of interviews, post-test tasks and workshops. They interviewed a convenience sample of 18 people in a variety of roles involved in the change management process who were asked "which potential issues are associated with performing impact analysis?". The post-test tasks were used to let participants determine their level in the organization (strategic, tactical, operational) and to let them prioritize issues from their own viewpoint and from the viewpoint of their organization. The prioritization was based on the cumulative voting technique which provided a quantitative measure of priority. Disagreements between individuals at their personal level and the organization level could therefore be quantified. The workshops provided a forum for participants to discuss the prioritization, to identify possible ways of mitigating important issues, and to understand each others' views.

Although there were substantial disagreements at both levels, and disagreements were correlated across levels, the researchers were able to identify the types of issues important at different levels, and a number of ideas for improving the process.

Zhang & Budgen (2012) report the results of a study aiming to investigate why three popular design patterns (specifically, *Facade*, *Singleton*, and *Visitor*) received conflicting opinions of their value in an earlier survey. Zhang & Budgen sent another questionnaire to the people who responded to their first survey. Part of the survey involved stating their agreement (based on a three-point response scale: agree, no opinion, disagree) to a set of statements about each pattern The statements included both positive and negative characteristics each pattern derived from comments reported in the original survey. Respondents were also asked to add:

- "Their own observation about that patterns and any thoughts about why it attracts conflicting views",

- "Provide any examples of good or bad use of the pattern based on their own experiences".

The quantitative analysis involved reporting the number of responses of each type for each question about each pattern. Across the three patterns, the positive questions achieved a large number of agree responses and relatively few disagree or no opinion responses. In sharp contrast, negative questions received a more mixed response: for *Visitor*, the number of agree, disagree and no opinion responses were very similar, for *Singleton* the majority of respondents disagreed although a substantial minority agreed, for *Facade* the number of agreements and disagreements was very similar while for the other question a large majority of respondents disagreed.

The qualitative analysis involved a four-step process of categorising and coding the six datasets 'comments' and 'experiences' for each of the three patterns. For each dataset:

1. Each of the co-authors read the responses and wrote a list of common issues.

2. The co-authors met and merged their lists generating a short agreed list of categories.

3. Each co-author independently coded the responses using the agreed categories.

4. They met and reviewed the codes, and discussed and resolved any disagreements.

5. They produced a set of categories and frequency counts for the comments and the positive and negative experiences for each pattern.

The authors used the Cohen's κ statistic to assess the agreement among coders. They discussed the implications for each of the patterns considering both the quantitative and qualitative results. They also identified threats to validity inherent in the survey method and the analysis. They conclude that:

- *Visitor* should "carry a 'health' warning, in that used outside of a well-constraint context, it is likely to increase complexity".

- *Singleton* appears to polarize opinion, and should only be used by experts. They note that it is often used in teaching because it is relatively simple, but their results suggest this practice should be avoided.

- *Facade* has some benefits in specific circumstances, but needs to be used with care since the consequences of misuse "may be more significant than for many other patterns".

19.4.2 Fully qualitative studies

Sharp & Robinson (2008) report an *ethnologically-informed* study involving three different companies with experience of eXtreme Programming (XP). They investigated the use of *story cards*, which are 3×5 inch index cards on which engineers write self-contained user functionality, and the *Wall* which is a physical place used to organise and display story cards. Each study was conducted at the XP teams' offices and lasted between 5 and 8 working days, plus one day at a later stage to report the findings. The data gathered included extensive field notes, photographs, and copies of work artefacts

Although there were some differences in working practices, they found substantial similarities among the three companies. The concepts of the story card and the Wall are simple but provide a "sophisticated and disciplined way to support levels of co-ordination and collaboration". Cards are annotated as they pass through the development process but but describe chunks of work that are to small to stand alone. The Wall provides an overview of the development plans and progress and supports various co-ordination activities:

- The daily meeting takes place around the Wall.

- Taking a card down acts as a means to signal that the work item will be addressed by the person taking the card, and prevents anyone else from working on it.

- Colour coding and annotating cards provide progress-tracking information.

- The structure of the Wall shows the current status of an iteration and it is continually updated.

- During a meeting, holding a card signifies an engineer has something to say.

- Cards can be moved around to help group reasoning

Sharp & Robinson also discuss why the physical Wall and cards are better than automated alternatives. They suggest it may be because using physical artefacts "relies on trust and a highly disciplined team".

It is not often the case that "raw" data from a qualitative study is compact enough to present in full. However, Kitchenham, Brereton & Budgen (2010) report a small study investigating the educational benefits of mapping studies, in which all the raw data was reported. The data was obtained from a self-reported questionnaire completed by six participants who had just finished a mapping study project. Based on personal knowledge of mapping studies and systematic reviews, we developed five propositions:

- *P1: Mapping studies teach students how to search the literature systematically and organize the results of such searches.*

- *P2: Postgraduate PhD students will find a mapping study a valuable means of initiating their research activities.*

- *P3: Postgraduate students and undergraduate students will find undertaking a mapping study provides them with transferable research skills.*

- *P4: Problems students find with mapping studies will primarily be concerned with the search and study classification processes.*

- *P5: Students should find mapping studies relatively easy to document and report.*

The questionnaire included questions about the students and their own experiences of undertaking mapping studies including:

- Was this a satisfactory educational experience?: Yes/Mostly/Somewhat/No. Please explain your answer?

- What (if anything) have you learnt from the experience?

- Did you find any aspect(s) of the projects particularly troublesome? Yes/No. If Yes please specify.

The free-format responses for each student for each question were read and linked to the propositions and other recurring themes. The responses of the students were then used to assess the validity of the propositions. The results found good support for propositions P1, P2 and P3, some support for P4 but little support for P5. Additional comments of interest were that some students found the work challenging and/or enjoyable and gave a good overview of the topic, but that it was time-consuming.

Since we performed the study ourselves, and can hardly be regarded as neutral commentators on systematic reviews, we were careful to discuss our personal bias and how we attempted to minimize its impact.

19.5 Reporting qualitative studies

If a qualitative study is going to contribute to a systematic review, it is helpful if the study reports clearly the methods it used and its findings. The meta-analyst needs to be clear about:

- Whether the study is of relevance to his/her systematic review.

- Whether the study results are trustworthy given the methodology reported by the author(s).

- What the actual findings of the study are.

Based on the discussion by Greenhalgh (2010) and the findings from CASP[1], authors of qualitative studies should ensure they report clearly the following information:

---

[1]The Critical Appraisal Skill Programme, see www.cap-uk.net

- The aims of the research.

- A justification for their choice of research method and an explanation of why it is appropriate given the aims of the study.

- An explanation of how they recruited appropriate participants, provide context and demographic details about the participants and information about the settings in which the participants were encountered.

- The methods they used to collect and analyse data. These activities often take place in parallel in a qualitative study, so authors might consider using a flow diagram to describe their research process.

- Data analysis showing how the raw data was converted into the study findings, for example, initial codes with examples of quotes from participants, and how initial codes relate to higher level codes

- The relationship between the researchers and the participants and any methodological issues arising from that relationship, for example, any limitations or caveats as a result of using a participant-observer design

- Any limitations or caveats related to the research design or the conduct of the research.

- A clear statement of ethical issues and how they were addressed. This is a particular issue when using student participants.

- A statement of the findings from the study, clearly separated from any more speculative discussion or discussion of other related research.

- A final summary of what the findings imply for software engineering practice.

Further reading

There are many books that discuss qualitative research methods. In the book "*Qualitative Data Analysis. A Methods Sourcebook*", Miles et al. (2014) provide detailed guidance on how to collect, display, and interpret qualitative data. In her book "*Researching Information Systems and Computing*", Oates (2006) provides a general overview of qualitative methods from an Information Systems viewpoint.

Other texts discuss specific methods in detail. In "*Grounded Theory. A reader for Researchers, Students, Faculty and Others*" Remenyi (2014) provides a general overview of grounded theory. There are, of course, books written by the original developers of grounded theory but unfortunately their ideas have diverged over the years. In her paper "A synthesis technique for grounded theory data analysis", Eaves (2001) provides flow diagrams showing the different versions of grounded theory and presents an integrated approach she used for her research.

Hammersley & Atkinson (1983) discuss the principles of ethnography. In the context of software engineering research, Robinson et al. (2007) describe the *ethnographically-informed* approach they have used to study software engineering practices in their paper "Ethnographically-informed empirical studies of software practice". They note the need both to base research on real practice not official procedures reported in company manuals, and to minimize the impact of the researcher's own background prejudices and assumptions. They also identify two major challenges: ensuring a good working relationship between researchers and participants, and adopting a rigorous method to avoid bias.

Chapter 20

Data Mining Studies

This chapter discusses the use of data mining techniques in software engineering and how data mining studies should be reported for purposes of future meta-analysis. Software engineering researchers have a long history of undertaking post-hoc analysis of industrial software project data to investigate issues such as:

- Factors that influence the costs of software projects.

- Factors that influence the quality of software components.

- Factors that affect the probability of a software project being judged a success.

This type of research has expanded even more with the availability of large datasets produced by the configuration management and change monitoring tools adopted by Open Source Software (OSS) projects.

In contrast to other types of empirical study, the analysis of industrial datasets, which we shall refer to as data mining, is one of the few areas of software engineering research that has a large volume of empirical studies. These studies have already led to interesting systematic reviews and mapping studies, some of which we discuss in this chapter, but there are also problems with the quality of primary studies that make meta-analysis difficult, and need to be addressed to facilitate aggregation of data mining research findings.

Readers should note that data mining techniques are used in many different disciplines for many different purposes. In this chapter, we discuss only the uses of data mining in the context of software engineering, which do not utilize the more complex machine learning methods needed for problems such as face recognition, speech recognition, or language translation.

20.1   Characteristics of data mining studies

Data mining is about the organizing and searching of large amounts of data with the aim of extracting important patterns and trends. It is used in many fields for many different purposes. In the context of software engineering, it is usually used to develop prediction models aimed at identifying some important characteristic of a project or of a software component that is currently unknown.

Data mining analyses can be categorized into two broad types: *supervised* or *unsupervised*. In *supervised learning* the goal is to predict the value of some outcome measure (for example, the expected effort required to develop a software project) given a number of input variables (for example, the estimated size of the project, the experience of the developers). In *unsupervised learning* there is no outcome measure, and the goal is to find patterns within the data (for example, groups of items that share similar properties).

The "data" in a software engineering data mining study is usually represented as one or more $n \times m$ data matrices with n rows specifying the number of data points (referred to as *instances* or *cases*) in a data matrix and m identifying the number of attributes associated with each data point. Thus, a data point is a *vector* with m elements. In the context of data describing software engineering projects, n would denote the number of projects in the dataset and m would denote the number of *attributes* (referred to as *variables* by statisticians and as *features* by data mining researchers) containing data about each case.

20.2   Conducting data mining research in software engineering

Data mining-based research involves:

- Identifying one or more datasets that can be used to answer the research questions. This may involves constructing a dataset from data held in a number of different sources.

- Pre-processing the data which involves

  - cleaning the data to remove or correct untrustworthy observations,
  - removing data points of attributes with missing values, or imputing (that is, estimating the value of) the missing values,
  - transforming the raw data values into values that have useful properties (such as having an approximately Normal distribution). For

example, in studies concerned with predicting the effort needed to complete a software project, effort values are often transformed into the logarithmic scale.

- Applying a computer-based algorithm to summarize and/or analyse the data in order to answer the research questions.

Many datasets related to software engineering activities are relatively small, however, we use the term data mining to cover the analysis of historic datasets that were not (usually) collected specifically to address the research questions of interest to the analyst.

Methods of analysis are usually based either on *statistical methods* or *machine learning methods*:

1. Statistical methods for data mining usually include:

   - For unsupervised learning problems, *cluster analysis* (which can be used to identify subsets of the data points that have similar properties) or *principal component analysis* (which can be used to identify subsets of the variables that exhibit similar patterns),

   - For supervised learning problems, *regression analysis* of various types (including *logistic regression* if the outcome variable is a binary variable and *least squares regression* if the outcome value is an ordinal or ratio-scale measure of some kind).

2. Machine learning methods usually use a *k-nearest neighbour* method for finding groups of k cases that are most similar to one another with respect to the measured attributes. Similarity measures are constructed between each pair of cases based on the attribute values and the resulting $n \times n$ matrix is analysed to find subsets of similar cases. This basic approach can be used to find clusters of similar cases for an unsupervised learning problem. For supervised learning, the algorithm can find the set of k cases that are most similar to a new case (based on a similarity measure, such as Euclidean distance). From this, the value of an outcome variable for the new case can be estimated (for example, by a weighted mean of the k similar data points with a weight based on the value of the similarity measure of the kth case to the new case).

   For a supervised learning problem, it is usual to separate the dataset into a training set (that is, a random subset of cases) and a validation dataset (that is, the other cases). The training set usually comprises between two-thirds and 90% of the dataset and is used to construct the predictors. Predictor construction may involve screening out redundant attributes and, for machine learning nearest neighbour methods, determining the optimum value of k to use when making predictions. The effectiveness of the predictor is then assessed by comparing the predictions of the outcome variable for the validation dataset with the actual values of the validation cases.

Hastie, Tibshirani & Friedman (2009) discuss and compare least squares and nearest neighbour methods and point out that:

- A regression predictor is relatively stable in the sense that the estimates increase smoothly as the input variables increase, but depends critically on the assumption that a linear relation is appropriate and can, therefore, result in very inaccurate predictions. Thus, regression methods have low variance but potentially high bias.

- A nearest neighbour predictor does not rely on any stringent assumptions but each estimate depends on a small number of cases and their specific attribute values, so estimates of the outcome attributes of different cases can fluctuate unpredictably. Thus, nearest neighbour methods have high variance but low bias.

20.3   Research questions that can be answered by data mining

Software engineering research questions addressed by data mining studies are usually supervised learning research questions and include:

- Which of several different predictors produces the most accurate estimates of the staff effort to develop a software product?

- Which of several different predictors provides the most accurate estimate of the number of remaining defects in a software component, or is best able to identify components likely to have defects?

- Which factors influence software effort predictions?

- Which is the best method of several different methods of identifying the likely success or otherwise of a development project?

- Which factors influence software project success?

Cost estimation and fault-proneness studies are extremely popular in software engineering research. In a mapping study, Jorgensen & Shepperd (2007) identified more than 300 refereed journal papers published up to 2004 and a recent update to this has located a further 268 journal papers in the decade 2004-2013 (Sigweni, Shepperd & Jørgensen 2014). Hall et al. (2012) identified 208 fault prediction studies published between January 2000 and December 2010.

20.4   Examples of data mining studies

There are two different types of data mining study that occur in software engineering research. Some studies use data mining approaches to address a specific software engineering problem, whereas other studies view software engineering data as a means of investigating different data mining methods. In practice, however, many studies address both issues.

Briand, Melo & Wust (2002) investigated whether fault-proneness models built on data obtained from one system would work well on data from another system, which is an important issue if models are to be used in software engineering practice. At the same time they compared a new exploratory analysis method MARS (Multivariate Adaptive Regression Splines) with multivariate logistic regression. They concluded that a model based on one system could be used to rank modules from another system in terms of the likelihood of fault-proneness. They also concluded that MARS was a more cost effective predictor than logistic regression.

Studies that are more interested in data mining usually analyse many different data sets to assess data mining methods rather than investigating a specific data set that addresses specific research questions. For example, Kocaguneli, Menzies, Keung, Cok & Madachy (2013) evaluate a method called *QUICK* that aims to find the "least number of features and instances required to capture information within software engineering datasets". They evaluated QUICK on 18 different data sets and concluded that a k=1 nearest neighbour predictor on reduced data sets worked as well as the Correlation and Regression Tree method (CART) when using the full dataset.

From a more practical software engineering viewpoint Kitchenham, Pfleeger, McColl & Eagan (2002) investigated whether regression models based on a function point size metric would produce more accurate project effort estimates than the estimates produced by company estimation experts working with the nominated project manager. From a dataset based on 145 projects they found expert estimates out-performed a regression-based function point model, even when the model was restricted to a homogeneous subset of the data and the model was re-calculated for separate time periods.

More recently Jørgensen (2014a) analysed a dataset of 785,325 small-scale global outsourcing software projects, in order to identify when and why such projects fail. He constructed a binary logistic regression model using information known at the time of project start-up as input variables that correctly predicted 74% of project failures and 64% of non-failures. The two factors that most strongly reduced the risk of project failure were previous collaboration between a client and a provider, and a low failure rate of previous projects by the provider.

20.5 Problems with data mining studies in software engineering

There are several significant problems with the quality of many data mining studies in software engineering:

- Shepperd, Song, Sun & Mair (2013) have identified a variety of quality problems with the NASA defect datasets that are frequently used in data mining studies. They point out that there are two very different versions of the datasets and both versions exhibit numerous quality problems. These include problems with features, such as identical or constant features, features with missing or implausible values, and problem with individual cases, such as identical cases, inconsistent cases, and cases with missing, conflicting or implausible values.

- Many studies use invalid evaluation metrics to compare the accuracy of different cost estimation methods. In particular, relative error metrics such as the mean magnitude relative error (MMRE) are biased metrics but are still in common use, both for assessing prediction accuracy and for use as a fitness function in machine learning situations (Foss, Stensrud, Myrtveit & Kitchenham 2003, Myrtveit, Stensrud & Shepperd 2005, Myrtveit & Stensrud 2012). One "excuse" is that they are commonly-used metrics, so continuing to use them allows for comparisons with previous studies. For example, two recent papers published in IEEE Transactions on Software Engineering used the MMRE statistic and other related relative error metrics as their evaluation metrics (Kocaguneli, Menzies & Keung 2012, Kocaguneli et al. 2013).

- A meta-analysis undertaken by Shepperd et al. (2014) analysed 600 experimental results from primary studies that compared methods for predicting fault-proneness. They found that the moderator factor that explained most of the variation among study results was the *research group that performed the study*. This factor accounted for over 30% of the variation. This can be contrasted with differences among estimation methods which was the topic being investigated by most of the studies, and which accounted for only 1.3% of the variation among studies. They conclude that "it matters more who does the work than what is done". Furthermore they observe that "Until this can be satisfactorily addressed there seems little point in conducting further primary studies".

20.6   Reporting data mining studies

There are no agreed standards for reporting data mining studies nor for what constitutes a good quality data mining study. However, based on the issues raised by studies criticising current data mining studies, we suggest that authors of data mining studies should consider the following guidelines, if they want their results to be incorporated into meta-analyses:

- Ensure that they define the source of their data, and justify why they have selected the specific ones they used from among the large numbers of available datasets identified by Mair et al. (2005).

- Explain clearly what they have done to validate and clean the data, including issues such as transforming any of the attributes and handling missing or implausible values.

- Define clearly how their prediction method(s) work with appropriate citations, identifying any study or dataset specific variants of the basic methods.

- Assess the accuracy of prediction methods based on an independent validation dataset.

- Avoid the use of accuracy metrics or fitness functions based on relative error statistics. For cost estimation prediction Shepperd & MacDonell (2012) recommend accuracy metrics based on the mean absolute residual (MAR)[1], which can be converted into an accuracy statistic by comparing MAR_{Pi} for predication method i with MAR_{P0} based on random guessing. This gives a standardised accuracy measure SA_i for prediction method i of:

$$SA_i = \left(1 - \frac{MAR_{Pi}}{MAR_{P0}}\right) \times 100 \qquad (20.1)$$

MAR can also be used to construct an effect size given an estimate of the standard deviation (s_{P0}) of the random guessing metric (based on simulation):

$$\Delta = \frac{MAR_{Pi} - MAR_{P0}}{s_{P0}} \qquad (20.2)$$

- Report the 2×2 confusion matrix for fault-proneness studies that identifies the number of components that exhibited, or did not exhibit, faults in a validation dataset and the number of components predicted as fault-prone or not by a prediction algorithm. In their recent meta-analysis, Shepperd et al. used the Matthews correlation coefficient (MCC) as an

---

[1]There is an argument for using the *median* absolute residual because it is more robust than the mean absolute residual.

effect size for fault proneness studies. Unlike many metrics based on the confusion matrix, MCC is based on all four quadrants of the confusion matrix, and, in addition, as a correlation coefficient, it is relatively easy to interpret.

- Clearly specify the study findings.

- Explain the implications of the study for researchers and practitioners.

Further reading

In their book *The Elements of Statistical Learning, Data Mining, Inference, and Prediction*, Hastie et al. (2009) provide a detailed discussion of the statistical methods underlying data mining, including comparisons of nearest-neighbour and regression-based estimation methods that are particularly relevant to software engineering data.

Chapter 21

Replicated and Distributed Studies

At the end of Chapter 15 we observed that the general acceptance of any new knowledge that has been derived from the outcomes of empirical studies first needs the outcomes to be confirmed by being replicated by other experimenters. Indeed, such replication is generally considered to be "a crucial aspect of the scientific method" (Lindsay & Ehrenberg 1993). We also noted that achieving consistent results from replicated studies has sometimes proved to be rather problematical for software engineering, an issue that we discuss a little more fully in this chapter.

When considered from the perspective of systematic reviews, replications of primary studies can provide two specific contributions to a review. Firstly they provide additional inputs, and secondly a replication can provide a degree of quality assessment for the outcomes from the original primary study.

So in this chapter, we examine the meaning of replication both as a general concept, and as employed in software engineering. We discuss why this has proved a challenging and sometimes contentious issue, and identify some of the work that has been done in this area. We also discuss the potentially useful concept of a *distributed* study, organised along lines that are in many ways quite similar to a replicated study, while addressing a different, but relevant, purpose.

21.1   What is a replication study?

An empirical study may be *replicated*, by the original researchers or by others, for a number of different reasons. As indicated above, a replication may be performed to *verify* the effects detected by the 'original' study. A replication may also be performed to investigate how far different changes made to the conditions of the study may alter the outcomes (the *scope* of the effects).

Both purposes require quite close control and measurement of the conditions of the study, so for software engineering, the concept of replication tends to be largely associated with experiments and quasi-experiments, where the environment, tasks and measures can be at least partly controlled. However, even here, the human-centric nature of so many software engineering experiments does introduce a substantial element of variability that the experimenter cannot easily control, with consequences for any attempts at replication.

Case studies, particularly multiple-case forms, may implicitly or explicitly involve a degree of replication. So, repeating a case study with a case from (say) another organisation might well be considered as being a replication, or can just be viewed as another case in the set.

A survey can also be replicated, for example, by varying the sample or the sampling frame, or by re-sampling periodically. However, the benefits of doing so are not evident for surveys that address software engineering questions. Although there might be good reason to repeat a survey periodically in social science, perhaps to study how the profile of responses changes over time, we are not aware of any examples of surveys from software engineering that have involved such an approach.

An important influence on thinking about the meaning of replication in such a context (in this case, the social sciences) has been the paper by Lindsay & Ehrenberg (1993). In this, they make a number of key points about the design of replicated studies.

- Achieving an *identical* repetition of a study is impractical. Indeed, as they point out, even if this were achievable, it would be of no real value in a human-centric context where we want to know whether the result remains the same *despite* any differences. A result that only applies to the specific group of people involved in the original study is likely to be of little general value.

- Replications can therefore be categorised as being *close* or *differentiated* in nature.

  - A *close* replication seeks to "keep almost all the known conditions of the study much the same or at least very similar". Lindsay & Ehrenberg suggest that at least one repeated study of this form is "particularly suitable early in a program of research to establish quickly and relatively easily and cheaply whether a new result can be repeated at all". So, if there is some factor that differs, but is (wrongly) assumed to have no effect (such as the time of year when the study is carried out), then if this does matter it will mean that the study will not replicate successfully. The problem with extremely close replications is that if the outcome is an artefact of the specific experimental design, the results will be successfully replicated, but the replication process itself will be invalid.

- A *differentiated* replication involves "deliberate or at least known, variations in fairly major aspects of the conditions of the study". Such a replication then helps to determine the boundaries within which particular results may occur, and possibly, how much the results might be changed as a result of other factors.

- The role of close replication is therefore one of *confirming* the existence of an effect, while the role of differentiated replication is to determine the *scope* within which the effect occurs.

Figure 21.1 provides an illustration of this 'spectrum of replication' in an abstract form. The vertical axis represents some measured 'effect', to which we arbitrarily assign a range of possible values between 0.0 and 1.0, while the horizontal axis represents the 'degree' of differentiation. Obviously the latter is not really one axis, but for illustrative purposes we have assumed that we have a single measure of this. The star represents the value measured in the original study, the squares are close replications (1 and 2), while the circles (3–6) are a set of differentiated replications.

So, our interpretation of this is that the two close replications produce effect values that are fairly near to the value observed in the original study, while the outcomes from three of the differentiated replications do not differ very much. However, one of them (5) does differ considerably, suggesting that one or more of the factors being adjusted in that study does probably have some influence upon the measured effect.

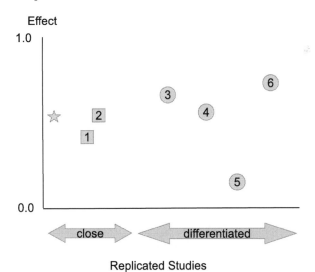

FIGURE 21.1: Illustration of replications.

A key issue for close replication is the question of exactly how close to the original this should be. In general, repeating a study while using a different group of participants drawn from the same "sampling frame" as the original participants, or using similar but different experimental material, can be considered to constitute a close replication, for which we don't expect to find a significant difference in the outcomes. From the viewpoint of synthesis for a systematic review however, close replications do not constitute a useful contribution, since they add "no useful information to the aggregation process" (Kitchenham 2008)[1]. Hence anyone conducting a systematic review is generally likely to be more interested in differentiated replications—and therefore needs to be able to distinguish between the two forms.

21.2   Replications in software engineering

As indicated earlier, for software engineering to date, *replication* has formed the major strategy for providing increased confidence in the outcomes from primary studies, (Miller 2005, Kitchenham 2008, Gómez, Juristo & Vegas 2010, Juristo & Vegas 2011). However, both Sjøberg, and later Juristo and Vegas have observed that, in software engineering, replications that were conducted by the same experimenters as those who conducted the original study usually obtain the same results as the original studies, whereas those performed by other researchers do not always lead to similar results (Sjøberg et al. 2005, Juristo & Vegas 2011).

21.2.1   Categorising replication forms

Although the model of Lindsay & Ehrenberg based on close and differentiated forms has been used quite widely in software engineering, other categorisations are in use within other disciplines. Gómez et al. (2010) have examined these, and sought to translate them to a software engineering context.

Their study identified some 18 different categorisation schemes (including that of Lindsay & Ehrenberg), which together included 79 'types' of replication. Most schemes came from Social Science (61%) and Business (33%). Unfortunately, they found that there was no consistency in the naming of replication types—with different categorisation schemes sometimes using the same term to mean different types of replication. Their analysis suggested a slightly extended form of categorisation, using three types.

---

[1]Kitchenham should perhaps have qualified her statement to say that concentrating on extremely close replications to the exclusion of differentiated replications is problematic.

- Replications which have little or no variation of conditions when compared with the original study (the 'close' category fitted into this type).

- Those that do vary, but where the same experimental method is used as in the original study.

- Those that use different experimental methods.

In effect, the last two sub-divide the *differentiated* form according to whether it differs from the original study by using different experimental materials or a different experimental process.

Within software engineering they found little evidence of methodological research into replication, apart from the concept of *families of experiments* (FoE) (Basili, Shull & Lanubile 1999). The goal for an *FoE* is to create and employ "a framework that makes explicit the different models used in the family of experiments". The motivation for developing the concept of an FoE is to enable better use of replication, as a step towards building a body of software engineering knowledge. One aspect of the framework is a set of three categorisations containing six replication types, summarised in Table 21.1.

TABLE 21.1: Replication Types Used in Families of Experiments

Category	Types
Replications that do not vary any research hypothesis (the dependent and independent variables are unchanged).	1. Strict replications. In effect these are 'close' replications that "duplicate as accurately as possible the original experiment".
	2. Replications that vary the way the experiment is run. This might involve using different group sizes, allocation criteria etc.
Replications that vary the research hypotheses (and hence vary attributes of the process, product and context models).	3. Replications that vary variables intrinsic to the object of study (that is, independent variables). Used to determine "what aspects of the process are important".
	4. Replications that vary variables intrinsic to the focus of the evaluation (dependent variables). Essentially, this is using different "effectiveness measures".
	5. Replications that vary context variables in the environment in which the solution is evaluated. For example, using practitioners rather than students as participants.
Replications that extend the theory.	6. These are a form of 'differentiated' replication making large changes to determine the bounds of an effect.

Based upon their study, Gómez et al. identified five elements of experimental practice that may usefully be varied in a replication study that is performed in software engineering.

1. The *site*. A replication can be conducted at the same location as the original study, or at a different one.

2. The *experimenters*. These might be the team who performed the original study, or a different team, or a mix of the original team plus others.

3. The *apparatus*. By this they mean the design, instruments, experimental material and procedures used for an experiment.

4. The *operationalizations*. These are the way that a construct is interpreted. They suggest that there are *cause operationalizations* in the form of the treatments that are to be evaluated, and *effect operationalizations* that represent the dependent variables used to measure the effect of the treatments.

5. The *population properties*. They suggest that there are two forms of population. One is the participants themselves, characterised by their role, experience etc. There are also the experimental objects such as the design documents, code, programs or any other artefact involved.

Clearly the choices made for each of these factors will be strongly influenced by the *purpose* of a replication. For the rest of this chapter we will continue to use the broad categorisations of *close* and *differentiated* replication, while recognising that there are many factors that might be involved in creating the most appropriate form of differentiation. However, one question that does arise is how many factors should be altered when performing a differentiated replication, and related to this, what degree of alteration is necessary to distinguish between a close and a differentiated replication.

21.2.2   How widely are replications performed?

To answer this question, we examine the outcomes from a systematic mapping study that charted the use of replication in software engineering up to the end of 2010. In particular, the study sought to identify which topics were most likely to be the subject of replicated studies (da Silva, Suassuna, França, Grubb, Gouveia, Monteiro & dos Santos 2014).

In summary, based upon a search of over 16,000 articles, they selected 96 articles that reported on 133 replications of 72 original studies. All were published between 1994 and 2010, with most being published after 2004, and with 70% of the studies (94) being "internal replications" (that is, performed by the same researchers who performed the original study). In this case, a study was classified as being 'internal' if there were one or more authors common between the articles describing the original study and the replication, otherwise as 'external'. An interesting observation was that, from their analysis of which authors were involved in a replication, there appeared to be two distinct research cultures present: one group preferring to perform internal replications and a second preferring to perform external ones.

Replications were mainly of quasi-experiments, and together with controlled experiments these formed 88% of the studies. Interestingly though, they found one survey and 15 case studies that were replications, with the latter being a fairly even mix of internal and external replications.

Topics where most replications occurred (in descending order) were: software requirements; software quality; software construction; software engineering management; and software maintenance. As the authors observe, this may simply be a reflection of there being more empirical studies on these topics that can be replicated, but they had no way to confirm that this was so. Only for software construction was the number of external replications close to the count of internal ones.

The study does not report how many replications were close and how many were differentiated. Indeed, the authors observe that the poor quality of reporting, particularly about the context of the original study, made it difficult to distinguish this aspect with any confidence.

One interesting discussion provided in the study is how well the replications confirmed the results of the original studies. We have already noted that previous studies have suggested a much greater likelihood that an internal replication will confirm the original outcomes than an external replication (Sjøberg et al. 2005, Juristo & Vegas 2011). For this review, the results of a replication were classified as being *confirmatory, partial* or *non-confirmatory*. Since four studies did not provide clear comparisons, the reviewers were left with 129 studies for this purpose.

Some 82% of internal replications confirmed results, with a further 9% providing partial confirmation. In contrast 46% of the external replications did not confirm the original results, reinforcing the trends previous observed. However, as the authors of the review note, there could be many factors influencing this, with one being publication bias. In particular, they observe that:

- Researchers are unlikely to publish non-confirmatory results for their own work;

- Negative results are probably less likely to be accepted for publication;

- Negative results might be easier to publish when related to the work of others.

There is also the likelihood that unintended variations are probably more likely to occur for external replications, and also it is possible that effective replication may well need access to tacit information about how the original study was conducted. Whatever the cause, as the authors of the review observe "the bulk of the replications analyzed in our review are isolated confirmatory internal or non-confirmatory external, and do not contribute to substantial knowledge building in our field". So while awareness of the value of replication studies does seem to be increasing, these are still making only a limited contribution.

21.2.3   Reporting replicated studies

The workshop paper by Carver (2010) provides some useful aggregated ideas about reporting of replications in the form of a small set of suggested guidelines. (Appropriately, these were also partly derived by performing a systematic review.)

The guidelines themselves address four key aspects of a replication.

1. *Information about the original study.* Key information about this is identified as being: the research question(s); the number and characteristics of the participants; the experimental design; details of any artifacts used; any important context variables that might have affected either the design or the analysis of the results; and a summary of the major findings from the study.

2. *Information about the replication.* The guidelines identify three important pieces of information that need to be provided:

   (a) The *motivation* for conducting the replication, and hence implicitly, the type of replication being reported (close/differentiated).

   (b) The degree of *interaction* with the original researchers, assuming that the replication is an external one of course.

   (c) What *changes* were made to the original experiment in terms of design, participants, procedures followed, artifacts used, data collection and analysis. It is also suggested that the reasons for any changes are also explained.

3. *Comparison of results with the original study.* This should highlight differences and similarities, and the guidelines suggest that this really merits a section of its own. (The paper also notes that in a few cases, this has been done very effectively by using a summary table.)

4. *Drawing conclusions across the studies.* For a replication to provide useful knowledge, it should really augment the conclusions drawn from the original in some way. So, the guidelines very sensibly suggest that a report should discuss what the studies indicate when viewed as a whole.

All of this information is of course very relevant for anyone performing a systematic review of any form, and particularly where this involves any form of aggregation.

21.3   Including replications in systematic reviews

From the perspective of a systematic review, *close* replications are of little direct value, since they are not sufficiently independent to be included in a meta-analysis—and indeed, should not be. However, they do potentially offer a *quality* measure for the original study, although as the discussion about publication bias in the previous section indicated, this might require to be treated with some care, depending upon who conducted the replication.

In contrast, most *differentiated* studies are likely to be a useful addition, assuming that their reporting is adequate. So, distinguishing whether a replication is close or differentiated is an important task, although as da Silva et al. (2014) have noted, this is not always made clear when researchers are reporting a replicated study.

So an important need is to be able to distinguish which type of study is being reported. Madeyski & Kitchenham (2014) suggest that pairs of studies where most or all of the factors below apply should be treated as being close replications.

- Both studies were run by the same experimenters.

- The same subject types were used (such as students, or staff from the same organisation).

- The same experimental materials were used.

- The same experimental design and analysis method were used.

- The studies took place in the same setting, such as the same university or within the same organisation.

For this purpose, the "same experimenters" might not be exactly the same group, but at least one, and probably more, of the authors will be common to both studies.

We might note too that, although the concept of *families of experiments* (FoE) is a potentially useful one, it does not specify a procedural approach to be followed by researchers (in contrast to the proposals of Juristo & Vegas (2011)). So, when viewed in terms of the needs of secondary studies, a major downside is that the individual experiments comprising the family can potentially violate the assumption that individual primary studies are independent, a characteristic which is essential to meta-analysis procedures. It is not clear whether studies comprising a family of experiments should count as independent for the purpose of meta-analysis, nor is it clear what the impact of violating the assumption would be on the outcomes from a meta-analysis. This therefore makes aggregating the results from a family of experiments problematic, and so the above criteria from Madeyski & Kitchenham should be applied quite strictly.

21.4 Distributed studies

One of the challenges for anyone conducting an experiment or quasi-experiment is to recruit an adequate number of participants. One reason for this is that software engineering experiments tend to require an element of technical knowledge or skill, and even when training is provided for this aspect, this still assumes a basic level of computing knowledge. So this usually forms an upper bound on the number of participants who can be recruited at any one site. As a result, software engineering experiments often have results with poor statistical power (Dybå et al. 2006). (As a reminder, the *statistical power* is the probability that a given test will correctly reject the null hypothesis.)

A second factor, which usually acts to reduce the number of available participants, is that students and practitioners who work in software engineering are usually unfamiliar with the idea of acting as participants. In many cases, they will have been exposed to little or no experience of empirical studies, and so be unaware of the importance these have in providing knowledge. Overcoming this can be quite challenging, as it is important to avoid applying undue and inappropriate pressure on people to take part, for both ethical reasons and also those of potentially biasing the outcomes.

One way of increasing the number of participants is to spread the study across a number of sites, so giving access to an increased number of potential participants. This is quite often done in clinical medicine, where such studies are termed "multi-site trials", and where the situation is helped by the way that participants are usually recipients of the 'treatment', rather than actively using their skills, as in software engineering. For the software engineering context, we prefer to use the terms *distributed (quasi-)experiment* or *distributed study*, which emphasises the need for suitable organisation.

In some ways this is similar to replication, although with the results from the different sites being aggregated for the analysis, rather than treated separately. An example of this is the study reported in (Gorschek, Svahnberg, Borg, Loconsole, Börstler, Sandahl & Eriksson 2007), which reports on a family of experiments undertaken at a number of sites, but presents the outcomes as a single entity, although this was not actually planned as a distributed study.

We ourselves have been involved in a trial distributed quasi-experiment, and describe our experiences of this in (Budgen, Kitchenham, Charters, Gibbs, Pohthong, Keung & Brereton 2013). We used a methodological topic (assessing completeness of structured abstracts when compared with conventional ones) as this was one that did not require any skill training. We chose this to limit the complexity of the experimental task, allowing us to concentrate on the distributed aspects. The core lessons from our study were the need for comprehensive documentation and effective communication in order to ensure that the five sites involved did conduct sufficiently similar elements of the quasi-experiment.

So, while distributed studies do provide the potential to address the limitations of single-site studies, they do need to be planned and conducted with considerable care. Both the recruitment of participants who have generally similar skill levels, and also the provision of training (or skill assessment) also needs to be considered. As such, this is still an area of research, but is mentioned here because the concept of the distributed study does offer a different perspective, while also involving many organisational issues that are common with replicated studies.

Further reading

The organisation and conduct of both replicated and distributed empirical studies are still very much issues of active research, and hence there is relatively little in the way of authoritative material that can be consulted. Two papers that we suggest are useful as further reading are as follows.

- *The Design of Replicated Studies* by Lindsay & Ehrenberg (1993) is highly readable and sets out many key issues of replication in a very clear form. As we note, other forms of category do exist, but the software engineering community seems to have largely adopted their terminology and while relatively simple, it seems to capture the key issues.

- *Towards Reporting Guidelines for Experimental Replications: A Proposal.* The value of this very short paper by Carver (2010) is that it sets out how a replication should be reported, and why. As such, it provides a useful baseline for anyone writing a paper that describes a replicated study.

Part III

Guidelines for Systematic Reviews

Chapter 22

Systematic Review and Mapping Study Procedures

22.1   Introduction

This rather long chapter is a revision of our previous guidelines for systematic reviews for software engineering research (Kitchenham & Charters 2007). There are deliberate overlaps with information provided in the preceding chapters of this book, so that this chapter can be used as a self-standing set of guidelines. However, we do cross-reference sections of the book where more detailed information about specific topics can be found.

Compared with the previous version of the guidelines, we have included more detailed advice for mapping studies and for the procedures needed by lone researchers including PhD students. We also include more guidelines based on our experiences of performing systematic reviews and the experiences of other software engineering researchers (Kitchenham & Brereton 2013).

Diagrams used in this chapter follow one of two standards:

1. When simple yes/no decisions are involved, we use a simple flowchart with the decisions shown as diamonds and actions shown in rectangles. Links (lines) between diamonds and rectangles have arrows which show the direction of flow, see Figure 22.1.

2. If decisions are more complex, as is often the case when planning a part of the systematic review process, we specify the decision using a rectangle with double lines. If decisions are inherently sequential, they are linked with lines using arrows to show the direction of flow. Simple rectangles are used to specify factors that affect a decision, or options available for a decision, or factors that influence the choice of options. Factors that influence a decision or an option are linked to the respective decision or option with a broken line with arrows at each end. Decisions are linked to options using a line with an arrow pointing at the option. An example of this form of diagram is shown in Figure 22.2.

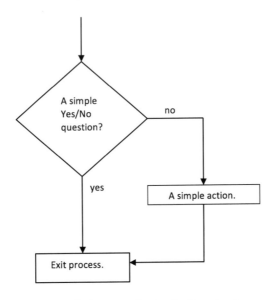

FIGURE 22.1: A simple flowchart.

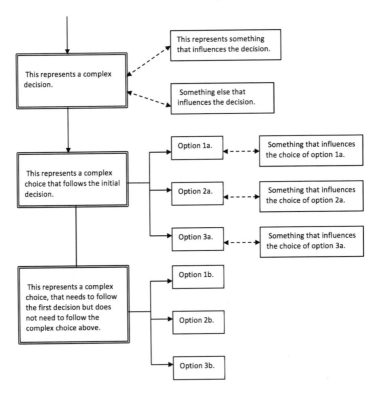

FIGURE 22.2: A complex planning process diagram.

22.2 Preliminaries

Before starting on a systematic review or mapping study, consider whether you have adequate background knowledge of the proposed topic area to be able to make decisions about the various choices involved. If not, you should begin by first reading about the topic (see Figure 22.3).

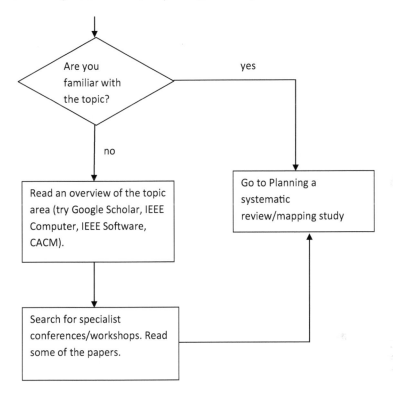

FIGURE 22.3: Initial considerations.

Point to remember

If you don't have any knowledge about the topic area, do not start planning your systematic review or mapping study yet. You need to read around the topic before you start.

22.3   Review management

A review is usually conducted by two or more researchers who comprise the review team. One researcher must act as the review manager or team leader in order to ensure all task activities are properly coordinated.

In the context of developing the protocol, the team leader is responsible for:

- Producing the protocol.

- Specifying the time scales for the review.

- Assigning the tasks specified in the protocol to named individuals.

- Obtaining any tools required to manage the review process and conduct individual tasks.

- Deciding how the protocol will be validated.

- Overseeing the protocol validation.

- Signing off the protocol and any subsequent changes to the protocol.

During the conduct of the review the team leader is responsible for monitoring the review progress, ensuring that team members complete their assigned tasks and managing any contingencies that arise during the review (such as disagreements about such aspects as primary study selection, quality evaluation, and data extraction).

Once the review is complete, the team leader is also responsible for signing off the final report.

Point to remember

The more researchers there are in a team, the more critical the role of the team leader becomes.

22.4 Planning a systematic review

Planning involves four main processes:

1. Justifying the need for a systematic review or mapping study.

2. Specifying the research questions.

3. Developing the protocol.

4. Validating the protocol

However, since reviews are usually done by a research team, planning also involves undertaking the basic project management actions such as task assignment, review coordination and monitoring, as discussed in the previous section.

22.4.1 The need for a systematic review or mapping study

An overview of the process of justifying the need for a systematic review is shown in Figure 22.4.

You should begin by checking whether any systematic reviews or mapping studies already exist in the topic area you want to study. If there are some, you may not need to do a review. Don't forget it is correct to use other researchers' work as the foundations for your own research. A major goal of systematic reviews in general, and mapping studies in particular, is to facilitate future research in a specific topic area.

However, if the existing reviews(s) do not cover the specific area you are interested in, or those that do exist are out of date, continue with planning the review. If there are already some existing secondary studies (that is, any literature reviews or state of the art surveys, whether systematic or not), you need to read these studies and decide whether you can use their results *as-is*, or you need to update one the review(s), or whether you need to undertake a new more focussed systematic review. In the case of an out of date systematic review, a sensible choice is to base your study on the protocol used by the initial study, while also amending the process if you can identify any limitations with the initial protocol.

If you decide to undertake a new more focussed review, the existing review(s) will have a significant impact in your research:

- You should read the papers describing the reviews and summarize their results. This will be the basis of the "related work" section in your final report. It should also allow you to specify the baseline of existing knowledge about the topic area and explain clearly how your results add to existing knowledge in the discussion part of your final report.

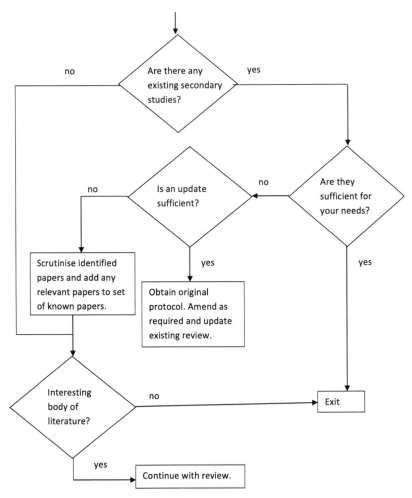

FIGURE 22.4: Justification for a systematic review.

- You should extract a list of any primary studies found by the previous reviews that are relevant to your topic area. These will be the basis for your set of known papers that can be used to construct and refine search strings for digital libraries and to help assess the completeness of your search (see Figure 22.5.1.3).

If there are no previous reviews, you need to be sure that there are likely to be sufficient relevant papers to make a systematic review or mapping study worthwhile. One way is to undertake a quick informal search using Google Scholar or a digital indexing system to look for relevant studies. In some cases a limited form of mapping study, called a scoping review, can be performed to determine whether there are sufficient empirical primary studies to justify a systematic review.

Finally, justifying the need for a review is about making a case that the topic is of interest and that it is an appropriate time to engage in an activities aimed at organising the literature or answering questions raised by the literature. For example, the reasons for performing a systematic review include:

1. There is a new development or testing method and practitioners would like to know if it is better than existing methods.

2. There are disagreements among researchers about the efficacy of a new method and the current empirical evidence needs to be collated.

3. There are field reports about a new software development method, or an international standard, and practitioners would like to understand what is known about the method or standard in terms of benefits and risks.

Reasons for doing mapping studies include:

1. To help assess the extent of research available in a topic area in order to identify sub-areas suitable for systematic reviews and sub-areas that need more basic research.

2. To organise a large number of independent research papers into a structured body of knowledge.

Points to remember

1. You need a genuine reason for undertaking a review, such as the likely existence of a large number of independent studies that have not previously been organised.

2. The existence of a substantial body of literature is not by itself a justification for a review. The topic for the review needs to be important to researchers and/or practitioners and the review needs to be timely.

3. Previous literature reviews (systematic or not) are extremely valuable for identifying known primary studies and validating your search process.

22.4.2   Specifying research questions

Research questions (RQs) are related to the justification for doing the review and the type of review being proposed.

22.4.2.1   Research questions for systematic reviews

If you are concerned with evaluating a technology, you should be planning to do a systematic review and the research questions should specify the type of evaluation you propose. For systematic reviews, the research question defines much of the search process. It is important to make sure the research questions(s) are properly formulated and stable, since changes to the RQ(s) will propagate other changes throughout the systematic review protocol.

If you are comparing two alternative techniques your research question will be of the form: Is technique A better than technique B? This basic question may need to be refined to determine what is meant by *better*, for example, more cost effective. Furthermore, you may want to qualify any answer in terms of any limitations or constraints on the answer, for example, does the answer apply to students or professionals, or to particular types of tasks, leading to questions of the type:

> Under what conditions, if any, is technique A more cost effective than technique B?

In software engineering, many empirical studies consider the impact and effectiveness of paradigm, method or standard A in an industry setting. Systematic reviews of such studies have research questions of the type:

- What are the risks or benefits associated with adopting paradigm, method or standard A?

- What factors motivate or de-motivate adoption of paradigm, method or standard A?

- How best should an organisation plan the adoption of paradigm, method or standard A?

22.4.2.2   Research questions for mapping studies

For mapping studies, research questions are often quite high level. This is because the characteristics of interest in the specific topic area may be hard to specify in advance. Thus, there is more likelihood of research questions being amended as result of identifying interesting aspects of the topic during data extraction.

Mapping studies usually have the overall goal of categorising the research literature for a specific topic in some way. This leads to research goals of the type: What trends can be observed among research studies discussing topic B?. The problem with a mapping study is deciding what trends will be of interest.

In practice, most software engineering mapping studies consider issues such as:

- The number of publications per year over the time period of the review, which gives an indication of the interest in the topic.

- The number of papers reporting studies of different types, often using the requirements engineering classification developed by Wieringa et al. (2006), which indicates the type of research being undertaken.

- The main researchers and research groups which identifies groups that interested researchers or practitioners might want to keep up with.

- The sources which published papers on the topic which identifies sources interested researchers or practitioners might want to monitor for future research.

This may be sufficient for the purposes of a student mapping study but is unlikely to be sufficient for a conference or a journal publication.

Mapping studies are far more interesting and beneficial to other researchers (and more likely to be published) if they also identify interesting subsets of the literature for example, the main subtopics, the different approaches/methods reported in the topic area and the extent to which they have been evaluated empirically, any significant limitations in existing research, as well as any significant controversies.

Points to remember

- For systematic reviews, research questions need to be well-defined and agreed to before the protocol is developed.

- For mapping studies, research questions are usually fairly high level and may be refined as the mapping study progresses.

- Student mapping studies are not always suitable for publication.

22.4.3   Developing the protocol

The research protocol defines and justifies what technical processes will be used to conduct and report the review and identifies which individuals will be assigned to which tasks. A template for a systematic review protocol is shown in Figure 22.5. The main technical issues that have not already been considered (that is, points 3 to 9 of Figure 22.5) will be discussed in later sections.

Search strings, quality extraction, data extraction, and data synthesis procedures need to be trialled as the protocol is developed.

Points to remember

- Sections of the protocol need to be tried out to ensure that the process is feasible and understood by all.

- The team leader is responsible for developing the protocol although some aspects can be delegated to other team members.

22.4.4   Validating the protocol

The protocol is a critical element of any systematic review. Researchers must agree to a procedure for validating the protocol. Where possible, you should try to find an independent reviewer.

Research teams should walk through the protocol and ensure that each researcher understands exactly what tasks he or she is scheduled to perform and the process he or she needs to follow to perform their allocated tasks. PhD students should present their protocol to their supervisors for review and criticism.

Since a systematic review aims to address specific research questions, the protocol should explain how those questions will be answered. Thus, a reviewer of a protocol needs to confirm that:

- The search strings are appropriately derived from the research questions.

- The data to be extracted will properly address the research question(s).

- The data analysis procedure is appropriate to answer the research questions.

The systematic review team leader is responsible for coordinating all the changes to the draft protocol and the final decision that the protocol is sufficiently complete for the systematic review to formally get under way.

Template for a Systematic Review Protocol

1. Change Record

This should be a list or table summarizing the main updates and changes embodied in each version of the protocol and (where appropriate), the reasons for these.

2. Background

a) explain why there is a need for a study on this topic

b) specify the main research question being addressed by this study

c) specify any additional research questions that will be addressed

d) if extending previous research on the topic, explain why a new study is needed

3. Search Process

a) specify and justify basic strategy: manual search, automated search, or mixed

b) for automated searches, specify search terms and compounds of these and record results of any prototyping of the search strings

c) for automated searches, identify resources to be used (specifying the digital libraries and search engines)

d) for manual searches, identify the journals and conferences to be searched

e) specify the time period to be covered by the review and any reasons for your choice

f) identify any ancillary search procedures, for example, asking leading researchers or research groups, or accessing their web sites; or checking reference lists of primary studies

g) specify how the search process is to be evaluated (for example, against a known subset of papers; or against the results from a previous systematic review)

4. Primary Study Selection Process

a) identify the *inclusion* criteria for primary studies

b) identify the *exclusion* criteria

c) define how selection will be undertaken (roles of reviewers)

d) define how agreement among reviewers will be evaluated

e) define how any differences between reviewers will be resolved

5. Study Quality Assessment Process

a) specify the quality checklists to be used

b) specify how the checklist will be evaluated (if a new checklist has been developed)

c) define how agreement among data extractors will be evaluated

d) define how any differences between data extractors will be resolved

e) identify the procedures to use for applying the checklists, such as details inclusion/exclusion, partitioning the primary studies during aggregation or meta-analysis, and explaining the results of primary studies

6. Data Extraction Process

a) design data extraction form (and check via a dry run)

b) specify the strategy for extracting and recording the data (for example, paper form, on-line. Form or database)

c) identify how the data extraction process is to be undertaken and validated, particularly any data that require numerical calculations, or are subjective

7. Data Synthesis Process

a) specify the form of analysis/synthesis to be used (for example, narrative, tabulation, meta-analysis)

b) discuss how the synthesis will be validated

8. Study Limitations

a) assess the threats to validity (construct, internal, external), particularly constraints on the search process and deviations from standard practice

b) specify residual validity issues including potential conflicts of interest that are inherent in the context of the study, rather than arising from the plan

9. Reporting

a) identify target audience, relationship to other studies, planned publications, authors of the publications

b} agree in advance who will be included in the list of authors and whose assistance will reported in the acknowledgements section.

10. Schedule

Provide time estimates for all of the major steps.

FIGURE 22.5: Template for a systematic review protocol

Points to remember

- Protocols will change throughout the conduct of a study.

- The team leader should take responsibility for keeping the protocol up to date and the team notified of all changes.

22.5   The search process

Planning the search process begins by defining a search strategy. After deciding the basic scope of the search strategy, you will need to determine the specific sources that will be searched and the search strings that will be used for automated searches and the sources that will be searched manually.

The final element of the search process is to integrate the set of candidate primary study references, remove duplicate copies of the same paper found in different sources, and store the references in the agreed storage tool (which can be a reference manager system, a spreadsheet or a database).

22.5.1   The search strategy

The factors that influence your search strategy are shown in Figure 22.6. There are three main decisions:

1. To decide whether completeness is critical or not.

2. To decide how to validate your search process.

3. To decide on an appropriate mix of search methods.

22.5.1.1   Is completeness critical?

The first issue to be decided is whether completeness is critical or not. In the case of a systematic review comparing SE technologies, completeness is a critical issue. In the case of a mapping study looking at the high level research trends in a broad topic area, completeness might be less critical; however, having an unbiased search strategy remains crucial. Nonetheless, there are occasions where even a mapping study may have a requirement for completeness. In particular, the more detailed the topic area, the more likely it is that completeness will be important. If you are in doubt consider your research questions. Can they be answered adequately if some relevant papers are not found by your search process?

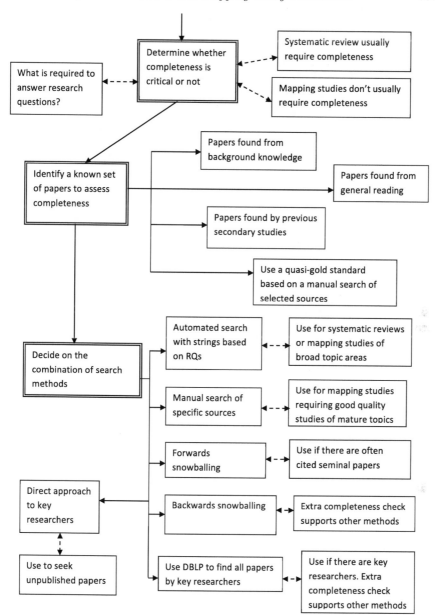

FIGURE 22.6: How to devise a search strategy.

22.5.1.2 Validating the search strategy

The next issue involves how you intend to refine and validate your search process. In the context of a systematic review or mapping study, validating the search process means quantifying, in some sense, the level of completeness achieved. The best way of doing this is to compare the primary studies identified by your search process with a known set of studies. To obtain a known set of studies several processes are possible:

1. If you have done some preliminary reading, you should be able to identify a set of papers that ought to be included in your review.

2. If you (or one of your research team) is an expert in the topic area, you (or your expert colleague) should make a list of the papers that are already known to be relevant.

3. If there are other related literature reviews (systematic or not) in the same topic area, review the papers which were included by them, and identify those papers that should also be included in your review.

4. For mapping studies, if none of these options is possible or the number of known papers is insufficient, you will need to construct a quasi-gold standard as proposed by Zhang et al. (2011). A quasi-gold standard is explained in Chapter 5 and involves conducting a systematic manual search of several defined sources including important journals and specialist conferences to identify a set of primary studies that is treated as a set of known papers. Deciding how many known papers is sufficient is clearly a subjective assessment, and depends upon the expected number of primary studies (which is obviously unknown at the start of a review, but may be clarified as you try out some of your search procedures). We suggest that 10 papers or fewer are insufficient for an assessment of completeness for a mapping study, whereas 30 papers would be enough (since completeness is judged by the percentage of known papers found).

Option: If you have a large set of known papers, select half the papers at random to use for constructing and refining automated search strings, and put aside the other half of the papers to help measure completeness.

In the case of mapping studies you should set an acceptable level of completeness which should be larger than 80%.

In the case of a systematic review there are likely to be fewer available papers and you may only have one or two known papers, so a numerical measure of completeness may be inappropriate. In this case, your search process needs to be as stringent as possible (that is, covering all options, automated and manual), and completeness may only be assessed against all the elements of your process rather than a numerical figure attached to the outcome of the process.

If the search process does not reach the required level of completeness, you need to specify a contingency plan in the protocol. For example:

1. Adding other search methods such as backwards snowballing.

2. Refining your search strings until the required completeness level is obtained.

22.5.1.3   Deciding which search methods to use

Finally you need to specify the detailed processes you will adopt. Methods include:

- Automated searches of digital libraries using search strings derived from the research questions. An automated search is usually required if you are doing a systematic review (and require completeness) or performing a mapping study of a broad topic area.

- Manual search of a restricted set of sources (that is, specific journals and conference proceedings). A manual search process aimed at specific journals is likely to find good quality research papers on mature topics. A manual search of specialist conferences is usually needed as an auxiliary method for reviews of new Software Engineering topics.

- Backwards and forwards snowballing, that is, searches based on extracting information from reference lists (backwards snowballing) and citation information (forwards snowballing). Backwards snowballing is based on searching the citations in each candidate paper to look for additional candidate studies. It is an ancillary method which is mainly used to support string-based automated searches. Forwards snowballing is based on finding all the papers that have cited a specific paper and searching that list for candidate primary studies. It is particularly useful if there are one or two seminal papers that first introduced the topic and are therefore cited by most subsequent papers. Both types of snowballing are supported by general indexing systems such as Scopus and Web of Science.

- Direct approach to active researchers or searching DBLP Computer Science Bibliography[1] for papers published by a specific author. A direct approach to active researchers is usually an ancillary method and is used to find out whether there are any related studies that have not yet been published. If there are specific authors who are known to contribute to the topic area, you can use the DBLP database to list all papers by those authors. This can be used as a completeness check.

A good search strategy will use a combination of these methods, although in most cases, one option is selected as the main search method, and then supported by other method(s).

When checking the references found by primary studies (that is, doing backwards snowballing), there are two main approaches:

1. If you have relatively few primary studies (for example, < 10), you may decide to use a manual approach. Two researchers should read the "Introduction","Related Work" and "Discussion" sections of each paper

---

[1]http://dblp.uni-trier.de/db/

and identify candidate studies. The candidate studies are the union of the set of studies found for each paper by each researcher.

2. Another approach is to use a general indexing system such as Scopus. Find each selected research paper in turn and extract all the references for that paper. The set of candidate studies is the union of the references extracted from each research paper.

The manual approach results in fewer candidate studies, since some screening of references takes place when the papers are read. However, both approaches can have some difficulties identifying duplicate reports because authors do not report their references in the same format and some authors make mistakes in their citations (for example, putting in the wrong date or leaving out "The" or "A" in the title).

Points to remember

- Systematic reviews usually require completeness. Mapping studies usually don't.

- Have a set of known studies to help assess completeness.

- Specify an appropriate completeness level.

- Have a contingency plan if your search does not reach the required completeness level.

- You will almost certainly need to do an automated search either using search strings or using citation analysis.

- You will usually need to consider ancillary search processes to achieve required completeness levels.

22.5.2 Automated searches

There are two main decisions:

- Decide on the sources that will be searched.

- Specify the search strings that will be used (unless the search is to be based on snowballing of some sort).

22.5.2.1 Sources to search for an automated search

Appropriate sources include publisher specific sources and general indexing systems. A mix of the two types of sources is best. In particular, the IEEE Digital library and the ACM digital library together cover important international journals such as IEEE Transactions on Software Engineering and most of the important computing-related conferences. These seem to be the best combination of publisher specific libraries. In addition, Springer publish a large number of conference proceedings and for new topics you may want to use SpringerLink as an additional source.

General indexing systems find many publisher specific sources (including ACM and IEEE papers) but may not index conference proceedings as quickly as ACM and IEEE. The Scopus, Web of Science and EI Compendex indexing systems are all possibilities and index papers published by Elsevier, Wiley and Springer which together with the IEEE publish most of the main internationally recognised software engineering journals that regularly publish empirical studies (which are of particular importance for systematic reviews). These include:

- *Empirical Software Engineering Journal* (Springer)

- *Journal of Systems and Software* (Elsevier)

- *Information and Software Technology* (Elsevier)

- *Software Quality Journal* (Springer).

- *Journal of Software Maintenance and Evolution: Research and Practice* (Wiley).

If completeness is critical, use several different indexing systems. The general indexing systems often provide mechanisms for extracting the references of papers they index and/or lists of papers that have cited a specific paper. These features are essential for efficient snowballing. It is possible to do backwards snowballing manually, although it is easier to extract references using facilities in an indexing system. It is not possible to perform citation analysis (forwards snowballing) without an automated system.

Standards in other domains emphasise the need to search for unpublished material such as Masters and PhD theses, technical reports, or industry "white papers". This is to ensure completeness of systematic reviews and minimise the possibility of publication bias (which occurs if negative results are less likely to be accepted by journals or conferences). It is unlikely to be necessary for mapping studies, but does need to be considered in the context of systematic reviews. Some researchers suggest using Google Scholar to search for unpublished searches but our experience of Google Scholar is that although it does identify unpublished material, it is often not possible to find a reliable source document that can be properly cited and guaranteed to remain publicly available. An alternative procedure is to approach key researchers directly and ask them if they have any relevant unpublished studies (including Masters or PhD theses or technical reports) that are publicly available.

22.5.2.2 Constructing search strings

PLEASE NOTE. *Previous versions of systematic review guidelines for software engineering researchers suggested using struc-*

tured questions to construct search strings. However, this approach has not proved to be very useful for software engineering reviews. Terminology in software engineering is neither well-defined nor stable, making it difficult to identify reliable keywords. Digital sources have limitations on the complexity of search strings and these are different for different libraries. Complex search strings are intended to identify a small number of highly relevant papers; however, in software engineering, they usually deliver large numbers of false positives.

Although in some rare cases a structured research question may help specify appropriate search strings, we recommend using fairly simple search strings based on the main topic of interest. Simple strings are more likely to work on a variety of different digital libraries without extensive refinement. To determine appropriate keywords:

- Review your research questions (RQs) and identify important concepts or terms used in the RQs.

- Review the terms used in the abstracts, keywords and title of your known set of papers. Match the frequently used terms to those found from your RQs.

- Try out your search strings on one of your selected digital indexing systems and identify the percentage of known papers you find. If the percentage of known papers found is low ($< 50\%$) (excluding, of course, any papers that could not have been found, such as papers not indexed by the specific digital indexing system or papers published in very recent conferences), review the papers that were not found. Refine your search strings by replacing existing keywords (for example, using more general terms) or adding new keywords (for example, adding qualifiers to make terms more specific).

If you are undertaking a systematic review aimed at aggregating comparative studies, you will need to specify some keywords to restrict your search to empirical studies, for example, "empirical" or "experiment".

Points to remember

- If automated searching is your main search strategy, you should search a variety of sources including IEEE, ACM and general indexing systems.

- General indexing systems have useful facilities for supporting the use of snowballing.

- Derive search strings from your research questions and terms used in known studies.

- Keep your search strings fairly simple.

- Try out your search strings on a general indexing system and refine them if they do not find the majority of known papers.

22.5.3   Selecting sources for a manual search

If you are using a manual search as the main strategy for a mapping study, there are several approaches you can take. If you are interested in high quality studies in a relatively mature topic (particularly if you are interested in empirical studies), the following sources are likely to be suitable for general software engineering topics:

- *IEEE Transactions on Software Engineering*

- *ACM Transactions on Software Engineering Methodology (TOSEM)*

- *Empirical Software Engineering Journal*

- *Journal of Systems and Software*

- *Information and Software Technology*

- *Proceedings of the International Conference on Software Engineering (ICSE)*

- *Empirical Software Engineering and Metrics Conference (ESEM).*

For a new topic area, you will need to review specialist conference and workshop proceedings. Whatever the circumstances, you should check the sources that published your known papers.

Points to remember

- Look for sources that are particularly likely to publish papers on your topic of interest.

- New topics are most likely to be reported in specialist workshops and conferences.

- If a topic is new, terminology may not be well-defined, complicating automated searches.

- Use your known papers to help with identifying possible sources.

22.5.4   Problems with the search process

A major search process problem is searching for topics that are unlikely to be the main research topic of research papers. For example, if you are interested in the use of some specific automated tools in a particular topic area, there may be many papers that report the use of the tools to support their validation or evaluation activities but do not mention the name of the tool in the title, abstract or keywords. Alternatively, if you are interested in the use of specific research practices, there will be difficulties because not only are specific experimental methods seldom identified in the title, abstract and keywords of primary studies, but it is also the case that software engineers are extremely poor at correctly specifying the empirical methods they used.

Searching for detailed aspects of a research process or a topic requires searches of the full research papers, which is not supported by indexing services nor by all the publishers' digital libraries. In such cases, you will probably need to do a relatively broad search and prepare to manage a large number of candidate primary studies including many false positives. In some cases, you might be able base your set of candidate papers on a randomly selected subset of the papers within the broad topic area.

Two other problems with the search process are finding too many (for example, many thousands) or too few candidate primary studies (for example, one or two).

You may find yourself with a very large number of primary studies if you are doing a mapping study of a topic with a very broad scope. In this case, you need to consider revising any automated search strings. However, before changing any search strings, you may need to reconsider your research questions. Are the research questions too broad in scope? Are any research questions unnecessary for your main research goals?

If you are doing a more focused systematic search you may find yourself with very few studies. In this case you have several options:

- Check whether your search parameters are too stringent. It might be possible that broadening the search would find additional relevant studies.

- If you have a well-designed search process, for example, your initial set of known papers was small and all those papers were found by the search, it may be that there is insufficient research for a systematic review. This can be recorded as an outcome of your systematic review, and you should, perhaps, plan to undertake your own primary study.

Points to remember

- In some cases, the information needed to answer your research question(s) won't be found in the title, abstract or keywords. This raises problems both for the search process and the selection process.

- If you have too many candidate papers about a topic likely to be mentioned in the title or abstract, are your research questions too broad?

- If you have too few candidate papers about a topic likely to be mentioned in the title or abstract, are your research questions too narrow, or are more primary studies needed?

22.6 Primary study selection process

Study selection is a multi-stage screening process by which irrelevant papers are removed from the set of candidate primary study papers. The selection process needs to be documented in the review protocol.

22.6.1 A team-based selection process

For a team-based systematic review or mapping study, the process is shown in Figure 22.7. At least two researchers should assess each candidate paper. The team leader is responsible for assigning researchers to individual papers and for collating the result of their evaluation.

Stage 1 selection is usually based on title and abstract. However, if the number of papers found by the search process is very large (for example, > 500), it may be appropriate to base a preliminary screening on title alone. Any paper that is considered irrelevant by all the researchers who assess it, based on the inclusion and exclusion criteria that can be evaluated from the title alone, is removed from the set of candidate papers.

The main Stage 1 selection activity is based on assessing the title, abstract and keywords of the remaining candidate papers. Again the process is to remove any paper that is considered irrelevant by all researchers that assess it, based on the inclusion and exclusion criteria that can be evaluated from the title, abstract and keywords alone.

At this point the team leader should review the agreement among the researchers and calculate an appropriate agreement statistic such as Cohen's kappa (Cohen 1960) or Krippendorff's alpha (Krippendorff 1978). If the agreement is poor, it is possible that some members of the research team are unclear about the interpretation of the research questions or the inclusion and exclusion criteria. The research team should meet to discuss possible reasons for poor agreement (which in our experience of software engineering papers, is sometimes due to poor quality abstracts).

After Stage 1 selection, the Stage 2 selection activity is based on the full text of the paper and all the defined inclusion and exclusion criteria. The goal of this screening activity is to positively include relevant papers as well as to exclude irrelevant papers.

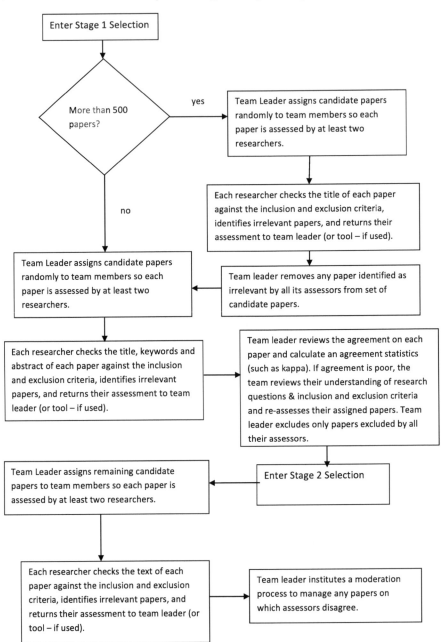

FIGURE 22.7: The team-based primary study selection process.

For a mapping study, it may be appropriate to apply the detailed inclusion/exclusion process only to papers where researchers disagreed about their relevance during Stage 1. That is, if all researchers agreed that a paper was relevant during Stage 1, it is not necessary to check the paper again against all the inclusion and exclusion criteria. However, for a systematic review it is usually better to apply the inclusion and exclusion criteria explicitly to each paper that passes the initial screening.

After Stage 2 screening, there may be some disagreements among researchers about the inclusion of specific papers. At this point, it is important to record the agreement among researchers in terms of an agreement statistic. If agreement is very poor, the team leader may need to initiate a procedure to investigate whether there is a systematic problem with the selection process. For example:

- If all the researchers show poor agreement, the team leader could call a team meeting to discuss the inclusion and exclusion criteria, in case there are previously overlooked ambiguities or other problems with the criteria.

- If the problem appears to involve a particular researcher, the team leader might initiate some additional training before asking the researcher to reassess their inclusion/exclusion decisions.

After identifying all disagreements, the team leader needs to institute a moderation process to gain an agreement on the relevance of each disputed paper. The moderation process can involve:

- Discussion among the researchers who assessed the paper.

- Assessment of the paper by another researcher.

- A trial data extraction to confirm whether or not the required data can be obtained from the paper.

It is sometimes necessary for the team leader to make a final decision but it is preferable for the researchers to come to a mutually agreed decision.

Points to remember

- The main study selection process usually involves two or three stages.

- If you have a very large number of papers, base initial inclusion/exclusion assessments on title alone, then assess the retained papers based on the keywords and abstract of the remaining papers, and finally assess the retained papers based on their contents.

- With a relatively small number of papers, base the initial inclusion/exclusion assessment on the title, abstract and keywords and then assess the retained papers based on their contents.

- Ancillary searches usually require a separate selection process.

22.6.2 Selection processes for lone researchers

The team-based process cannot be followed by a lone researcher or a PhD student. If you are a lone researcher, you should adopt a test-retest approach whereby you assess the papers once and then, at a later time, assess them again (preferably not in exactly the same order). A substantial disagreement should prompt you to review your research questions and inclusion and exclusion criteria.

If you are a PhD student, you can involve your supervisors by asking them to assess a random selection of papers. This also gives your supervisors an opportunity to provide you with feedback. Again a substantial disagreement would be an indication that you (or your supervisor) misunderstand some aspects of your research question(s) or inclusion and exclusion criteria. If you and your supervisors plan to publish the results of the systematic review or mapping study, it is appropriate for the supervisors to act as members of the research team to ensure the selection process is of an appropriately high quality.

Points to remember

- Lone researchers should use test-retest to validate their inclusion/exclusion decisions.

- PhD students should ask their supervisors to apply inclusion/exclusion criteria to a proportion of the candidate papers and assess the level of agreement.

22.6.3 Selection process problems

If you are doing a broad mapping study rather than a focussed systematic review you may find you have a very large number of candidate primary studies (that is, many thousands of articles rather than a few hundreds). If, in addition, you are a lone researcher or the leader of a small review team, you may find this number of studies impossible to screen in the time available for the review.

Assuming that the search process has been performed correctly, you have two main options for the preliminary stage in the process:

- Recruiting more members to your review team, but bear in mind any new recruits will require time to get up to speed on the planned review procedures.

- Using a text mining tool to identify the set of papers that are most likely to be relevant to your research questions and excluding papers with a low probability of relevance.

If you are at the end of your selection process and still have a very large number of primary studies (for example, many hundreds of papers) that are now *confirmed* as being relevant to your research questions, you may anticipate a potential problem with managing the primary study analysis and synthesis process. In this, case options include:

- Recruiting more review team members, but bear in mind that, the later in the process that you decide to recruit more team members, the more difficult it will be for them to get up to speed on the planned review procedures.

- Revising your research questions, which is possible if your research questions are answered by different subsets of the primary study. You may be able to use a text mining tool to look for primary study clusters.

- Basing selection on a random sample of primary studies, using stratified sampling if the primary studies cluster by research question, domain or topic. Several systematic studies of research methods in software engineering have been based on a sampling strategy (for example, Glass et al. (2004) and Zelkowitz & Wallace (1998))

Points to remember

- Managing an extremely large number of candidate primary studies is difficult and time consuming.

- Consider contingencies for managing large numbers of primary studies during the planning process.

22.6.4 Papers versus studies

An important issue for a systematic review is the relationship between papers and individual studies. Software engineering papers often exhibit overlaps:

- There may be several different papers reporting the same study. This can occur if there is a conference version of the paper followed by an extended journal version of the paper.

- It is possible that the results of a large study may be published in a series of different papers.

- It is also possible that a single paper may report the results of several independent studies.

It is important that a systematic review does not double-count study results, particularly if some form of statistical meta-analysis is to be performed.

Thus, after the completion of the primary study selection, a research team must review papers that have similar titles and authors and assess the relationship between the papers and individual studies. We advise you to keep a record of papers related to a specific study (for completeness and auditability), but ensure your results are reported against the individual study.

Furthermore, all papers should be scanned for the possibility of multiple studies. This is by no means a straightforward procedure since authors may regard some studies as independent that you consider to be a single study. Issues that occur are:

- Researchers may report both a pilot experiment and a main experiment. In some cases, it may be appropriate to ignore the pilot experiment. This is likely to be appropriate if the authors report many changes to the research methods as a result of the pilot experiment, or the pilot experiment was based on a very small sample. In other cases, it may be appropriate to treat the pilot study and main study as one study with two blocks.

- Researchers may report several case studies, and it is unclear whether their design is a multi-case case study or several independent case studies. If the case studies have the same research questions and used the same methodology then our advice is to treat the study as one multi-case case study. If the case study methodologies are very different, for example, ethnography in one case and semi-structured interview in another, we would suggest treating the paper as reporting two independent case studies.

Often there is no obvious "right" answer to the number of independent studies. You need to report your basic approach (for example, two researchers will discuss each case) and the decisions you make in each individual case.

In most cases, this is less of a problem for mapping studies, since they are usually able to work at the paper level irrespective of the relationship between papers and individual studies. However, it is often useful to identify duplicate reports of the same piece of research, particularly if the aim of the mapping study is to identify whether sufficient research is available for a detailed systematic review. Furthermore, if a mapping study aims to assess the quality of research, you may need to consider quality at the study level rather than the paper level.

Points to remember

- The relationship between studies or pieces of research and published paper is many-to-many.

- For systematic reviews, it is important not to over-count (or under-count) the number of independent studies.

22.6.5 The interaction between the search and selection processes

Although the search and selection activities are different, the process of undertaking those activities is often entwined. If you intend to use backwards snowballing, you cannot do such a search until you have selected a set of primary studies from the set of candidate primary studies you found during your main search process. A similar issue arises if you want to write to authors who have published a number of primary studies to seek other as yet unpublished results.

Thus, after performing your main selection process be it automated or manual or a mixture, you need to suspend the search process and enter the selection process to screen the current set of candidate papers. After the screening process is complete, you will need to reactivate the search process in order to check the references of the current set of primary studies or identify the most frequently cited authors.

Clearly if you find more candidate primary studies, the selection process needs to be re-activated.

Point to remember

Ancillary searches often depend on having an existing list of candidate studies, which means they cannot be started until an initial round of the search and selection process has been completed

---

22.7 Validating the search and selection process

The team leader should assess the validity of the search process against the criteria specified in the protocol. This information should be reported in the methods section of the final report.

The team leader should expect to:

- Justify the comprehensiveness of the search process given the type of review, that is, whether it is a quantitative or qualitative systematic review or a mapping study.

- Report the agreement achieved during the Stage 2 selection process, prior to any moderation process (see Figure 22.7).

- Confirm that all papers that were known before the start of the selection process were found by the search process and selected during the selection process.

- *Optional.* Consider validating the search process using textual analysis tools (see Chapter 13). Such tools can be used to check the frequency of the use of main keywords to investigate whether included papers that and excluded papers that differed with respect to usage of keywords and if they did, whether any papers might have been misidentified. Tools can also be used to review the extent to which included and excluded papers cross-reference one another. Again, such an analysis can be used to identify any papers that might have been misclassified. Any papers that may have been misclassified can be reviewed again and their classification revised if necessary. This approach is particularly useful for single researchers.

- If any known papers were kept separate for validation purposes, the team leader must report the coverage of these papers. For a systematic review, coverage should be 100% if the study involves a comparison of two technologies.

- If any previous systematic reviews or mapping studies were kept separate for validation purposes, the team leader must identify the primary studies reported by the previous reviews that should have been found by the current review. The team leader should report the number of such primary studies that were missed by the current review.

Of course, if the final two validation exercises discover missing papers they must be added to the set of primary studies.

Points to remember

- You will need to justify your overall search and selection process.

- You will need to provide evidence that your search process was effective.

- You should report the values of agreement statistics to confirm selection process was effective.

22.8   Quality assessment

The main decisions that need to be made during quality assessment are:

1. Deciding whether or not a quality assessment is necessary.

2. Deciding appropriate quality assessment criteria.

3. Deciding how the quality assessment will be used to support the goals of the review.

4. Deciding how the quality assessment will be managed.

The results of these decisions should be documented in the protocol. Each of these issues is discussed below.

22.8.1   Is quality assessment necessary?

For any systematic review, quality assessment should be considered mandatory. It is important to ensure that the results of any aggregation are based on *best* available evidence. This means either simply excluding poor quality studies from the aggregation, or investigating the impact of excluding such studies.

For mapping studies, quality assessment is not required, unless one of the aims of the mapping study is to assess the quality of existing studies. This can happen in the case of tertiary studies investigating the methodology used in systematic reviews.

22.8.2   Quality assessment criteria

There are two aspects to quality assessment:

- Assessing the quality of individual primary studies.

- Assessing the overall strength of evidence of the review findings.

These are discussed in the following sections.

22.8.2.1   Primary study quality

Quality assessment of primary studies is usually done by means of a quality instrument comprising a number of questions related to the goals, design, conduct and results of each study. The questions are referred to as *quality criteria*. The quality instrument is often referred to as a quality checklist. A checklist for a particular study type is usually made up of questions related to:

- The goals, research questions, hypotheses and outcome measures.

- The study design and the extent to which it is appropriate to the study type.

- Study data collection and analysis and the extent to which they are appropriate given the study design.

- Study findings, the strength of evidence supporting those findings, the extent to which the findings answer the research questions, and their value to researchers and practitioners.

Quality assessment criteria usually depend on the *type* of primary study being evaluated since factors that determine a good example of one type of study may be irrelevant for a different type of study. For example, factors that identify a good quality experiment such as random allocation to treatment, and sufficient experimental units to achieve a reasonably high power are different from the factors that determine a good case study such as an appropriate choice of *case*, and consideration of alternative explanations for the case study findings.

Quality assessment criteria also depend on the *subject type*. For studies that compare human-intensive methods or techniques, the subjects will be human beings and checklists can be adapted from the many recommendations available in the medical and healthcare domain. We advise you to look at some of the published checklists, choose the one(s) most appropriate for your systematic review, and adapt it (if necessary) to your own study. For example, checklists for randomized controlled trials (which are field experiments), qualitative studies, and systematic reviews can be found at the Critical Appraisal Skills Programme (CASP) website[2] or the SURE Critical Appraisal Checklists[3]. A version of the randomised trials checklist suitable for software engineering experiments is shown in Figure 22.9. In addition, Runeson et al. (2012) provide checklists specifically designed to help researchers undertaking and reading software engineering case studies.

In contrast, for studies that compare or evaluate algorithms or tools which are *technology-intensive* studies, specialised checklists will need to be constructed, see for example Figure 22.8 which is adapted from a checklist developed by Kitchenham, Burn & Li (2009), and includes suggestions for scoring each question.

For non-comparative or qualitative systematic reviews, or if a large variety of study types are found, a more general quality assessment instrument may be appropriate. The quality criteria proposed by Dybå & Dingsøyr (2008*a*) based on the CASP checklist for qualitative studies has been used in several software engineering systematic reviews. However, if you have a large number of different study types but only one quality checklist, you need to consider the study type as well as answers to the checklist questions when using the quality assessment (see Section 22.8.2.2 below).

22.8.2.2 Strength of evidence supporting review findings

Dybå & Dingsøyr (2008*b*) recommend using the GRADE approach (Guyatt, Oxman, Vist, Kunz, Falck-Ytter, Alonso-Coello & Schünemann

---

[2]http://www.casp-uk.net/find-appraise-act/appraising-the-evidence/
[3]http://www.cardiff.ac.uk/insrv/libraries/sure/checklists.html

Question No	Question	Scoring
1	Are the goals of the experiment clear	No=0, Partly=0.5, Yes=1
2	Were the research questions and hypotheses defined?	No=0, Partly=0.5, Yes=1
3	Was there any replication, for example, multiple test objects, multiple test sets?	Yes=1, No=0 – If No this is not an experiment and should be considered a case study, feasibility study or example.
4	Are the study measures valid?	None=0 / Some (0.33) / Most (0.75) / All (1)
5	If test cases were required by the Test Treatment, how were the test cases generated?	Not applicable (reduce number of questions by 1) By the experimenters (Yes=0) By an independent third party (Yes=0.5) Automatically (Yes=0.75) By industry practitioners when the test object was created (Yes=1)
6	How were Test Objects generated?	Small programs (Yes=0) Derived from industrial programs but simplified (Yes=0.5) Real industrial programs but small. (Yes=0.75) Real industry programs of various sizes including large programs (Yes=1)
7	How were the faults/modifications found?	Not applicable (reduce number of questions by 1 and go to question 8) Naturally occurring Yes=1, go to question 8 If No go to questions 7a
7a	For seeded faults/modifications, how were the faults identified?	Faults introduced by the experimenters (Yes=0), Independent third party (Yes=0.25) Generated automatically (Yes=0.5)
7b	For seeded faults/modifications, were the type and number of faults/modifications introduced justified?	Type & Number: Yes (0.5) Type or Number (Yes=0.25) No=0
8	Did the statistical analysis match the study design?	No=(0), somewhat (0.33) Mostly (0.66), Completely (1)
9	Was any sensitivity analysis done to assess whether results were due to a specific test object or a specific type of fault/modification?	Yes=1 / Somewhat=0.5 / No=0
10	Were limitations of the study reported either during the explanation of the study design or during the discussion of the study results?	No=0, Somewhat=0.5, Extensively=1
11	Were the findings clearly reported?	No=0, Partly=0.5, Fully-1
12	Are the findings of value to industry or researchers?	No=0, Somewhat=0.5, Extensively=1

FIGURE 22.8: Quality criteria for studies of automated testing methods.

2008) to assess the strength of the evidence for recommendations. The GRADE approach is mainly used to assess the strength of *recommendations* when a decision has to be made concerning the *adoption* of a recommendation in a particular situation. However, it can also be considered for assessing the strength of evidence associated with individual findings.

GRADE defines strength of evidence in terms of the confidence we have that further research will or will not change the estimate of effect size:

1. High confidence means that further research is unlikely to change the estimate.

2. Moderate confidence means further research may change the estimate.

3. Low confidence means further research is likely to change the estimate.

Broad issue	Detailed Issue	Factors to consider
Are the results of the experiment valid? Initial screening questions	Did the experiment address a clearly focussed issue?	Is the population study defined? Are the methods (including the control) well defined? Are the outcomes appropriate?
	Was the assignment of participants to methods randomised?	How was assignment carried out? Was the allocation concealed from researchers and subjects?
	Were all participants who took part in the study accounted for at its conclusion?	Were participants analysed in the groups to which they were assigned? How were partial results from dropouts handled? Were dropout rates related to the method?
Validity – detailed questions	Were any of the experimenters "blind to treatment"?	Did the researchers running the experiment know who was in which treatment group? Were the outputs independent of the method and if so were markers/evaluators blind to the method used by each participant?
	Were the groups similar at the start of the experiment?	Have other factors been considered such as SE experience, knowledge of the different methods?
	Aside from the method, were groups treated equally?	Were both groups given appropriate training? Were the trainers equivalently skilled in the method(s) they taught? Did the trainers have a vested interest in the success of one of the methods?
What are the results?	How large was the treatment effect?	What outcomes were measured? Is the primary outcome clearly specified? What effect sizes were found for each outcome?
	How precise was the estimate of the treatment effect?	What are the confidence limits?
Will the results help locally?	Can the results be applied in your context?	Are the participants similar to the intended population, for example, being practitioners rather than students?
	Were all industrially important outcomes considered?	Is there other information you would like to have seen? If not does this affect the value of the experiment?
	Are the benefits worth the risks and costs?	Will likely cost savings outweigh adoption costs?

FIGURE 22.9: Quality criteria for randomised experiments.

4. Very Low confidence means the estimate is very uncertain.

GRADE also considers factors that may decrease or increase confidence in the strength of evidence. Factors that decrease the strength of evidence relate to poor methodological quality, inconsistent findings, sparse data, or reporting bias found in individual studies. Factors that increase the strength of evidence relate to very large effect sizes, only having confounders that would decrease the effect size, or evidence of a "dose response gradient" (which means that more of the treatment results in a better outcome).

Formulating GRADE evidence in terms of effect sizes and study bias indicates that the method is intended to apply primarily to randomised controlled trials (which are controlled field experiments) or systematic reviews of such studies. However, such studies are extremely rare in software engineering. We must often make do with much weaker forms of study and more varied types of empirical study

To apply the GRADE concept to findings from software engineering systematic reviews, specific findings from the review will need to be discussed in the light of the methodological quality of the related primary studies and their study type(s). For example, if a specific finding is supported only by feasibility studies, even if they are high quality feasibility studies, the evidence for that finding must be considered to be very weak.

Study types that provide very weak evidence are:

- Feasibility studies, including small experiments (that is, experiments with very few subjects), and small-scale examples.

- Lessons learned studies.

- Before-after within-subject quasi-experiments which are the weakest form of quasi-experiment.

Types of study that provide slightly stronger evidence (although still relatively weak) include:

- Post-hoc re-working of large-scale examples (often mistakenly called *case studies*).

- Post-hoc analyses of industry datasets (for example, correlation and regression studies).

Types of study that provide moderately strong evidence include:

- Laboratory-based experiments and quasi-experiments. Good quality studies of these types are likely to give a reliable indication of whether or not an effect size is significant and the *direction* of the effect size. However, they are likely to give biased estimates of the *magnitude* of the effect size (probably overestimates) because the studies did not take place in an industrial context where other factors such as time scale pressure, team dynamics, task complexity, task dependencies, and personal motivation influence outcomes. Also, if the studies involve students rather than professionals or the software engineering task is particularly straightforward, confidence in the evidence should be downgraded.

- Industry case studies, preferably using multiple cases, which can provide reasonably reliable evidence.

The most trustworthy form of evidence in software engineering comes from industry-based field studies including:

- Randomised field experiments. Such designs represent the most reliable form of empirical study but can seldom be performed in software engineering contexts.

- Field-based quasi-experiments based on cross-over designs, interrupted time-series, regression discontinuity, or differences-in-differences designs (Shadish et al. 2002) which can provide highly reliable evidence.

22.8.3   Using quality assessment results

There is little point in collecting data about primary study quality if you have no plan as to how such data will be used. There are several possibilities:

- Specific quality criteria may be used as part of the inclusion criteria to screen out low quality studies.

- The quality score (that is, the sum of numerical values assigned to the answer of each quality question) for each primary study may be used to identify poor quality studies. Then, the impact of the poor quality studies on the results of the review findings can be assessed to investigate whether poor quality studies are causing the results to be biased.

- The quality data may be assessed to see if there are systematic problems with primary study quality, for example it may be problematic if most or all of the studies use student participants.

- Specific quality criteria may be used as moderating factors (that is, factors that might explain the differences among study results) in meta-analysis, for example whether or not the empirical study was an experiment (formal or quasi) or a less rigorous form of study type.

Finally, as mentioned above, the quality score of studies can be used as part of an assessment of the strength of evidence supporting individual findings.

22.8.4   Managing the quality assessment process

Many quality criteria require subjective assessment, for example, any criterion that asks whether something was *appropriate*. To reduce the problem of bias associated with subjective assessments, it is customary for at least two researchers to assess the quality criteria of each paper and for disagreements to be moderated.

The general process used for managing quality assessment in a team-based systematic review is shown in Figure 22.10. The specific process you decide to adopt should be documented in the systematic review protocol.

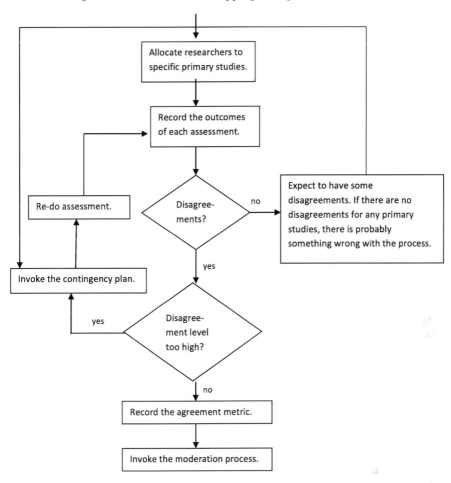

FIGURE 22.10: Process for managing team-based quality assessment.

22.8.4.1 A team-based quality assessment process

For team-based systematic reviews, the team leader should assign at least two researchers to assess the quality (and study type, if necessary) of each primary study. Each researcher should complete the quality evaluation form independently, then the results of the evaluations for a specific primary study should be compared and disagreements recorded (this may be done by the team leader, a systematic review management tool, or the two assigned researchers working together).

If there are disagreements, some form of moderation must take place as defined in the systematic review protocol. Options include:

- Assigning a third person to assess the quality of the primary study and discuss the assessments with the two original researchers.

- Asking the researchers assigned to the primary study to work together to arrive at an agreed assessment.

- If more than two assessments are available, aggregating the assessment into a combined score (for example, by taking the mean).

The team leader should expect to report initial agreement rates using an appropriate agreement measure. The agreement between independent assessors is used to assess the validity the evaluation process. If the agreement for individual primary studies is very low, the quality criteria may not be well understood, so there should be a contingency plan ready. Options for the contingency plan should include:

- Calling a team meeting to discuss the quality criteria and the reasons for disagreements.

- Additional training for specific members of the team.

The team leader is responsible for deciding which option should be chosen given the specific circumstances. It should also be noted that unusually high agreement can also be a sign of misunderstandings among team members. The team leader should be prepared to invoke the contingency plan if this condition arises.

The actual quality evaluation process may take place as part of the general data collection process or may precede the data collection phase. If some of the quality criteria are being used as inclusion/exclusion criteria, it is better to complete quality evaluation before beginning data extraction.

22.8.4.2 Quality assessment for lone researchers

If you are a lone researcher or a PhD student, the main problem you will have is validating your quality assessment.

If you are a PhD student, options include:

- Requesting your supervisors to assess a random selection of primary studies and comparing the results.

- Re-assessing of a random selection (or all) of the primary studies after a suitable elapsed time and calculating the test-retest agreement.

In both cases, you should specify in the protocol what constitutes a dangerous level of disagreement, and have ready a contingency plan to deal with this possibility. Any disagreements identified during this validation exercise need to be resolved. This should usually be done by discussing each case with a supervisor.

If you are a lone researcher you must also specify how to validate your quality assessment process. This will usually require a test-retest assessment based on all the primary studies and a justification of the process for reaching an agreed evaluation for any disagreements (which might be taking the mean score) or recording a justification for each revised score.

Points to remember

- The subjective nature of many quality criteria makes quality assessment far from simple.

- You should be clear about how the quality assessment will be used.

- If your primary studies include many different study types and you use a general-purpose quality checklist, keep a record of the study type as well. In this case you should use the quality assessment information and study type to assess the reliability of individual findings.

- You should report agreement statistics to indicate the reliability of the quality assessment process.

22.9 Data extraction

Data extraction and data synthesis are phases where the differences between quantitative systematic reviews, qualitative systematic reviews and mapping studies are most significant. You need to consider the type of review you are undertaking both when planning your data extraction process and when conducting data extraction.

22.9.1 Data extraction for quantitative systematic reviews

The data extraction process is most well-defined for quantitative systematic reviews. You should be in a position to define in advance the data you intend to extract from each paper in order to answer your research question(s). However, this presupposes that you know enough about the topic area and the available literature to determine whether a meta-analysis is feasible and if so what effect sizes are most appropriate. If this is not the case, you will need to defer formalising the data extraction and analysis processes until you have selected the primary studies and identified the statistical designs used in the studies and the nature of the outcome metrics they report.

22.9.1.1 Data extraction planning for quantitative systematic reviews

Once you have adequate knowledge of the primary studies, the decisions you need to make to identify the data you need to collect are shown in Figure 22.11.

You will need to decide whether you intend to undertake a formal meta-analysis or a more qualitative-style of analysis. Even with a quantitative systematic review, you will not be able to do a formal meta-analysis:

- If the primary studies use different treatment combinations, and there are insufficient studies that compare the same pair of treatments.

- If your outcome measures include many different incompatible metrics. For example, the quality of a program (or a maintenance change made to a program) might be evaluated using static complexity measures, number of residual errors, subjective quality assessments, or test coverage statistics. You may be able to identify whether differences are statistically significant which is all that is needed for vote-counting, but more detailed meta-analysis methods may not be possible.

Whether you are aiming for vote-counting or a full meta-analysis, data extraction will be based on:

- Basic information about the study, including the treatments being compared and the outcome metrics reported.

- The quantitative outcomes of the experiments being included in the review, as required for the type of meta-analysis being planned (see Chapter 11 and Table 22.1). This would include all the metrics needed to construct the specified effect size such as values of any test statistics, the probability level achieved by the test(s), sample sizes, mean values, standard deviations.

- Contextual information that can be used in any meta-analysis to investigate any heterogeneity among primary studies or support a detailed qualitative analysis (Chapter 11 and Table 22.2). Note, however, that some relevant contextual information may already have been specified in the quality criteria.

After defining the data to extract, you will need to construct a data collection form. This can be a paper form, a spreadsheet, or database form. The data collection form and any necessary associated documentation should define the data being extracted and provide clear guidelines for data extractors. To complete the planning process, the form should be trialled using some known primary studies.

All members of the review team who are expected to extract data should take part in the trial and report any problems with the data collection form to the team leader. Any problems with the form should be resolved prior to finalizing the protocol. A procedure both for checking individual extraction forms and for resolving any disagreements should be defined in the review protocol.

TABLE 22.1: Common Effect Sizes Used in Meta-Analysis

Type	Formula	Definition	Variants
Significance level	p-value	Probability obtained from a statistical test	
Point serial correlation for between groups design	$r = \dfrac{\sum \left(x_{ij} - \overline{x} \right)\left(y_{ij} - \overline{y} \right)}{\sqrt{\sum \left(x_{ij} - \overline{x} \right)^2 \sum \left(y_{ij} - \overline{y} \right)^2}}$	Pearson correlation where $x_{ij} = 0$ for group 1 and $x_{ij} = 1$ for group 2 and y_{ij} is the outcome value for observation j in group i.	R-squared for ANOVA designs. Standard Pearson correlation for regression or correlation studies.
Standardized mean difference for numerical outcome metrics	$d = \dfrac{(m_1 - m_2)}{stdev}$	Difference between the means of observations in each group divided by an appropriate standard deviation.	Cohen's g, Hedge's d, and Glass's Δ.
Odds ratio for counts and probability outcome metrics	$O = \dfrac{p_1(1 - p_2)}{p_2(1 - p_1)}$	Ratio of odds related to one group and odds related to a second group. Odds are the probability of an event divided by 1 minus the probability.	Log odds which is the logarithm of the Odds ratio.

TABLE 22.2:   Contextual Information Appropriate for Meta-Analysis

Type	Options	Value
Study type	Experiment, Case study, Quasi-experiment, Survey, Benchmarking, Data mining, Lessons learnt	Provides information about constraints on study rigour.
Participants	Students, Practitioners, Consultants, Academics	Indicates the population to which results apply.
Materials	Programs, Software specifications, Test cases	For benchmarking studies or testing studies, define type of systems to which results apply.
Settings	University course, Training course, Industry	Indicates realism of setting.
Task	Task time, Task complexity	Indicates realism of task.

22.9.1.2   Data extraction team process for quantitative systematic reviews

Once the data collection form has been designed and tested, the actual data extraction process should be fairly straightforward. For a team-based review, the team leader must assign two team members to each primary study and monitor the data extraction process (see Figure 22.9.1.1). Note that data extraction may be done at the same time as extracting quality data.

However, it is always possible that a primary study could be found that performed a novel analysis and presented its results in a manner that was not anticipated in the protocol. This should be reported to the team leader, who needs to halt further data extraction in case there are other examples of such analyses among the primary studies. Data extraction should only be restarted when it is clear how to deal with papers using the new type of analysis. This might involve amending the data collection form and/or providing additional training/guidelines for data extractors.

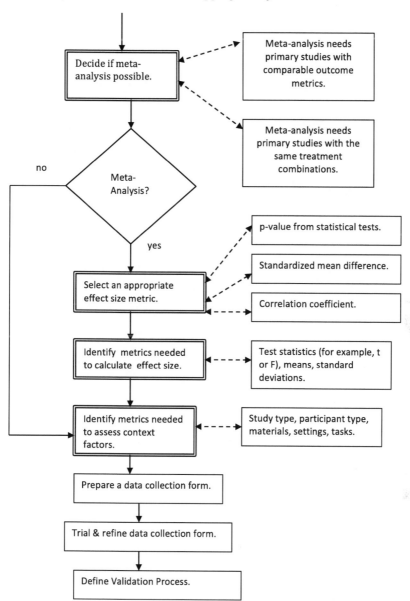

FIGURE 22.11: Initial planning decisions for quantitative systematic reviews.

22.9.1.3   Quantitative systematic reviews data extraction process for lone researchers

If you are a lone researcher, you should use a test-retest approach to validate that all the data was correctly extracted.

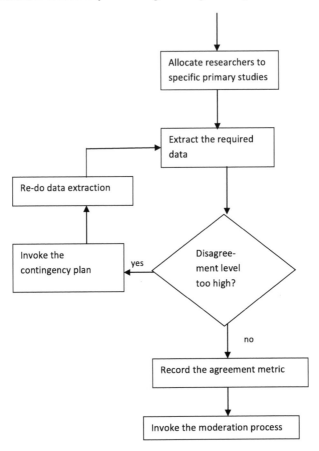

FIGURE 22.12: Quantitative systematic reviews data extraction process.

PhD students can either use test-retest, or ask their supervisors to act as independent extractors. Note, however, performing a full systematic review would be a major task for a PhD student and should be suitable for a journal publication. In such circumstances supervisors should be prepared to act as members of the review team and ensure all the extracted data is properly validated.

22.9.2   Data extraction for qualitative systematic reviews

Qualitative systematic reviews are usually based on data extracted from qualitative primary studies. Qualitative studies are primary studies that:

- used semi-structured or unstructured interviews, or

- were based on observations researchers made about software developers, software teams or managers and their work processes, or

- were based on subjective opinion surveys.

They are the most difficult type of review from the viewpoint of data extraction and data synthesis. This is because such reviews are looking for *textual* information provided by the study authors of issues such as risks, cost benefits, motivators, barriers to adoption, definitions of terminology and other themes or concepts related to the research topic. Problems arise because authors of primary studies may use different terms for the same concept or the same terms for different concepts. This means the textual information required from each primary study cannot usually be defined in advance. Furthermore, data extraction and data synthesis become inextricably linked as you attempt to identify and define core terms including all homonyms and synonyms.

22.9.2.1 Planning data extraction for qualitative systematic reviews

In most cases you should expect the data extraction process to centre on creating a database of evidence, similar in concept to a case study database (Yin 2014). You should specify, in the protocol, the tool that will be used to hold the data. The data itself would usually include textual information extracted from each primary study, a reference to the place in the primary study the text was found, and a comment indicating the relevance of the text (if you have several different research questions, you should identify which research question(s) it addresses.)

Some general points can be made:

- The more specific your research questions are, the easier data extraction and synthesis will be, since you will be able to specify relevant themes prior to starting data extraction. In contrast, if you have some general high level topic such as "Global Software Development" or "Cloud-Based software engineering" and intend looking for unspecified "themes" from the information reported by the primary study authors, your task will be much more difficult.

- The more familiar you are with the topic area, the more likely you are to be able to identify appropriate research questions and interesting themes in advance.

22.9.2.2 Data extraction process for qualitative systematic reviews

It is difficult to organise a team-based data extraction/synthesis process. Since any primary study could introduce a new homonym or synonym, none of the primary studies can be considered as independent for the purposes of data extraction and synthesis. Currently our personal experience and experiences reported by other researchers have consisted only of two-person teams where either one person does all the extraction and defines a set of terms which the other member of the team then checks, or both team members jointly read and extract data from each paper agreeing terminology together. In both cases, the extraction is likely to involve considerable iteration as new terms and concepts are identified and need to be reconciled with the data extracted from previously reviewed primary studies. There are several other options:

- You could use a text analysis tool such as *NVivo* to identify relevant areas of text across all the studies, but as yet there have not been any large-scale software engineering systematic reviews that have reported using such tools. If you are a lone researcher or post-graduate student, the use of a textual analysis tool is a particularly attractive option.

- All team members could read all the primary studies to get a sound overview of the topic area, and then work together to define appropriate themes and agreed terminology before undertaking a systematic data extraction and synthesis process.

However, as yet we have no definitive evidence as to which process is the most effective.

Whether you are a lone researcher or a member of a team, we advise you to read some of the papers and trial various data extraction and synthesis processes on some known studies before making any firm decisions about how to organise data extraction and synthesis.

22.9.3   Data extraction for mapping studies

Mapping studies are generally about finding and classifying the literature related to a specific topic area. Reviewers need to specify a set of characteristics that define the nature of the topic area. In the sense that characteristics might seem similar to themes, there may appear to be some overlap with qualitative systematic reviews, the differences are:

- In a mapping study, the type of study (for example, theoretical or empirical) is a means of classifying the primary study. In contrast, a qualitative systematic review will usually use the type of study as an inclusion/exclusion criterion (for example, including only empirically-based qualitative studies) or as part of a quality assessment of the primary studies.

- A mapping study is not usually concerned with the outcomes of empirical studies whereas a qualitative systematic review aims to aggregate information from the outcomes of qualitative primary studies.

22.9.3.1   Planning data extraction for mapping studies

Mapping studies aim to organise and classify the literature on a specific topic area. This process is done by identifying a set of *features* (sometimes referred to as *attributes*, or *characteristics*) that describe the research goals and methods employed in the topic area. A feature is often specified as a set of mutually exclusive categories to which a primary study can belong, for example, the *research type* might be one feature with categories *case study, experiment, quasi-experiment, opinion survey, lessons learnt, personal opinion*, etc. In this case, the feature is represented as a nominal scale metric. Features may also be ordinal scale, for example, the feature *concept definition* might have one of the values *fully defined, partially defined, undefined*. Other features may be or integer-valued or real-values. Features may relate to one another in hierarchies, for example the category *experiment* belonging to the feature *research type*, might be a sub-feature with categories *fully randomised, randomised block, latin square, n by m factorial*, etc. The features required for a mapping study are related to the specific topic area and the research questions.

The planning activities for a mapping study are shown in Figure 22.13. You need to specify in the protocol the features you will use to classify each primary study, as discussed in Section 22.4.2.2. The major problem with mapping studies is that it may be difficult to identify in advance all the features of interest. We suggest a multiple-phase data extraction process.

Firstly some information needed to answer some of your research questions will already be available from the primary study citation information, for example:

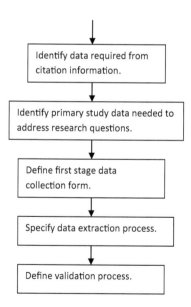

FIGURE 22.13: Planning mapping studies.

- Date of Publication

- Publication type (journal, conference, workshop, technical paper)

- Publication source (journal, conference, workshop name)

- Authors' names, affiliation and country

This information can be specified in the protocol and should be in a suitable format as the outcome of the primary study selection process.

Next, some information you need will be derived from your research questions and can be specified in the protocol. This information will be the basis of the first stage of data extraction. This will include some features and some free-format textual information such as:

- The type of study (using, for example, the categories proposed by Wieringa et al. (2006)).

- The goal(s) of the paper.

- The specific topic(s) or subtopic(s) being addressed in the paper.

- Any issues of interest raised in the paper.

Other information of interest may be identified during the data extraction process.

22.9.3.2 Data extraction process for mapping studies

The overall process for mapping study data extraction is shown in Figure 22.14. The first task is to ensure that all the citation information is held in a format suitable for analysis.

The first stage of data extraction uses the standard data extraction process with at least two members of the team manually extracting data from each primary study.

The team leader should then convene a team meeting to discuss whether there are any more trends or general topics of interest against which to classify your primary studies. If more features are identified, the team leader will need to amend the protocol to include some additional research questions and organise a second round of data extraction to classify the primary studies against the newly defined features. A further round of data extraction would follow the usual data extraction process (that is, two extractors and a process for moderating disagreement). This iterative process continues until no more interesting topics are identified.

Note in the case of mapping studies, the quality of individual primary studies is rarely evaluated and data extraction can usually begin as soon as study selection is completed.

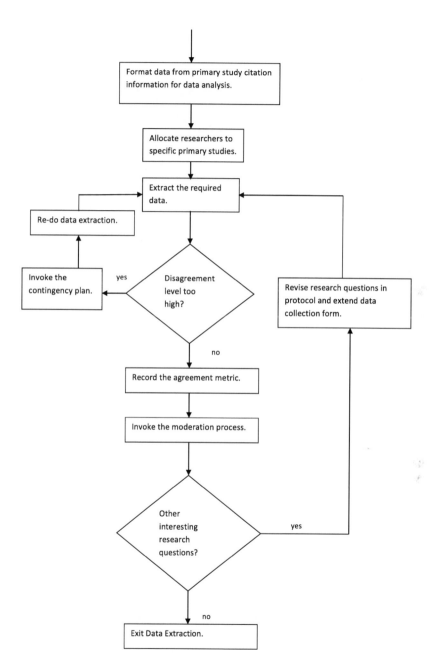

FIGURE 22.14: Mapping study data extraction process.

22.9.4   Validating the data extraction process

For quantitative systematic reviews, the team leader should expect to report initial agreement rates using an appropriate agreement measure. Similar to the quality evaluation process, the agreement between independent assessors is used to assess the validity of the data extraction process. If the agreement for individual primary studies is very low, the data form may not be well understood, so there should be a contingency plan ready. Options for the contingency plan should include:

- Calling a team meeting to discuss the data form and the reasons for disagreements.

- Additional training for specific members of the team.

The team leader is responsible for deciding which option should be chosen given the specific circumstances.

For mapping studies, the process is similar to that for quantitative systematic reviews but if the data extraction process requires two (or more) separate data extraction steps, the agreement measures should be evaluated for each step separately. Felizardo et al. (2010) have suggested using visual text mining tools to support the classification of primary studies in mapping studies. This may be a useful validation approach, particularly for lone researchers and postgraduate students.

It is not clear how data extraction for qualitative primary studies should be validated. It might be possible to use a tool such as *NVivo* to check whether it finds the same textual elements as the review team, but we are not aware of any systematic reviews that reported using this approach.

22.9.5   General data extraction issues

Whatever type of review you are doing, you are supposed to employ critical reasoning when reading primary studies. If you identify some interesting trend or characteristic common to many studies, but it is not mentioned in the protocol, do not ignore it. You should notify your team leader. The team leader needs to decide whether the data extraction process (and the protocol) need to be enhanced to collect information about the newly identified characteristic.

If you are a postgraduate student doing a mapping study as a starting point for your PhD, you need to recognise that the main aim is not to classify a set of primary studies. A mapping study should help you to find the most relevant studies to read, and classifying them may allow you to present a well-organised literature review in your thesis. However, the main aim is for you to read and understand the topic area you intend to study. The secondary aim is for you to understand how to conduct a systematic literature search.

Points to remember

- For quantitative systematic reviews the data extraction process should, in principle, be fully defined in the protocol. However, you need to be alert to any circumstances that indicate an omission in the protocol and be prepared to amend the protocol if necessary.

- For mapping studies, it is not always possible to define all the trends and topics of interest in the protocol. You should expect to iterate the data extraction process if new trends or topics of interest are identified during the data extraction process.

- For qualitative systematic reviews, it is difficult to define the data extraction process or the data synthesis process in advance. You are usually only able to decide whether or not to use a textual analysis tool, and specify the basic strategy that will be used.

- For qualitative systematic reviews, data extraction and data synthesis cannot be regarded as independent processes. In general, you should expect data extraction and synthesis to be iterative including re-reading papers and re-evaluating definitions of terms and themes.

22.10   Data aggregation and synthesis

Like data collection, data synthesis depends on the type of review you are undertaking. There is some disagreement among quantitative and qualitative researchers about the use of the terms "synthesis" and "aggregation". Qualitative researchers use the term "aggregation" to describe results obtained either from statistical analysis or simple counts, whereas "synthesis" is used to refer to analyses that *interpret* findings from qualitative studies. However, quantitative meta-analysts also refer to their statistical analyses as "synthesis". For the purposes of these guidelines we will refer to meta-analysis and qualitative meta-synthesis using the term "synthesis" and we use the term "aggregation" to refer to analysing results from a mapping study.

22.10.1   Data synthesis for quantitative systematic reviews

Your data synthesis process will depend on your data collection plan. This will either be vote counting with a qualitative analysis or a full meta-analysis. For more details on how to perform a meta-analysis consult Chapter 11; vote counting is discussed in Chapter 10.

22.10.1.1   Data synthesis using meta-analysis

Meta-analysis is a statistical method to synthesise the results from primary studies that have reported a statistical analysis of the same (or very similar) hypotheses. For a full meta-analysis, we recommend the process shown in Figure 22.15. It comprises the following steps:

1. Calculate the specified effect size for a specific outcome metric and its variance from the extracted data. If several different outcome metrics are reported in the primary studies, you will need to analyse each outcome metric separately.

2. Tabulate the effect size, its variance and the number of observations per study.

3. Check whether any of the primary studies should be considered dependent replications (that is, primary studies performed by the same researchers, using the same subjects types, and materials). If any of the studies are dependent, you will need to decide how to incorporate these studies. This will involve either using a multi-level analysis model or integrating the dependent studies into a single large-scale experiment.

4. Import the data to an appropriate statistical tool, for example, the R *metafor* package, see Viechtbauer (2010).

5. Perform a random effects analysis to identify whether there is significant heterogeneity among the primary studies, see Viechtbauer (2007). Note, a fixed effects analysis is only appropriate when you have a very small number of effect sizes to aggregate or you are sure that the effect sizes all come from very close replications.

6. If heterogeneity is not significant, data synthesis is completed (subject to appropriate sensitivity analysis, as discussed below) and the overall mean and variance derived from aggregating the primary study results provide the best estimate of the difference between the treatment outcomes.

7. If there is significant heterogeneity, you will need to investigate whether any of the context factors and/or specific quality criteria might have influenced the outcomes and could be used as explanatory factors in a *moderator analysis*. This will require a mixed-effects analysis. Synthesis will depend on whether statistically significant explanatory factors (often referred to as *moderators*) can be found.

8. You will need to consider sensitivity analysis for example, assessing the impact of removing each primary study turn, the impact of high influence studies, and the impact of removing low quality primary studies.

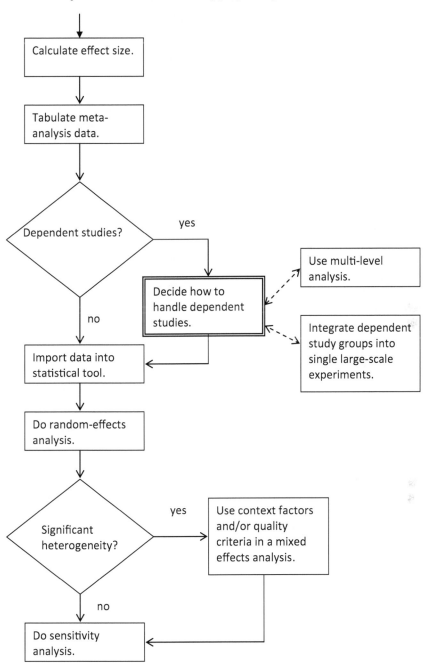

FIGURE 22.15: Meta-analysis process.

22.10.1.2   Reporting meta-analysis results

The R *metafor* package provides the standard meta-analysis graphics to report results (see Chapter 11). The most important graphics are:

- Forest plots to show the mean and variance of the effect size metric for each primary study as well as the overall mean and variance. An example of a forest plot is shown in Figure 22.16. Such plots can also be used to show the impact of significant moderating factors.

- Funnel plots to investigate publication bias (which is the tendency for only papers with significant results to be published). Funnel plots plot the effect size metric for each study against its standard error. The effect size of studies with large standard errors should be more varied than the effect size of studies with relatively small standard errors. The white funnel on the funnel plot shows the acceptable variation for the data points. The funnel plot application can also estimate the number of *missing* studies and display the funnel plot with estimated missing data points filled-in. An example of a funnel plot is shown Figure 22.17. It is based on the same data as the forest plot and includes three extra data points (shown by 3 white filled dots in the left-side of the plot) which correspond to estimated missing data points. The funnel plot also shows

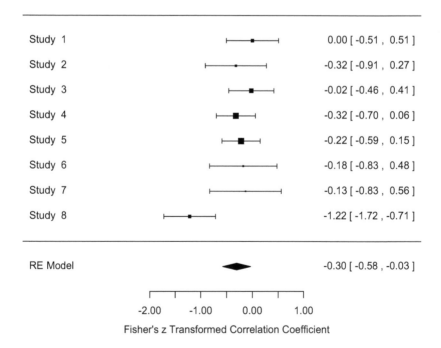

FIGURE 22.16: Forest plot example.

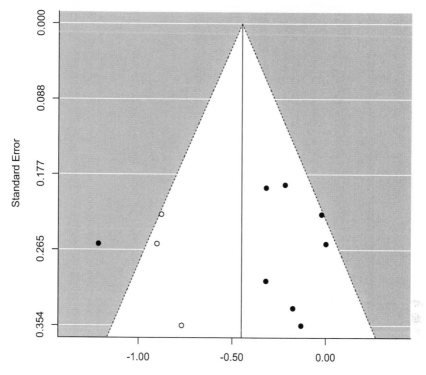

FIGURE 22.17: Funnel plot example.

that one data point is outside the acceptable range (that is, the black dot on the left-hand side of the plot).

Sensitivity analysis can be performed using the influential case diagnostic procedures. In addition Q-Q (Quantile-Quantile) plots can be used to investigate whether the distribution of the primary studies is normal.

22.10.1.3 Vote counting for quantitative systematic reviews

If you cannot do a full meta-analysis because there is too much diversity among the outcome measures or experimental methods, we recommend using a vote counting approach. Vote counting is based on counting the number of primary studies that found a significant affect, and if they constitute the majority, assuming that there is a true effect.

Most meta-analysts object to the use of vote counting for two reasons:

1. Assuming that a study is valid, a significant effect indicates a true effect. However, a non-significant effect does not indicate that there is no effect, because it can be due to low power.

2. Vote counting does not consider the size of effect, so it is not clear whether an effect is of practical importance as well as being significant.

However, in practice, the technique is useful for synthesizing software engineering studies, particularly when it is integrated with some form of moderator analysis. That is, you look for additional factors that might explain differences in primary study outcomes. This is similar to moderator analysis in a meta-analysis, but is qualitative rather than quantitative.

Before considering moderator factors, you need to define the outcome of each primary study. Popay et al. (2006) suggest a five-point scale to describe the outcome:

1. Significantly favours intervention

2. Trends towards intervention

3. No difference

4. Trends towards control

5. Significantly favours control.

They also recommend reporting any effect sizes that can be calculated, not the significance of the outcome.

Some of the synthesis methods used for qualitative primary studies support vote counting (see Chapter 10 and Table 22.3). In particular, the display methods used for Qualitative Cross-Case Analysis (Miles et al. 2014) can also be used to display the results of vote counting and qualitative moderator analysis. The outcomes of each primary study can be tabulated, together with the identified moderator factor values for the study. The display can be organised so that primary studies with the same outcome are kept together. Alternatively, if there is a good reason for displaying the results in a different order (for example, based on the date of the study), it is useful to colour-code the entries according to the outcome. You should then look for any trends among the moderator factors that seem consistent with favourable outcomes. In some cases, it may be possible to use more sophisticated methods such as Comparative Analysis (Ragin 1989) or Case Survey Analysis (Yin & Heald 1975) to analyse vote counting and moderator factor data.

22.10.2   Data synthesis for qualitative systematic reviews

As far as planning is concerned, you should specify in the protocol, the type of synthesis method you intend to use (see Chapter 10 and Table 22.3, for an overview of qualitative methods that have been used in software engineering studies).

In most cases, data synthesis will be integrated with data extraction. You should expect an iterative process whereby the results of reading, extracting and synthesizing data from some primary studies influence the data extraction

TABLE 22.3: Synthesis Methods for Qualitative Analysis

Type	Description
Narrative Synthesis (Popay et al. 2006)	The results and any trends are reported as a textual narrative. Narrative synthesis must be supported by tabulating results or it is very difficult to demonstrate traceability from research questions to the data, to aggregated results that answer the research questions.
Thematic Analysis (Thomas & Harden 2008)	Cruzes & Dybå (2011a) define a 5 stage process starting with reading the text and identifying specific segments of text. The segments of text are labelled and coded, then the codes are analysed to reduce overlaps and define themes. Themes are analysed to create higher-order themes and/or models of the phenomenon being studied. Note some themes are likely to be defined in advance as a result of the research questions, while others may arise as a result of reading the primary studies.
Comparative Analysis (Ragin 1989)	List and categorises cases and attempts to assess what inferences the data supports using boolean algebra. For example, looks for factors that are consistently associated with favourable outcomes AND not associated with unfavourable outcomes.
Meta-Ethnography (Noblit & Hare 1988)	7 stage process in which interpretations and explanations reported in the primary studies are translated into one another. Translations may result in agreement among studies, contradictions among studies, or may form parts of a coherent argument.
Case Survey (Yin & Heald 1975)	This is similar to Comparative Analysis but is appropriate when there are a large number primary studies. Individual primary study results and context information are classified and tabulated looking for commonalities and differences.
Qualitative Cross-case Analysis (Miles et al. 2014)	This uses matrices to report textual and quantitative information from each primary study. The matrices allow similarities and differences among primary studies to be identified.
Metasummary (Sandelowski et al. 2007)	This is a quantitatively oriented aggregation method for analysing thematic analysis papers and opinion surveys. Counts of themes such as risks, motivators, barriers to adoption are made on a primary study basis irrespective of the number of participants in each study.

and synthesis of subsequent primary studies and may initiate a re-assessment of some of the primary studies which have already been synthesised. If you are using a textual analysis tool you need to decide whether it will be used during the initial data extraction and synthesis process or as part of the validation process. The basic data extraction and qualitative synthesis process involves:

1. Identifying textual elements (which can be phrases, sentences, paragraphs, items in tables) in each primary study. The textual element should be stored in your data collection form (which can be a database, document or spreadsheet) with associated information identifying where in the document the element was found, and the research question(s) that it addresses.

2. Each textual element is coded, that is, allocated a single word or phrase that defines its content.

3. Codes are cross-checked for consistency across different primary studies and the data extracted by different team members.

4. Codes may be used for context analysis, that is, the frequency of occurrence of the individual codes are counted for research questions.

5. Codes may be used to create a model of the topic of interest by grouping codes together into related higher-level characteristics and themes. Then the relationships among those higher level characteristics and themes are investigated.

In these guidelines it is not possible to describe every approach to qualitative synthesis. You will find more detailed information in Chapter 10. We recommend reviewing existing software engineering systematic reviews that have used qualitative approaches. A good starting point is a paper written by Cruzes & Dybå (2011b) which cross-references software engineering systematic reviews to the qualitative synthesis methods they used. For specific qualitative methods, Cruzes & Dybå (2011a) provide a detailed explanation of thematic analysis, while Da Silva et al. (2013) present a worked example of using meta-ethnography to synthesise four primary studies. In addition, Cruzes et al. (2014) present an example of synthesising two case studies using three different methods: thematic analysis, qualitative cross case analysis and narrative analysis.

22.10.3   Data aggregation for mapping studies

Mapping study data collection involves identifying important features that describe the characteristics of the primary studies and identifying the appropriate metrics to measure those features. Mapping study aggregation involves tabulating the primary study features. In the case of nominal and ordinal scale features, you should count the number of primary studies in the different categories. For numerical features, you should use standard statistical measures of location and scale (for example, mean and standard deviations, and graphical representations such as box plots or histograms). It is often useful to represent the data in two-way tables that show the relationship between two different categorical features.

For features represented as nominal and ordinal scale metrics, two graphical representation are particularly useful:

1. Trend plots that have counts of primary studies in a specific category as the y-variable and a year as the x-variable. It is sometimes useful to have more than one y-variable. For example, trend plots might be used to show the number of primary studies of different types per year.

2. Bubble plots that are graphical representations which allow you to view information from two two-way tables on the same diagram when the tables share a nominal scale measure. An example is shown in Figure 22.18. In this diagram, the y-variable is called "Variability Context Factor" and has six categories. There are two x-variables, one called "Contribution Facet" which has five categories, and the other called "Research Facet" which has six categories. The bubble containing the value 21 identifies that 21 studies were categorised as having a y-category of "Requirement Variability" and a "Research Focus" category of "Solution". The diagram shows the total number of primary studies classified by "Research Facet" was 128 (which is the number next to the name of the x-variable) and that 21 primary studies corresponds to 16.4% of those studies. The fact that there are different numbers of primary studies classified against each variable indicates that either that some primary studies were not classified against one of the x-valuables or some studies were classified in more than one category for the same x-variable. Note this is not a three-dimensional table because it does not show the distribution of the third nominal variable conditional on the values of the two other variables.

More information about synthesis for mapping studies can be found in Chapter 9.

22.10.3.1 Tables versus graphics

Although graphical representations of the data are important for showing the distribution of primary studies, it is also important to tabulate the results. Without a table showing the values of the categories for each primary study (or a publicly available on-line supporting database), other researchers cannot make constructive use of the results of a mapping study.

22.10.4 Data synthesis validation

There are no standard validation procedures for data synthesis. You must aim to ensure that there is a clear link from the research questions to the data and then to the syntheses that answers those research questions.

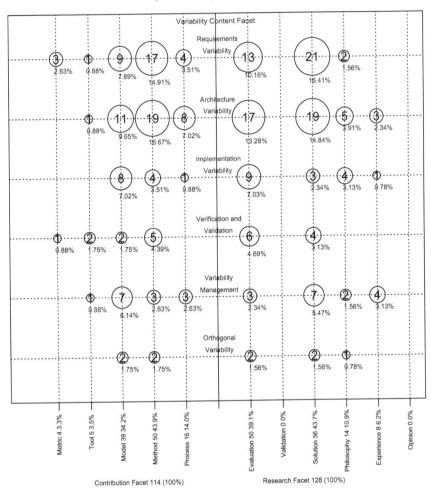

FIGURE 22.18: Bubbleplot example.

General points to remember

- Quantitative systematic reviews can sometimes be analysed using meta-analysis but it may still be necessary to use qualitative synthesis if individual primary studies use non-comparable measurements or treatments.

- Qualitative synthesis is difficult. You should aim to maintain traceability from the data to the synthesis.

- Mapping study aggregation is relatively straightforward.

- If a mapping study is intended for publication in a conference or journal papers, you must make sure that all the primary study data including the citations and classification information is available to the reader. For very large mapping studies, this may mean publishing an on-line database holding the citations and feature data for each primary study.

22.11   Reporting the systematic review

There are three things you need consider when reporting your results:

1. Who do you expect to be interested in the results of the systematic review and what format of the report(s) do they need?

2. The format of each type of report you need to write.

3. How you plan to validate the report.

22.11.1   Systematic review readership

Systematic reviews (in contrast to mapping studies) should consider two main types of reader: researchers and practitioners. A reader of an academic journal or conference paper will expect to see a full description of the methodology, as well as a report of the results of the study with traceability between the data and the analysis. Practitioners, however, will be more concerned about the implication of the result for software engineering practice. It is also the case that practitioners are more likely to read short magazine articles than ponderous academic papers.

Thus, there is an argument for writing both an academic paper and a separate article for practitioners. Remember, however, to reference any related journal paper in the practitioner article and vice versa. You also need to ensure that you do not violate any originality or copyright requirements of the publication outlets.

For mapping studies, the main readership will be researchers. It is important to include information about each primary study including the full citations, as well as how it was classified against each feature. For conference papers, it may be difficult to fit in all the data for each primary study. In such cases you need to consider providing ancillary data on-line (either an extended technical report or a database).

22.11.2   Report structure

With respect to the structure of a systematic review report, PRISMA, the current guideline for reporting systematic reviews and meta-analysis studies in health care, has been published in several open-access papers (one example being Liberati et al. (2009)).

The high-level structure for a PRISMA report is very similar to the standard scientific format which comprises the Title and Abstract followed by the IMRAD sections (Introduction, Methods, Results, and Discussion). In software engineering, we usually have a separate *Conclusions* section but PRISMA makes *Conclusions* a subset of the *Discussion* section and adds a final section called *Funding*.

We recommend using the basic PRISMA structure with a few minor changes:

1. Title: Identify the topic of the study and that it is a systematic review, meta-analysis or mapping study.

2. Abstract: Structured abstract, including headings for background, objectives, methods, results, conclusions.

3. Introduction: Justification for the review and the research question(s).

4. Background: Any information needed to understand the topic of the systematic review.

5. Methods: Indicate where the protocol can be accessed. Report search and selection process including databases searched, search terms and inclusion and exclusion criteria - consider what needs to be in the body of the paper and what can be put into appendices. Report data collection process and data items collected including quality data and the data items needed to answer the research questions. For meta-analysis, identify the principal summary measures and methods of aggregation. For other forms of review, describe the data analysis and synthesis process. Report how quality data will be used and any other means of identifying possible biases. Describe any additional analysis such as sensitivity analysis or subgroup analysis.

6. Results: Report the study selection process including number of studies excluded at each major stage, preferably with a flow diagram. Report the identified primary study characteristics. Discuss the quality of individual studies and any systematic biases. Present data syntheses using the graphical methods discussed in Section 22.10. Report the results of any additional analyses.

7. Discussion: provide a summary the evidence for each research question and any additional analyses. Discuss the limitations of the review at the primary study level and at the review level.

8. Conclusions: Provide a general interpretation of the result in the context of any other evidence. Provide recommendations for researchers and practitioners.

9. Acknowledgements: Identify the funding agency (if any). Thank any researchers who made useful contributions to the study but were not part of the research team, for example external reviewers of the protocol or final report.

10. Appendices: Report the search strings used for individual digital sources. Specify the data collection form. Provide a list of "near miss" papers, for example, candidate primary studies excluded from the review after the second screening process. Report any quality checklists used.

If you are intending to report your results in a practitioner magazine, you should expect to use a much less formal presentation. You should aim to explain the topic covered by the review and why it is important. It is unnecessary to provide any detailed description of the methodology. When presenting the findings you should explain what the implications are for practice and the confidence that can be placed in the findings.

22.11.3   Validating the report

All report authors have a responsibility to read and review the report, with the aim of ensuring:

- The research questions are clearly specified and fully answered.

- The research methodology is fully and correctly reported.

- There is traceability from the research questions to data collection, data synthesis and conclusions.

- All the tables and figures used to present the results are correct and internally consistent.

- In the case of systematic reviews, the conclusions are written clearly and are targeted both at researchers and practitioners.

If possible, for systematic reviews, you should find someone to act as an independent reviewer of the report. For example, within a research group or a university department, you can try to encourage colleagues to undertake independent reviews on a quid-pro-quo basis. An independent reviewer would probably find useful the following questions, which are based on review assessment questions suggested by Greenhalgh (2010):

- Q1: Does the review address an important software engineering problem?

- Q2: Was a thorough search done of the appropriate databases(s) and were other sources considered?

- Q3: Was methodological quality assessed and primary studies weighted accordingly?

- Q4: How sensitive are the results to the way the review was done?

- Q5: Have numerical results been interpreted sensibly in the context of the problem?

Points to remember

- Identify your target audience and write appropriately.

- Make sure to report your methodology fully.

- For reviews including a large number primary studies, consider providing an online database to hold the information collected about each primary study.

- Particularly for systematic reviews, consider the implications for practitioners.

- All authors need to review the final report carefully.

- For systematic reviews, an independent reviewer can be very helpful.

- For journal or conference papers reporting mapping studies, you may need to provide citation and classification information in a separate technical report or online database.

Appendix

Catalogue of Systematic Reviews Relevant to Education and Practice

with Sarah Drummond and Nikki Williams

In seeking to identify what evidence-based management can learn from evidence-based medicine, Barends & Briner (2014) observed that, in the field of medicine, "evidence-based practice started out as a teaching method, developed at McMaster University in Canada". Hence, it was looking for better ways to teach about clinical *practice* that formed the key driver for the emergence of evidence-based medicine. Later in their paper the distinction is made between a minority of "evidence practitioners" who "critique the literature" (by conducting systematic reviews) and the majority of "evidence users" who have an understanding of where the evidence comes from, and turn to those sources to advise their practices. So, while this book is largely concerned with the activities of the "evidence practitioners" in software engineering, this appendix concerns itself with the needs of our many potential "evidence users".

Perhaps because EBSE-related research has so far been mainly focused upon applying evidence-based research practices to specific topics, its potential role in support of the ways that we teach our subject has received less attention. However, the growing number of systematic reviews does provide scope to make our teaching more *evidence-informed* by being able to illustrate it with evidence about how well software engineering techniques and models work, as well as with data drawn from industrial settings. In (Budgen et al. 2012) we identified 43 systematic reviews, published up until mid-2011, that contained material that could be used to support *introductory* courses about software engineering practices and concepts. This is not to say that many other systematic reviews may not also be useful for teaching in some way, and obviously, more specialist courses might be able to use a rather different set of studies. For this appendix, we have extended the restricted search used in (Budgen et al. 2012) up to the end of 2014, and this has added a further 16 reviews. All of the studies have been categorised using the *Knowledge Areas* (KAs) and *Knowledge Units* (KUs) identified in the 2004 ACM/IEEE guidelines for undergraduate curricula in software engineering—although we should note that some studies do not easily fall into just one of these categorisations.

Obviously, not all the KUs have been covered by this set of systematic reviews, especially those concerned with 'fundamentals' and 'concepts'. However, from the viewpoint of the teacher, these reviews can help to provide the student with a realistic picture of the many complexities of software engineering practice, and by aiding evidence-informed teaching, they can help 'flesh out' our students' knowledge about the discipline. However, we would advise that before using any of these it is also worth checking if newer systematic reviews have been published on those topics.

Each of the following sections provides a summary of the studies categorised as belonging to one Knowledge Area, highlighting some examples for each KA and then listing the others. We should however emphasise that we have chosen highlights that are intended to be *illustrative* and that these do not form specific recommendations. Each person teaching about software engineering therefore needs to identify the set of studies that can provide the best support for their specific course.

Table A.1 summarises the distribution of studies across the different KAs and KUs, identifying both clusters and gaps as appropriate for a mapping study. (Indeed, this provides an example of using an existing set of categories to help identify any gaps.)

A.1 Professional Practice (PRF)

There are two studies that were categorised as addressing this Knowledge Unit.

Professionalism (pr) Useful insight into human aspects of the discipline is provided in Ghapanchi & Aurum (2011), which examines why IT professionals change their jobs, and identifies some of the factors that might help an organisation retain its technical staff.

Group dynamics/Psychology (psy) The review by Renger, Kolfschoten & de Vreede (2008) examines the challenges faced by teams of developers and stakeholders when working together to create graphical models for describing systems, and identifies some trade-offs in the processes such as

TABLE A.1: Distribution of Systematic Reviews across Knowledge Areas

Knowledge Area	No. KU	KUs covered	No. gaps	KUs not covered
Professional Practice	2	psy(1), pr(1)	1	com
Modelling & Analysis	4	tm(4), af(1), er(3), rsd(1)	3	md, rfd, rv
Design	2	str(1), ar(2)	4	con, hci, dd, ste
V & V	3	rev(2), tst(6), par(1)	1	fnd, hct
Evolution	1	ac(3)	1	pro
Software Process	2	imp(6), con(3)	0	
Software Quality	3	pro(3), pca(1), pda(2)	2	cc, std
Software Management	3	pp(11), per(2), ctl(2)	2	con, cm

"involving stakeholders can improve correctness, buy-in and complete-ness, but lead to conflict (with developers) due to different perspectives" and "including a skilled modeller can improve model quality but reduce the sense of ownership by stakeholders".

A.2   Modelling and Analysis (MAA)

We have highlighted three studies that provide a spectrum of the issues covered, and then listed the remaining ones in Table A.2.

Types of models (tm) One of the studies under this heading looks at experience with using model-driven engineering (MDE) in industry (Mohagheghi & Dehlen 2008). Perhaps in part because the primary studies were industry studies they found some difficulty in generalising their results, and they also noted that there was a lack of suitably mature tool support. However, they did comment that they "found reports of improvements in software quality and of both productivity gains and losses, but these reports were mainly from small-scale studies. There are a few reports on advantages of applying MDE in larger projects, how-ever, more empirical studies and detailed data are needed to strengthen the evidence." From the teaching perspective this has the benefit of being one of the few systematic reviews to have looked at what is happening in industry, and hence can provide a useful perspective upon this.

Eliciting requirements (er) The study by López, Nicolás & Toval (2009) is a fairly short report on the risks to requirements elicitation that can occur when moving from co-located development to a global software development context. While set in the context of exploring a particular approach to addressing the problems, this can help to provide the reader with an appreciation of the potential pitfalls of moving to what is an increasingly popular form of software development.

Eliciting requirements (er) We have discussed the study by Dieste & Juristo (2011) earlier in the context of knowledge translation (see Section 14.3.3). Its main value for teaching is that it compares different techniques for requirements elicitation and gives a number of recommendations, together with reasoning.

TABLE A.2: Other Studies Addressing MAA

KA	Title	Reference
tm	Comparing Local and Global Software Effort Estimation Models—Reflections on a Systematic Review	(MacDonell & Shepperd 2007)
	Does the technology acceptance model predict actual use? A systematic literature review	(Turner et al. 2010)
	A systematic literature review on fault prediction performance in software engineering	(Hall et al. 2012)
af	A systematic review of domain analysis tools	(Lisboa, Garcia, Lucrédio, de Almeida, de Lemos Meira & de Mattos Fortes 2010)
rsd	On the generation of requirements specifications from software engineering models: A systematic literature review	(Nicolás & Toval 2009)
er	A systematic literature review of stakeholder identification methods in requirements elicitation	(Pacheco & Garcia 2012)

A.3   Software Design (DES)

Only three papers were categorised as coming under this heading.

Design strategies (str) The paper by Ali, Babar, Chen & Stol (2010) addresses an important if challenging topic, which is comparing the strengths and weaknesses of aspect-oriented programming with those of other design approaches. The authors identified a range of generally positive effects, particularly for performance, memory consumption, modularity etc. However, some studies did report negative effects and many of the primary studies included were rated as being 'poor' for the credibility of any evidence provided.

Architectural design (ar) The paper by Shahin, Liang & Babar (2014) looks at the different forms used to visualise software architecture, how extensively they are used and for what purposes, and the forms most widely used in industry. While there are no guidelines as to which form to choose, it does provide a useful report of what is happening in industry.

The third paper, listed in Table A.3 has already been described earlier in the book.

TABLE A.3: Other Studies Addressing DES

KA	Title	Reference
dd	What do We Know about the Effectiveness of Software Design Patterns?	(Zhang & Budgen 2012)

A.4   Validation and Verification (VAV)

This topic is fairly well covered, with a number of systematic reviews, some on fairly specialist topics. We have identified two that may be particularly useful for general teaching, summarising the remaining papers in Table A.4.

Reviews (rev) The paper by Kollanus & Koskinen (2009) looks at software inspection research and profiles what is known about inspections, based on 153 articles and considering both technical and management aspects.

Testing (tst) The topic addressed in (Haugset & Hanssen 2008) is that of automation of acceptance testing, and in particular its application in agile development. The authors identify eight different effects and discuss the benefits of these in terms of encouraging developers to reflect on design and system behaviour.

TABLE A.4: Other Studies Addressing VAV

KA	Title	Reference
rev	Capture-recapture in software inspections after 10 years research—theory, evaluation and application	(Petersson, Thelin, Runeson & Wohlin 2004)
tst	In Search of What We Experimentally Know about Unit Testing	(Juristo, Moreno, Vegas & Solari 2006)
	A systematic review of search-based testing for non-functional system properties	(Azfal, Torkar & Feldt 2009)
	A systematic review on regression test selection techniques	(Engström, Runeson & Skoglund 2010)
	On evaluating commercial Cloud services: A systematic review	(Li, Zhang, O'Brien, Cai & Flint 2013)
	A systematic review on the functional testing of semantic web services	(Tahir, Tosi & Morasca 2013)
par	What Do We Know about Defect Detection Methods?	(Runeson, Andersson, Thelin, Andrews & Berling 2006)

A.5   Software Evolution (EVO)

The three papers categorised under this heading demonstrate some very relevant aspects of software engineering knowledge in an area where students will usually have little first-hand experience.

Evolution activities (ac) The first paper, by Williams & Carver (2010) sought to identify the architectural characteristics that link changes in software to the resulting effects in a system. From this, they created a software architecture change characterisation scheme (SACCS) mapping high-level changes to lower-level characteristics, together with an assessment of likely impact. So here, we see how a systematic review led to the development of a categorisation model.

The second paper, by (Arias, van der Spek & Avgeriou 2011) is concerned with *dependency analysis* concerned with making explicit the details and form of interconnections between program units, with particular concern about their effects upon the evolution of software-intensive systems. It provides an assessment of how usable the different techniques are, and assesses some of their limitations.

The third paper, by Hordijk, Ponisio & Wieringa (2009), examined whether duplication of code (cloning) affected the ease with which a system could be changed. They found that it was not possible to demonstrate or reject the existence of any direct link between duplication of code and changeability, so demonstrating the limitations of a secondary study when lacking enough strong primary studies. However, their study did identify some useful guidelines for handling systems containing cloned code, based on the studies that they reviewed.

A.6 Software Process (PRO)

All of the papers under this heading are of relevance to teaching, and most have a bias towards practice (although in turn, this may limit the number of primary studies available). However, we have identified three as particularly relevant for providing insight into industry practices, with details of the other two being provided in Table A.5.

Process implementation (imp) All three papers are in this category, although addressing very different issues.

The study by Hanssen, Bjørnson & Westerheim (2007) addresses the way that the Rational Unified Process (RUP) can be tailored to meet the needs of individual organisations. A key conclusion from a teaching perspective is that the RUP is too complex and that software development needs more agile practices.

Regarding agile practices, the review by Hossain, Babar & Paik (2009) does address the use of agile practice, although in a rather specific context, examining how Scrum can be used in global software development (GSD) projects. The authors conclude that Scrum practices may be constrained by GSD contextual factors affecting communication, coordination and collaboration.

Finally the study by Mohagheghi & Conradi (2007) addresses the issues of software reuse and its effect upon quality and productivity as well as any economic benefits it provides, finding a positive and significant effect for both software quality and productivity.

TABLE A.5: Other Studies Addressing PRO

KA	Title	Reference
con	A systematic literature review on the industrial use of software process simulation	(Ali, Peterson & Wohlin 2014)
	Software fault prediction metrics: A systematic literature review	(Radjenović, Heričko, Torkar & Živkovič 2013)
	The Effects of Test Driven Development on External Quality and Productivity: A Meta-analysis	(Rafique & Misic 2013)
imp	The effectiveness of pair programming: A meta-analysis	(Hannay et al. 2009)
	Challenges and Improvements in Distributed Software Development: A systematic review	(Jiménez, Piattini & Vizcaíno 2009)
	Reconciling software development models: A quasi-systematic review	(Magdaleno, Werner & de Araujo 2012)
	Considering rigor and relevance when evaluating test driven development: A systematic review	(Munir, Moayyed & Peterson 2014)
	Agile product line engineering: A systematic literature review	(Díaz, Pérez, Alarcón & Garbajosa 2011)

A.7   Software Quality (QUA)

There are several useful studies that address issues such as process improvement and the assessment of processes and products. For illustration, we have summarised two studies that address these process and product aspects, with the remaining papers being listed in Table A.6.

Software quality processes (pro) Galin & Avrahami (2006) investigated the value for an organisation in investing in a CMM program for software process improvement. Their results were expressed using seven common performance metrics and the authors argue that these indicate that CMM programs can lead to improved software development and maintenance.

Product assurance (pda) Riaz, Mendes & Tempero (2009) examined techniques and models commonly used for predicting the maintainability of software. Their study provides classification of techniques and a list of metrics that were successful for predicting maintainability, as well as providing a review of definitions of maintainability.

TABLE A.6: Other Studies Addressing QUA

KA	Title	Reference
pro	Software process improvement in small and medium software enterprises: a systematic review	(Pino, Garcia & Piattini 2008)
	Systematic review of organizational motivations for adopting CMM-based SPI	(Staples & Niazi 2008)
pac	Towards a Defect Prevention Based Process Improvement Approach	(Kalinowski, Travassos & Card 2008)
pda	Coupling Metrics for Aspect-Oriented Programming: A Systematic Review of Maintainability Studies	(Burrows, Garcia & Taïani 2009)

A.8 Software Management (MGT)

The large number of studies in the 'pp' KA (project planning) is in part because of the importance of estimation for software engineering projects, and also because some of the pioneering EBSE researchers were particularly interested in this. Again, we have highlighted two studies to provide examples and then listed the remaining ones in Table A.7 This is in part because it is quite difficult to know how these topics will be addressed in a course, and so the highlighted studies are again essentially indicative.

Project planning (pp) Our example for this is (Jørgensen 2007). While there has tended to be the expectation that model-based estimation will be 'better' than expert judgement, this paper observes that this is not necessarily so, and explains some reasons for this.

Project personnel & organisation (per) Understanding how programmers and programming teams are motivated, and the effect that motivation has is important knowledge for project management. The study by Sharp, Baddoo, Beecham, Hall & Robinson (2009) identifies key motivation factors and creates a new model of motivation.

TABLE A.7: Other Studies Addressing MGT

KA	Title	Reference
pp	A review of studies on expert estimation of software development effort	(Jørgensen 2004)
	The Consistency of Empirical Comparisons of Regression and Analogy-based Software Project Cost Prediction	(Mair et al. 2005)
	A Survey on Software Estimation in the Norwegian Industry	(Moløkken-Østvold, Tanilkan, Gallis, Lien & Hove 2004)
	The Clients' Impact on Effort Estimation Accuracy in Software Development Projects	(Grimstad, Jørgensen & Moløkken-Østvold 2005)
	Evidence-Based Guidelines for Assessment of Software Development Cost Uncertainty	(Jørgensen 2005)
	A Comparison of Software Cost, Duration, and Quality for Waterfall vs. Iterative and Incremental Development: A Systematic Review	(Mitchell & Seaman 2009)
	Managing Risks in Distributed Software Projects: An Integrative Framework	(Persson, Mathiassen, Boeg, Madsen & Steinson 2009)
	Empirical evidence in global software engineering: a systematic review	(Smite, Wohlin, Gorschek & Feldt 2010)
	Barriers in the selection of offshore software development outsourcing vendors: An exploratory study using a systematic literature review	(Khan et al. 2011)
	Software development in start-up companies: A systematic mapping study	(Paternoster, Giardino, Unterkalmsteiner & Gorschek 2014)
per	Motivation in Software Engineering: A systematic literature review	(Beecham et al. 2008)
	Empirical studies on the use of social software in global software development–A systematic mapping study	(Giuffrida & Dittrich 2013)
ctl	Factors Influencing Software Development Productivity — State-of-the-Art and Industrial Experiences	(Trendowicz & Münch 2009)
	Measuring and predicting software productivity: A systematic map and review	(Peterson 2011)

Bibliography

AGREE (2009), 'Appraisal of Guidelines for Research and Evaluation II (AGREE II)', AGREE Next Steps Consortium Report.

Ali, M. S., Babar, M. A., Chen, L. & Stol, K.-J. (2010), 'A systematic review of comparative evidence of aspect-oriented programming', *Information and Software Technology* **52**(9), 871–887.

Ali, N. B., Peterson, K. & Wohlin, C. (2014), 'A systematic literature review on the industrial use of software process simulation', *Journal of Systems & Software* **97**, 65–85.

Alves, V., Niu, N., Alves, C. & Valença, G. (2010), 'Requirements engineering for software product lines: A systematic literature review', *Information and Software Technology* **52**(8), 806–820.

Ampatzoglou, A. & Stamelos, I. (2010), 'Software engineering research for computer games: A systematic review', *Information and Software Technology* **52**(9), 888–901.

Anjum, M. & Budgen, D. (2012), A mapping study of the definitions used for Service Oriented Architecture, *in* 'Proceedings of 16th EASE Conference', IET Press, pp. 1–5.

Arias, T. B. C., van der Spek, P. & Avgeriou, P. (2011), 'A practice-driven systematic review of dependency analysis solutions', *Empirical Software Engineering* **16**, 544–586.

Atkins, S., Lewin, S., Smith, H., Engel, M., Fretheim, A. & Volmink, J. (2008), 'Conducting a meta-ethnography of qualitative literature: Lessons learnt', *BMC Medical Research Methodology* **8**(21).

Azfal, W., Torkar, R. & Feldt, R. (2009), 'A systematic review of search-based testing for non-functional system properties', *Information and Software Technology* **51**, 957–976.

Babar, M. A. & Zhang, H. (2009), Systematic literature reviews in software engineering: Preliminary results from interviews with researchers, *in* 'Proceedings of the 2009 3rd International Symposium on Empirical Software Engineering and Measurement', ESEM '09, IEEE Computer Society, Washington, DC, USA, pp. 346–355.

Barends, E. G. R. & Briner, R. B. (2014), 'Teaching evidence-based practice: Lessons from the pioneers—An interview with Amanda Burls and Gordon Guyatt', *Academy of Management Learning & Education* **13**(3), 476–483.

Barnett-Page, E. & Thomas, J. (2009), 'Methods for the synthesis of qualitative research: a critical review', *BMC Medical Research Methodology* **9**(59).

Basili, V., Green, S., Laitenberger, O., Lanubile, F., Shull, F., Sorumgard, S. & Zelkowitz, M. (1996), 'The empirical investigation of perspective-based reading', *Empirical Software Engineering* **1**(2), 133–164.

Basili, V. R., Shull, F. & Lanubile, F. (1999), 'Building Knowledge through Families of Experiments', *IEEE Transactions on Software Engineering* **25**(4), 456–473.

Beecham, S., Baddoo, N., Hall, T., Robinson, H. & Sharp, H. (2006), *Protocol for a Systematic Literature Review of Motivation in Software Engineering*, University of Hertfordshire.

Beecham, S., Baddoo, N., Hall, T., Robinson, H. & Sharp, H. (2008), 'Motivation in software engineering: A systematic literature review', *Information and Software Technology* **50**(9–10), 860–878.

Benbasat, I., Goldstein, D. K. & Mead, M. (1987), 'The case research strategy in studies of information systems', *MIS Quarterly* **11**(3), 369–386.

Boehm, B. W. (1981), *Software Engineering Economics*, Prentice-Hall.

Booth, A., Papaioannou, D. & Sutton, A. (2012), *Systematic Approaches to a Successful Literature Review*, Sage Publications Ltd.

Borenstein, M., Hedges, L. V., Higgins, J. P. T. & Rothstein, H. T. (2009), *Introduction to Meta-Analysis*, John Wiley and Sons Ltd.

Bowes, D., Hall, T. & Beecham, S. (2012), Slurp: A tool to help large complex systematic reviews deliver valid and rigorous results, in '*Proceedings 2nd International Workshop on Evidential Assessment of Software Technologies (EAST'12)*', ACM Press, pp. 33–36.

Bratthall, L. & Jørgensen, M. (2002), 'Can you trust a single data source exploratory software engineering case study?', *Empirical Software Engineering* **7**(1), 9–26.

Briand, L. C., Melo, W. L. & Wust, J. (2002), 'Assessing the applicability of object-oriented software projects fault-proneness models across', *IEEE Transactions on Software Engineering* **28**, 706–720.

Brooks Jr., F. P. (1987), 'No silver bullet: essences and accidents of software engineering', *IEEE Computer* **20**(4), 10–19.

Budgen, D. (1971), 'A $KN \to \Lambda\pi$ partial-wave analysis in the region of the $\Sigma(1670)$', *Lettere al Nuovo Cimento* **2**(3), 85–89.

Budgen, D., Burn, A., Brereton, P., Kitchenham, B. & Pretorius, R. (2011), 'Empirical evidence about the UML: A systematic literature review', *Software — Practice and Experience* **41**(4), 363–392.

Budgen, D., Burn, A. & Kitchenham, B. (2011), 'Reporting student projects through structured abstracts: A quasi-experiment', *Empirical Software Engineering* **16**(2), 244–277.

Budgen, D., Drummond, S., Brereton, P. & Holland, N. (2012), What scope is there for adopting evidence-informed teaching in software engineering?, in 'Proceedings of 34th International Conference on Software Engineering (ICSE 2012)', IEEE Computer Society Press, pp. 1205–1214.

Budgen, D., Kitchenham, B. A., Charters, S., Turner, M., Brereton, P. & Linkman, S. (2008), 'Presenting software engineering results using structured abstracts: A randomised experiment', *Empirical Software Engineering* **13**(4), 435–468.

Budgen, D., Kitchenham, B. & Brereton, P. (2013), The Case for Knowledge Translation, in 'Proceedings of 2013 International Symposium on Empirical Software Engineering & Measurement', IEEE Computer Society Press, pp. 263–266.

Budgen, D., Kitchenham, B., Charters, S., Gibbs, S., Pohthong, A., Keung, J. & Brereton, P. (2013), Lessons from conducting a distributed quasi-experiment, in 'Proceedings of 2013 International Symposium on Empirical Software Engineering & Measurement', IEEE Computer Society Press, pp. 143–152.

Burgers, J. S., Grol, R., Klazinga, N. S., Mäkelä, M. & Zaat, J. (2003), 'Towards evidence-based clinical practice: an international survey of 18 clinical guidelines programs', *International Journal for Quality in Health Care* **15**(1), 31–45.

Burrows, R., Garcia, A. & Taïani, F. (2009), Coupling metrics for aspect-oriented programming: A systematic review of maintainability studies, in 'ENASE 2009 - Proceedings of the 4th International Conference on Evaluation of Novel Approaches to Software Engineering, Milan, Italy, May 2009', pp. 191–202.

Canfora, G., Cimitile, A., Garcia, F., Piattini, M. & Visaggio, C. A. (2007), 'Evaluating performances of pair designing in industry', *Journal of Systems & Software* **80**, 1317–1327.

Carifio, J. & Perla, R. J. (2007), 'Ten common misunderstandings, misconceptions, persistent myths and urban legends about likert scales and likert response formats and their antidotes', *Journal of Social Science* **3**(3), 106–116.

Carver, J. C. (2010), Towards reporting guidelines for experimental replications: A proposal, *in* 'Proceedings of the 1st International Workshop on Replication in Empirical Software Engineering Research (RESER 2010)', ACM Press, pp. 1–4.

Casey, V. & Richardson, I. (2008), The impact of fear on the operation of virtual teams, *in* 'Proceedings of IEEE International Conference on Global Software Engineering', IEEE Computer Society Press.

Chen, L. & Babar, M. A. (2011), 'A systematic review of evaluation of variability management approaches in software product lines', *Information and Software Technology* **53**(4), 344–362. Special section: Software Engineering track of the 24th Annual Symposium on Applied Computing Software Engineering track of the 24th Annual Symposium on Applied Computing.

Ciolkowski, M. (2009), What do we know about perspective-based reading? an approach for quantitative aggregation in software engineering, *in* 'Proceedings 3rd International Symposium on Empirical Software Engineering & Measurement (ESEM)', pp. 133–144.

Cochran, W. (1954), 'The combination of estimates from different experiments.', *Biometrics* **10**(101-129).

Cochrane, A. L. (1971), *Effectiveness and Efficiency: Random Reflections on Health Services*, The Nuffield Provincial Hospitals Trust.

Cohen, J. (1960), 'A coefficient of agreement for nominal scales', *Educational and Psychological Measurement* **20**(1), 37–46.

CRD (2009), 'Systematic reviews crd's guidance for undertaking reviews in health care', Centre for Review and Dissemination.

Cruzes, D. S. & Dybå, T. (2011*a*), Recommended steps for thematic synthesis in software engineering, *in* 'Proceedings ESEM 2011'.

Cruzes, D. S. & Dybå, T. (2011*b*), 'Research synthesis in software engineering: A tertiary study', *Information and Software Technology* **53**(5), 440–455.

Cruzes, D. S., Dybå, T., Runeson, P. & Höst, M. (2014), 'Case studies synthesis: a thematic, cross-case, and narrative synthesis worked example', *Empirical Software Engineering* .

Cruzes, D. S., Mendonca, M., Basili, V., Shull, F. & Jino, M. (2007*a*), Automated information extraction from empirical software engineering literature, in 'Proceedings of First International Symposium on Empirical Software Engineering & Measurement (ESEM 2007)', pp. 491–493.

Cruzes, D. S., Mendonca, M., Basili, V., Shull, F. & Jino, M. (2007*b*), Using context distance measurement to analyze results across studies, *in* 'Proceedings of First International Symposium on Empirical Software Engineering & Measurement (ESEM 2007)', pp. 235–244.

Cumming, G. (2012), *Understanding the New Statistics. Effect Sizes, Confidence Intervals, and Meta-Analysis*, Routledge Taylor & Francis Group, New York, London.

Da Silva, F. Q. B.; Cruz, S. S. J. O.; Gouveia, T. B.; & Capretz, L. F. (2013), 'Using meta-ethnography to synthesize research: A worked example of the relations between personality and software team process', *2013 ACM/IEEE International Symposium on Empirical Software Engineering and Measurement*, pp. 153–162 .

da Silva, F. Q., Santos, A. L., Soares, S., França, A. C. C., Monteiro, C. V. & Maciel, F. F. (2011), 'Six years of systematic literature reviews in software engineering: An updated tertiary study', *Information and Software Technology* **53**(9), 899–913.

da Silva, F. Q., Suassuna, M., França, A. C. C., Grubb, A. M., Gouveia, T. B., Monteiro, C. V. & dos Santos, I. E. (2014), 'Replication of empirical studies in software engineering research: A systematic mapping study', *Empirical Software Engineering* **19**, 501–557.

Davis, D., Evans, M., Jadad, A., Perrier, L., Rath, D., Ryan, D., Sibbald, G., Straus, S., Rappolt, S., Wowk, M. & Zwarenstein, M. (2003), 'The case for knowledge translation: shortening the journey from evidence to effect', *BMJ* **327**, 33–35.

Díaz, J., Pérez, J., Alarcón, P. P. & Garbajosa, J. (2011), 'Agile product line engineering—A systematic literature review', *Software: Practice and Experience* **41**, 921–941.

Dickinson, T. L. & McIntyre, R. M. (1997), A conceptual framework of teamwork measurement, *in* M. T. Brannick, E. Salas & C. Prince, eds, *Team Performance Assessment and Measurement: Theory, Methods and Applications*, Psychology Press, NJ, USA, pp. 19–43.

Dieste, O., Grimán, A. & Juristo, N. (2009), 'Developing search strategies for detecting relevant experiments', *Empirical Software Engineering* **14**(5), 513–539.

Dieste, O., Griman, A., Juristo, N. & Saxena, H. (2011), Quantitative determination of the relationship between internal validity and bias in software engineering experiments: Consequences for systematic literature reviews, *in* 'Empirical Software Engineering and Measurement (ESEM), 2011 International Symposium on', pp. 285–294.

Dieste, O. & Juristo, N. (2011), 'Systematic review and aggregation of empirical studies on elicitation techniques', *IEEE Transactions on Software Engineering* **37**(2), 283–304.

Dieste, O., Juristo, N. & Martinez, M. D. (2014), Software industry experiments: A systematic literature review, in *'Proceedings of the 1st International Workshop on Conducting Empirical Studies in Industry (CSEI'14)'*, IEEE Computer Society Press, pp. 2–8.

Dieste, O. & Padua, O. (2007), Developing search strategies for detecting relevant experiments for systematic reviews, in *'Empirical Software Engineering and Measurement, 2007. ESEM 2007. First International Symposium on'*, pp. 215–224.

Dixon-Woods, M., Sutton, A., Shaw, R., Miller, T., Smith, J., Young, B., Bonas, S., Booth, A. & Jones, D. (2007), 'Appraising qualitative research for inclusion in systematic reviews: a quantitative and qualitative comparison of three methods', *Journal of Health Services Research and Policy* **12**(1), 42–47.

Dybå, T. & Dingsøyr, T. (2008a), 'Empirical studies of agile software development: A systematic review', *Information & Software Technology* **50**, 833–859.

Dybå, T. & Dingsøyr, T. (2008b), Strength of evidence in systematic reviews in software engineering, in *'Proceedings of International Symposium on Empirical Software Engineering and Metrics (ESEM)'*, pp. 178–187.

Dybå, T., Kampenes, V. & Sjøberg, D. (2006), 'A systematic review of statistical power in software engineering experiments', *Information & Software Technology* **48**(8), 745–755.

Eaves, Y. D. (2001), 'A synthesis technique for grounded theory data analysis', *Journal of Advanced Nursing* **35**(5), 654–663.

Eisenhardt, K. M. (1989), 'Building theories from case study research', *Academy of Management Review* **14**, 532–550.

Elamin, M. B., Flynn, D. N., Bassler, D., Briel, M., Alonso-Coello, P., Karanicolas, P. J., Guyatt, G., Malaga, G., Furukawa, T. A., Kunz, R., Schnemann, H., Murad, M. H., Barbui, C., Cipriani, A. & Montori, V. M. (2009), 'Choice of data extraction tools for systematic reviews depends on resources and review complexity', *Journal of Clinical Epidemiology* **62**(5), 506–510.

Elberzhager, F., Rosbach, A., Münch, J. & Eschbach, R. (2012), 'Reducing test effort: A systematic mapping study on existing approaches', *Information and Software Technology* **54**, 1092–1106.

Engström, E., Runeson, P. & Skoglund, M. (2010), 'A systematic review on regression test selection techniques', *Information and Software Technology* **52**, 14–30.

Ericsson, K. & Simon, H. (1993), *Protocol Analysis: Verbal Reports as Data*, MIT Press.

Felizardo, K. R., Andery, G. F., Paulovich, F. V., Minghim, R. & Maldonado, J. C. (2012), 'A visual analysis approach to validate the selection review of primary studies in systematic reviews', *Information and Software Technology* **54**(10), 1079 – 1091.

Felizardo, K. R., Nakagawa, E. Y., Feitosa, D., Minghim, R. & Maldonado, J. C. (2010), An approach based on visual text mining to support categorization and classification in the systematic mapping, *in* 'Proceedings EASE '10', British Computer Society.

Fenton, N. E. & Pfleeger, S. L. (1997), *Software Metrics: A Rigorous and Practical Approach*, 2nd edn, PWS Publishing.

Fernández-Sáez, A. M., Bocco, M. G. & Romero, F. P. (2010), SLR-Tool—a tool for performing systematic literature reviews, *in* J. A. M. Cordeiro, M. Virvou & B. Shishkov, eds, 'Proceedings of ICSOFT (2)', SciTePress, pp. 157–166.

Fichman, R. G. & Kemerer, C. F. (1997), 'Object technology and reuse: Lessons from early adopters', *IEEE Computer* **30**, 47–59. (Reports on four Case Studies).

Fink, A. (2003), *The Survey Handbook*, 2 edn, Sage Books. Volume 1 of the Survey Kit.

Foss, T., Stensrud, E., Myrtveit, I. & Kitchenham, B. (2003), 'A simulation study of the model evaluation criterion MMRE', *IEEE Transactions on Software Engineering* **29**(11), 985–995.

Galin, D. & Avrahami, M. (2006), 'Are CMM program investments beneficial? Analysing past studies', *IEEE Software* pp. 81–87.

Gamma, E., Helm, R., Johnson, R. & Vlissides, J. (1995), *Design Patterns: Elements of Reusable Object-Oriented Software*, Addison-Wesley.

Ghapanchi, A. H. & Aurum, A. (2011), 'Antecedents to IT personnel's intentions to leave: A systematic literature review', *Journal of Systems & Software* **84**, 238–249.

Giuffrida, R. & Dittrich, Y. (2013), 'Empirical studies on the use of social software in global software development–A systematic mapping study', *Information & Software Technology* **55**, 1143–1164.

Glaser, B. & Strauss, A. (1967), *The Discovery of Grounded Theory: Strategies for Qualitative Research*, Aldine Publishing Company.

Glass, R., Ramesh, V. & Vessey, I. (2004), 'An Analysis of Research in Computing Disciplines', *Communications of the ACM* **47**, 89–94.

Goldacre, B. (2009), *Bad Science*, Harper Perennial.

Gómez, O. S., Juristo, N. & Vegas, S. (2010), Replications types in experimental disciplines, *in* 'Proceedings of Empirical Software Engineering & Measurement (ESEM)', pp. 1–10.

Gorschek, T., Svahnberg, M., Borg, A., Loconsole, A., Börstler, J., Sandahl, K. & Eriksson, M. (2007), 'A controlled empirical evaluation of a requirements abstraction model', *Information & Software Technology* **49**(7), 790–805.

Gotterbarn, D. (1999), 'How the new Software Engineering Code of Ethics affects you', *IEEE Software* **16**(6), 58–64.

GRADE Working Group (2004), 'Grading quality of evidence and strength of recommendations', *British Medical Journal (BMJ)* **328**(1490).

Graham, I. D., Logan, J., Harrison, M. B., Straus, S. E., Tetroe, J., Caswell, W. & Robinson, N. (2006), 'Lost in knowledge translation: Time for a map?', *Journal of Continuing Education in the Health Professions* **26**(1), 13–24.

Greenhalgh, T. (2010), *How to read a paper The basics of evidence-based medicine*, 4th edn, Wiley-Blackwell BMJIBooks.

Greenhalgh, T., Robert, G., MacFarlane, F., Bate, P. & Kyriakidou, O. (2004), 'Diffusion of Innovations in Service Organisations: Systematic Review and Recommendations', *The Milbank Quarterly* **82**(4), 581–629.

Greenhalgh, T. & Wieringa, S. (2013), 'Is it time to drop the "knowledge translation" metaphor? a critical literature review', *Journal of the Royal Society of Medicine* **104**, 501–509.

Grimstad, S. & Jørgensen, M. (2007), 'Inconsistency of expert judgment-based estimates of software development effort', *Journal of Systems & Software* **80**, 1770–1777.

Grimstad, S., Jørgensen, M. & Moløkken-Østvold, K. (2005), The clients' impact on effort estimation accuracy in software development projects, *in* 'Proceedings of 11th IEEE International Software Metrics Symposium (METRICS 2005)', pp. 1–10.

Gu, Q. & Lago, P. (2009), 'Exploring service-oriented system engineering challenges: a systematic literature review', *Service Oriented Computing and Applications* **3**(3), 171–188.

Guyatt, G. H., Oxman, A. D., Vist, G. E., Kunz, R., Falck-Ytter, Y., Alonso-Coello, P. & Schünemann, H. J. (2008), 'Grade: an emerging consensus on rating quality of evidence and strength of recommendations', *British Medical Journal* **336**, 924–926.

Guzmán, L., Lampasona, C., Seaman, C. & Rombach, D. (2014), Survey on research synthesis in software engineering, in *'18th International Conference on Eavluation and Assessment in Software Engineering'*, ACM, New York, USA.

Hall, T., Beecham, S., Bowes, D., Gray, D. & Counsell, S. (2012), 'A systematic literature review on fault prediction performance in software engineering', *IEEE Transactions on Software Engineering* **38**(6), 1276–1304.

Hammersley, M. (2005), 'Is the evidence-based practice movement doing more good than harm? Reflections on Iain Chalmers' case for research-based policy making and practice', *Evidence & Policy* **1**(1), 85–100.

Hammersley, M. & Atkinson, P. (1983), *Ethnography, Principles in Practice*, Tavistock.

Hannay, J., Dybå, T., Arisholm, E. & Sjøberg, D. (2009), 'The effectiveness of pair programming. A meta analysis', *Information & Software Technology* **51**(7), 1110–1122.

Hannes, K., Lockwood, C. & Pearson, A. (2010), 'A comparative analysis of three online appraisal instruments' ability to assess validity in qualitative research', *Qualitative Health Research* **20**(12), 1736–1743.

Hanssen, G. K., Bjørnson, F. O. & Westerheim, H. (2007), Tailoring and introduction of the rational unified process, in *'Software Process Improvement (EuroSPI 2007)'*, Vol. LNCS 4764/2007, Springer, pp. 7–18.

Hastie, T., Tibshirani, R. & Friedman, J. (2009), *The Elements of Statistical Learning Data Mining, Inference, Prediction*, 2nd edn, Springer.

Haugset, B. & Hanssen, G. K. (2008), Automated acceptance testing: A literature review and an industral case study, in *'Proceedings of Agile 2008'*, IEEE Computer Society Press, pp. 27–38.

Hedges, L. V. & Olkin, I. (1985), *Statistical Methods for Meta-Analysis*, Academic Press.

Hernandes, E., Zamboni, A., Fabbri, S. & Thommazo, A. A. D. (2012), 'Using GQM and TAM to evaluate StArt–a tool that supports systematic review', *CLEI Electronic Journal* **15**, 3.

Higgins, J. P. T. & Thompson, S. G. (2002), 'Quantifying heterogeneity in a meta-analysis.', *Statistics in Medicine* **21**(11), 1539–1558.

Higgins, J. P. T., Thompson, S. G., Deeks, J. J. & Altman, D. G. (2003), 'Measuring inconsistency in meta-analyses', *BMJ* **327**(7414), 557–560.

Hordijk, W., Ponisio, M. L. & Wieringa, R. (2009), Harmfulness of code duplication–a structured review of the evidence, in *'Proceedings of 13th International Conference on Evaluation and Assessment in Software Engineering (EASE 2009)'*, pp. 1–10.

Hossain, E., Babar, M. A. & Paik, H. (2009), Using scrum in global software development: A systematic literature review, in *'Proceedings of 4th International Conference on Global Software Engineering'*, IEEE Computer Society Press, pp. 175–184.

IEEE-CS/ACM (1999), ACM/IEEE-CS software Engineering Code of Ethics and Professional Practice. (Version 5.2).
URL: *http://www.acm.org/about/se-code/*

Jalali, S. & Wohlin, C. (2012), Systematic literature studies: Database searches vs. backward snowballing, in *'Empirical Software Engineering and Measurement (ESEM)*, 2012 ACM-IEEE International Symposium on', pp. 29–38.

Jedlitschka, A. & Pfahl, D. (2005), Reporting guidelines for controlled experiments in software engineering, in *'Proc. ACM/IEEE International Symposium on Empirical Software Engineering (ISESE) 2005'*, IEEE Computer Society Press, pp. 95–195.

Jiménez, M., Piattini, M. & Vizcaíno, A. (2009), Challenges and improvements in distributed software development: A systematic review, in *'Advances in Software Engineering'*, pp. 1–14.

Jørgensen, M. (2004), 'A review of studies on expert estimation of software development effort', *Journal of Systems & Software* **70**(1–2), 37–60.

Jørgensen, M. (2005), 'Evidence-based guidelines for assessment of software development cost uncertainty', *IEEE Transactions on Software Engineering* **31**(11), 942–954.

Jørgensen, M. (2007), 'Forecasting of software development work effort: Evidence on expert judgement and formal models', *Int. Journal of Forecasting* **23**(3), 449–462.

Jørgensen, M. (2014*a*), 'Failure factors of small software projects at a global outsourcingmarketplace', *The Journal of Systems and Software* **92**, 157–169.

Jørgensen, M. (2014*b*), 'What we do and don't know about software development effort estimation', *IEEE Software* pp. 37–40.

Jorgensen, M. & Shepperd, M. (2007), 'A systematic review of software development cost estimation studies', *Software Engineering, IEEE Transactions on* **33**(1), 33–53.

Juristo, N., Moreno, A. M., Vegas, S. & Solari, M. (2006), 'In search of what we experimentally know about unit testing', *IEEE Software* pp. 72–80.

Juristo, N. & Vegas, S. (2011), 'The role of non-exact replications in software engineering experiments', *Empirical Software Engineering* **16**, 295–324.

Kabacoff, R. I. (2011), *R in Action*, Manning.

Kakarla, S., Momotaz, S. & Namim, A. (2011), An evaluation of mutation and data-flow testing: A meta analysis, in *'Fourth International Conference on Software Testing, Verification and Validation Workshops'*, ICSTW, IEEE Computer Society, Washington, DC, USA, pp. 366–375.

Kalinowski, M., Travassos, G. H. & Card, D. N. (2008), Towards a defect prevention based process improvement approach, in *'Proceedings of the 34th Euromicro Conference on Software Engineering and Advanced Applications'*, IEEE Computer Society Press, pp. 199–206.

Kampenes, V. B., Dybå, T., Hannay, J. E. & K. Sjøberg, D. I. (2009), 'A systematic review of quasi-experiments in software engineering', *Inf. Softw. Technol.* **51**(1), 71–82.

Kampenes, V. B., Dybå, T., Hannay, J. E. & Sjøberg, D. I. K. (2007), 'Systematic review: A systematic review of effect size in software engineering experiments', *Information and Software Technology* **49**(11-12), 1073–1086.
URL: *http://dx.doi.org/10.1016/j.infsof.2007.02.015*

Kasoju, A., Peterson, K. & Mäntylä, M. (2013), 'Analyzing an automotive testing process with evidence-based software engineering', *Information & Software Technology* **55**, 1237–1259.

Kearney, M. H. (1998), 'Ready-to-wear: Discovering grounded formal theory', *Research in Nursing and Health* **21**(2), 179–186.

Khan, K. S., Kunz, R., Kleijnen, J. & Antes, G. (2011), *Systematic Reviews to Support Evidence-Based Medicine*, 2nd edn, Hodder Arnold.

Kitchenham, B. (2008), 'The role of replications in empirical software engineering—a word of warning', *Empirical Software Engineering* **13**, 219–221.

Kitchenham, B. A., Budgen, D. & Brereton, O. P. (2011), 'Using mapping studies as the basis for further research—a participant-observer case study', *Information & Software Technology* **53**(6), 638–651. Special section from EASE 2010.

Kitchenham, B. A., Fry, J. & Linkman, S. (2003), The case against cross-over designs in software engineering, in *'Proceedings of Eleventh Annual International Workshop on Software Technology & Engineering (STEP 2003)'*, IEEE Computer Society Press, pp. 65–67.

Kitchenham, B. A., Li, Z. & Burn, A. (2011), Validating search processes in systematic literature reviews, in *'Proceeding of the 1st International Workshop on Evidential Assessment of Software Technologies'*, pp. 3–9.

Kitchenham, B. A. & Pfleeger, S. L. (2002a), 'Principles of survey research part 2: Designing a survey', *ACM Software Engineering Notes* **21**(1), 18–20. (For Part 1, see under Pfleeger).

Kitchenham, B. A. & Pfleeger, S. L. (2002b), 'Principles of survey research part 4: Questionnaire evaluation', *ACM Software Engineering Notes* **27**(3), 20–23.

Kitchenham, B. A. & Pfleeger, S. L. (2008), Personal opinion surveys, in F. Shull, J. Singer & D. I. Sjøberg, eds, *'Guide to Advanced Empirical Software Engineering'*, Springer-Verlag London, chapter 3.

Kitchenham, B. & Brereton, P. (2013), 'A systematic review of systematic review process research in software engineering', *Information and Software Technology* **55**(12), 2049–2075.

Kitchenham, B., Brereton, P. & Budgen, D. (2010), The educational value of mapping studies of software engineering literature, in *'Proceedings ICSE'10'*, ACM.

Kitchenham, B., Brereton, P. & Budgen, D. (2012), Mapping study completeness and reliability—a case study, in *'Proceedings of 16th EASE Conference'*, IET Press, pp. 1–10.

Kitchenham, B., Brereton, P., Budgen, D., Turner, M., Bailey, J. & Linkman, S. (2009), 'Systematic literature reviews in software engineering — a systematic literature review', *Information & Software Technology* **51**(1), 7–15.

Kitchenham, B., Burn, A. & Li, Z. (2009), A quality checklist for technology-centred testing studies, in *'Proceedings EASE '09'*.

Kitchenham, B. & Charters, S. (2007), Guidelines for performing systematic literature reviews in software engineering, Technical report, Keele University and Durham University Joint Report.

Kitchenham, B., Dybå, T. & Jørgensen, M. (2004), Evidence-based software engineering, in *'Proceedings of ICSE 2004'*, IEEE Computer Society Press, pp. 273–281.

Kitchenham, B., Mendes, E. & Travassos, G. H. (2007), 'Cross versus within-company cost estimation studies: A systematic review', *IEEE Transactions on Software Engineering* **33**(5), 316–329.

Kitchenham, B., Pfleeger, S. L., McColl, B. & Eagan, S. (2002), 'An empirical study of maintenance and development estimation accuracy', *Journal of Systems and Software* **64**, 57–77.

Kitchenham, B., Pfleeger, S. L., Pickard, L., Jones, P., Hoaglin, D., Emam, K. E. & J.Rosenberg (2002), 'Preliminary Guidelines for Empirical Research in Software Engineering', *IEEE Transactions on Software Engineering* **28**, 721–734.

Kitchenham, B., Pfleeger, S., Pickard, L., Jones, P., Hoaglin, D., El Emam, K. & Rosenberg, J. (2002), 'Preliminary guidelines for empirical research in software engineering', *IEEE Transactions on Software Engineering* **28**(8), 721–734.

Kitchenham, B., Pretorius, R., Budgen, D., Brereton, P., Turner, M., Niazi, M. & Linkman, S. (2010), 'Systematic literature reviews in software engineering — a tertiary study', *Information & Software Technology* **52**, 792–805.

Kitchenham, B., Sjøberg, D. I., Dybå, T., Brereton, P., Budgen, D., Höst, M. & Runeson, P. (2013), 'Trends in the quality of human-intensive software engineering experiments–a quasi-experiment', *IEEE Transactions on Software Engineering* **39**(7), 1002–1017.

Kocaguneli, E., Menzies, T. & Keung, J. W. (2012), 'On the value of ensemble effort estimation', *IEEE Transactions on Software Engineering* **38**(6), 1403–1416.

Kocaguneli, E., Menzies, T., Keung, J. W., Cok, D. & Madachy, R. (2013), 'Active learning and effort estimation: Finding the essential content of software effort estimation data', *IEEE Transactions on Software Engineering* **39**(8), 1040–1053.

Kollanus, S. & Koskinen, J. (2009), 'Survey of software inspection research', *The Open Software Engineering Journal* **3**, 15–34.

Kothari, A. & Armstrong, R. (2011), 'Community-based knowledge translation: unexplored opportunities', *Implementation Science* **6**(59).

Krippendorff, K. (1978), 'Reliability of binary attribute data', *Biometrics* **34**(1), 142–144.

Laitenberger, O., Emam, K. E. & Harbich, T. G. (2001), 'An internally replicated quasi-experimental comparison of checklist and perspective-based reading of code documents', *IEEE Transactions on Software Engineering* **27**(5), 387–421.

Li, Z., Zhang, H., O'Brien, L., Cai, R. & Flint, S. (2013), 'On evaluating commercial cloud services: A systematic review', *Journal of Systems & Software* **86**, 2371–2393.

Liberati, A., Altman, D. G., Tetzlaff, J., Mulrow, C., Gøtzsche, P. C., Ioannidis, J. P. A., Clarke, M., Devereaux, P. J., Kleijnen, J. & Moher, D. (2009), 'The prisma statement for reporting systematic reviews and meta-analyses of studies that evaluate healthcare interventions: explanation and elaboration', *BMJ* **339**.

Lindsay, R. M. & Ehrenberg, A. S. C. (1993), 'The design of replicated studies', *The American Statistician* **47**(3), 217–228.

Lisboa, L. B., Garcia, V. C., Lucrédio, D., de Almeida, E. S., de Lemos Meira, S. R. & de Mattos Fortes, R. P. (2010), 'A systematic review of domain analysis tools', *Information and Software Technology* **52**(1), 1–13.

López, A., Nicolás, J. & Toval, A. (2009), Risks and safeguards for the requirements engineering process in global software development, in *'Proceedings of 4th International Conference on Global Software Engineering'*, IEEE Computer Society Press, pp. 394–399.

Lucia, A. D., Gravino, C., Oliveto, R. & Tortara, G. (2010), 'An experimental comparison of ER and UML class diagrams for data modelling', *Empirical Software Engineering* **15**, 455–492.

MacDonell, S. & Shepperd, M. (2007), Comparing local and global software effort estimation models – reflections on a systematic review, in *'Empirical Software Engineering and Measurement, 2007. ESEM 2007. First International Symposium on'*, pp. 401–409.

MacDonell, S., Shepperd, M., Kitchenham, B. & Mendes, E. (2010), 'How reliable are systematic reviews in empirical software engineering?', *IEEE Transactions on Software Engineering* **36**(5), 676–687.

Madeyski, L. & Kitchenham, B. (2014), How variations in experimental designs impact the construction of comparable effect sizes for meta-analysis. Available from Barbara Kitchenham.

Magdaleno, A. M., Werner, C. M. L. & de Araujo, R. M. (2012), 'Reconciling software development models: a quasi-systematic review', *Journal of Systems & Software* **85**, 351–369.

Mair, C., Shepperd, M. & Jørgensen, M. (2005), An analysis of data sets used to train and validate cost prediction systems, in *'Proceedings PROMISE '05'*, ACM.

Marques, A., Rodrigues, R. & Conte, T. (2012), Systematic literature reviews in distributed software development: A tertiary study, in *'Global Software*

Engineering (ICGSE), 2012 IEEE Seventh International Conference on', Global Software Engineering pp. 134–143.

Marshall, C., Brereton, O. P. & Kitchenham, B. A. (2014), Tools to support systematic literature reviews in software engineering: A feature analysis, *in* 'Proceedings of 18th International Conference on Evaluation and Assessment in Software Engineering (EASE'14)', ACM Press, pp. 13:1–13:10.

Marshall, C. & Brereton, P. (2013), Tools to support systematic literature reviews in software engineering: A mapping study, *in* 'Proceedings ACM/IEEE International Symposium on Empirical Software Engineering and Measurement (ESEM)', IEEE Computer Society Press, pp. 296–299.

McGarry, F., Burke, S. & Decker, B. (1998), Measuring the impacts individual process maturity attributes have on software products, *in* 'Proceedings 5th Software Metrics Symposium', pp. 52–60.

Mierzewski, P. (2001), Developing a methodology for drawing up guidelines on best medical practices, Technical report, Council of Europe.

Miles, M. B., Huberman, A. M. & Saldaña, J. (2014), *Qualitative Data Analysis A Methods Sourcebook*, 3rd edn, Sage Publications Inc.

Miller, J. (2005), 'Replicating software engineering experiments: a poisoned chalice or the holy grail', *Information & Software Technology* **47**, 233–244.

Mitchell, S. & Seaman, C. (2009), A comparison of software cost, duration, and quality for waterfall vs. iterative and incremental development: A systematic review, *in* 'Empirical Software Engineering and Measurement, 2009. ESEM 2009. 3rd International Symposium on', pp. 511–515.

Moe, N. B., Dingsøyr, T. & Dybå, T. (2010), 'A teamwork model for understanding an agile team: A case study of a Scrum project', *Information & Software Technology* **52**, 480–491.

Mohagheghi, P. & Conradi, R. (2007), 'Quality, productivity and economic benefits of software reuse: a review of industrial studies', *Empirical Software Engineering* **12**, 471–516.

Mohagheghi, P. & Dehlen, V. (2008), Where is the proof? – a review of experiences from appying MDE in industry, in '*Model Driven Architectures–Foundations & Application*', Vol. 5095/2008, Lecture Notes in Computer Science, Springer, pp. 432–443.

Moher, D., Liberati, A., Tetzlaff, J. & Group, D. G. A. T. P. (2009), 'Preferred reporting items for systematic reviews and meta-analyses: The prisma statement', *PLoS Med 6(7)* .

Moløkken-Østvold, K., Tanilkan, M. J. S. S., Gallis, H., Lien, A. C. & Hove, S. E. (2004), A survey on software estimation in the Norwegian industry, in *'Proceedings of 10th Internation Symposium on Software Metrics (METRICS'04)'*, pp. 1–12.

Morris, S. B. (2000), 'Distribution of the standardized mean change effect size for meta-analysis on repeated measures', *British Journal of Mathematical and Statistical Psychology* **53**, 17–29.

Morris, S. B. & DeShon, R. P. (2002), 'Combining effect size estimates in meta-analysis with repeated measures and independent-groups designs', *Psychological Methods* **7**(1), 105–125.

Munir, H., Moayyed, M. & Peterson, K. (2014), 'Considering rigor and relevance when evaluating test driven development: A systematic review', *Information & Software Technology* **56**, 375–394.

Myrtveit, I. & Stensrud, E. (2012), 'Validity and reliability of evaluation procedures in comparative studies of effort prediction models', *Empirical Software Engineering* **17**, 23–33.

Myrtveit, I., Stensrud, E. & Shepperd, M. (2005), 'Reliability and validity in comparative studies of software prediction models', *IEEE Transactions on Software Engineering* **31**(5), 380–391.

Nascimento, D., Cox, K., Almeida, T., Sampaio, W., Almeida Bittencourt, R., Souza, R. & Chavez, C. (2013), Using open source projects in software engineering education: A systematic mapping study, in *'Frontiers in Education Conference, 2013 IEEE'*, pp. 1837–1843.

NICE (2009), *The Guidelines Manual*, National Institute for Clinical Excellence (NICE).

Nicolás, J. & Toval, A. (2009), 'On the generation of requirements specifications from software engineering models: A systematic literature review', *Information and Software Technology* **51**(9), 1291–1307.

Noblit, G. & Hare, R. (1988), *Meta Ethnography: Synthesizing Qualitative Studies*, Sage Publications Ltd.

Noyes, J. & Lewin, S. (2011), Chapter 6: Supplemental guidance on selecting a method of qualitative evidence synthesis, and integrating qualitative evidence with cochrane intervention reviews, in J. Noyes, A. Booth, K. Hannes, A.Harden, J. Harris, S. Lewin & C. Lockwood, eds, *'Supplementary Guidance for Inclusion of Qualitative Research in Cochrane Systematic Reviews of Interventions.'*, version 1 (updated august 2011) edn, Cochrane Collaboration Qualitative Methods Group.

Oates, B. (2006), *Researching Information Systems and Computing*, SAGE.

Oza, N. V., Hall, T., Rainer, A. & Grey, S. (2006), 'Trust in software outsourcing relationships: An empirical investigation of Indian software companies', *Information and Software Technology* **48**, 345–354.

Pacheco, C. & Garcia, I. (2012), 'A systematic literature review of stakeholder identification methods in requirements elicitation', *Journal of Systems & Software* **85**, 2171–2181.

Paternoster, N., Giardino, C., Unterkalmsteiner, M. & Gorschek, T. (2014), 'Software development in startup companies: A systematic mapping study', *Information & Software Technology* **56**, 1200–1218.

Penzenstadler, B., Raturi, A., Richardson, D., Calero, C., Femmer, H. & Franch, X. (2014), Systematic mapping study on software engineering for sustainability (SE4S) — protocol and results, ISR Technical Report UCI-ISR-14-1, Institute for Software Research, University of California, Irvine.

Persson, J. S., Mathiassen, L., Boeg, J., Madsen, T. S. & Steinson, F. (2009), 'Managing risks in distributed software projects: An integrative framework', *IEEE Transactions on Engineering Management* **56**(3), 508–532.

Petersen, K., Feldt, R., Mujtaba, S. & Mattsson, M. (2008), Systematic mapping studies in software engineering, in *'Proceedings of the 12th International Conference on Evaluation and Assessment in Software Engineering'*, EASE'08, British Computer Society, Swinton, UK, pp. 68–77.

Peterson, K. (2011), 'Measuring and predicting software productivity: A systematic map and review', *Information & Software Technology* **53**, 317–343.

Petersson, H., Thelin, T., Runeson, P. & Wohlin, C. (2004), 'Capture-recapture in software inspections after 10 years research—theory, evaluation and application', *Journal of Systems and Software* **72**, 249–264.

Petre, M. (2013), UML in practice, in *'Proceedings of the 2013 International Conference on Software Engineering (ICSE)'*, IEEE Computer Society Press, pp. 722–731.

Petticrew, M. & Roberts, H. (2006), *Systematic Reviews in the Social Sciences A Practical Guide*, Blackwell Publishing.

Pfleeger, S. L. (1999), 'Understanding and improving technology transfer in software engineering', *Journal of Systems & Software* **47**, 111–124.

Phalp, K., Vincent, J. & Cox, K. (2007), 'Improving the quality of use case descriptions: empirical assessment of writing guidelines', *Software Quality Journal* **15**(4), 383–399.

Pino, F. J., Garcia, F. & Piattini, M. (2008), 'Software process improvement in small and medium software enterprises: a systematic review', *Software Quality Journal* **16**, 237–261.

Popay, J., Roberts, H., Sowden, A., Petticrew, M., Arai, L., Rodgers, M., Britten, N., Roen, K. & Duffy, S. (2006), Guidance on the conduct of narrative synthesis in systematic reviews, Technical report, Lancaster University, UK, Available from j.popay@lancaster.ac.uk.

Publication Manual of the American Psychological Association (2001), 5th edn, American Psychological Association, Washington, DC, USA.

Radjenović, D., Heričko, M., Torkar, R. & Živkovič, A. (2013), 'Software fault prediction metrics: A systematic literature review', *Information & Software Technology* **55**, 1397–1418.

Rafique, Y. & Misic, V. (2013), 'The effects of test-driven development on external quality and productivity: A meta-analysis', *IEEE Transactions on Software Engineering* **39**(6).

Ragin, C. C. (1989), *The Comparative Method*, University of California Press.

Remenyi, D. (2014), *Grounded Theory. A reader for Researchers, Studentts, Faculty and Others*, 2nd edn, Academic Conferences and Publishing International Ltd, Reading, UK.

Renger, M., Kolfschoten, G. L. & de Vreede, G.-J. (2008), Challenges in collaborative modelling: A literature review, in *'Proceedings of CIAO!* 2008 and EOMAS 2008', Vol. LNBIP 10, Springer-Verlag Berlin, pp. 61–77.

Riaz, M., Mendes, E. & Tempero, E. (2009), A systematic review of software maintainability prediction and metrics, in *'Proceedings of Third International Symposium on Empirical Software Engineering and Measurement (ESEM 2009)'*, pp. 367–377.

Robinson, H., Segal, J. & Sharp, H. (2007), 'Ethnographically-informed empirical studies of software practicel studies of software practice', *Information and Software Technology* **49**(540-551).

Robson, C. (2002), *Real World Research*, 2nd edn, Blackwell Publishing, Malden.

Rogers, E. M. (2003), *Diffusion of Innovations*, 5th edn, Free Press, New York.

Ropponen, J. & Lyytinen, K. (2000), 'Components of software development risk: How to address them. A project manager survey', *IEEE Transactions on Software Engineering* **26**(2), 98–111.

Rosenthal, R. & DiMatteo, M. (2001), 'Meta-analysis: Recent developments in quantitative methods for literature reviews', *Annual Review of Psychology* **52**, 59–82.

Rosenthal, R. & Rubin, D. B. (2003), 'r(equivalent): A simple effect size indicator', *Psychological Methods* **8**(4), 492–496.

Rosnow, R. L. & Rosenthal, R. (1997), *People Studying People Artifacts and Ethics in Behavioural Research*, W.H. Freeman & Co., New York.

Rovegard, P., Angelis, L. & Wohlin, C. (2008), 'An empirical study on views of importance of change impact analysis issues', *IEEE Transactions on Software Engineering* **34**(4), 516–530.

Runeson, P., Andersson, C., Thelin, T., Andrews, A. & Berling, T. (2006), 'What do we know about defect detection methods?', *IEEE Software* **23**(3), 82–86.

Runeson, P. & Höst, M. (2009), 'Guidelines for conducting and reporting case study research in software engineering', *Empirical Software Engineering* **14**(2), 131–164.

Runeson, P., Höst, M., Rainer, A. & Regnell, B. (2012), *Case Study Research in Software Engineering: Guidelines and Examples*, Wiley.

Sackett, D., Straus, S., Richardson, W., Rosenberg, W. & Haynes, R. (2000), *Evidence-based medicine: how to practice and teach EBM*, second edn, Churchill Livingstone.

Salleh, N., Mendes, E. & Grundy, J. (2009), 'Empirical studies of pair programming for CS/SE teaching in higher education: A systematic literature review', *IEEE Transactions on Software Engineering* **37**(4), 509–525.

Sandelowski, M. & Barroso, J. (2003), 'Creating metasummaries of qualitative findings', *Nursing Research* **52**(4), 226–233.

Sandelowski, M., Barroso, J. & Voils, C. I. (2007), 'Using qualitative metasummary to synthesize qualitative and quantitative descriptive findings', *Research in Nursing and Health* **30**(1), 99–111.

Sandelowski, M., Docherty, S. & Emden, C. (1997), 'Focus on qualitative methods qualitative metasynthesis: issues and techniques', *Research in Nursing and Health* **20**, 365–372.

Santos, R. E. S. & da Silva, F. Q. (2013), Motivation to perform systematic reviews and their impact on software engineering practice, in *'Proceedings ACM/IEEE International Symposium on Empirical Software Engineering and Measurement (ESEM)'*, IEEE Computer Society Press, pp. 292–295.

Schünemann, H. J., Fretheim, A. & Oxman, A. D. (2006), 'Improving the use of research evidence in guideline development: 1. guidelines for guidelines', *Health Research Policy and Systems* **4**(13).

Seaman, C. B. (1999), 'Qualitative methods in empirical studies of software engineering', *IEEE Transactions on Software Engineering* **25**(4), 557–572.

Seaman, C. B. & Basili, V. R. (1998), 'Communication and organization: An empirical study of discussion in inspection meetings', *IEEE Transactions on Software Engineering* **24**(6), 559–572.

Shadish, W., Cook, T. & Campbell, D. (2002), *Experimental and Quasi-Experimental Design for Generalized Causal Inference*, Houghton Mifflin Co.

Shahin, M., Liang, P. & Babar, M. A. (2014), 'A systematic review of software architecture visualization techniques', *Journal of Systems & Software* **94**, 161–185.

Shang, A., Huwiler-Müntener, K., Nartey, L., Jüni, P., Dörig, S., Sterne, J. A. C., Pewsner, D. & Egger, M. (2005), 'Are the clinical effects of homoeopathy placebo effects? comparative study of placebo-controlled trials of homoeopathy and allopathy', *The Lancet* **366**(9487), 726–732.

Sharp, H., Baddoo, N., Beecham, S., Hall, T. & Robinson, H. (2009), 'Models of motivation in software engineering', *Information and Software Technology* **51**, 219–233.

Sharp, H. & Robinson, H. (2008), 'Collaboration and co-ordination in mature extreme programming teams', *International Journal of Human-Computer Studies* **66**, 506–518.

Shaw, M. (2003), Writing good software engineering research papers (mini-tutorial), in *'Proceedings of 25th International Conference on Software Engineering (ICSE 2003)'*, IEEE Computer Society Press, p. 726.

Shepperd, M., Bowes, D. & Hall, T. (2014), 'Researcher bias: The use of machine learning in software defect prediction', *IEEE Transactions on Software Engineering* **40**(6), 603–616.

Shepperd, M. J. & MacDonell, S. G. (2012), 'Evaluating prediction systems in software project estimation', *Information and Software Technology* **54**(8), 820–827.

Shepperd, M., Song, Q., Sun, Z. & Mair, C. (2013), 'Data quality: Some comments on the NASA software defect datasets', *IEEE Transactions on Software Engineering* **39**(9), 1208–1215.

Sigweni, B., Shepperd, M. & Jørgensen, M. (2014), 'An extended mapping study of software development cost estimation studies', *IEEE Transactions on Software Engineering*. Under review.

Sjøberg, D., Hannay, J., Hansen, O., Kampenes, V., Karahasanović, A., Liborg, N.-K. & Rekdal, A. (2005), 'A survey of controlled experiments in software engineering', *IEEE Transactions on Software Engineering* **31**(9), 733–753.

Sjøberg, D. I. K., Dybå, T. & Jørgensen, M. (2007), The future of empirical methods in software engineering research, in *'Future of Software Engineering (FOSE'07)'*, Future of Software Engineering.

Skoglund, M. & Runeson, P. (2009), Reference-based search strategies in systematic reviews, in *'13th International Conference on Evaluation and Assessment in Software Engineering (EASE)'*.

Smite, D., Wohlin, C., Gorschek, T. & Feldt, R. (2010), 'Empirical evidence in global software engineering: a systematic review', *Empirical Software Engineering* **15**, 91–118.

Spencer, L., Ritchie, J., Lewis, J. & Dillon, L. (2003), *Quality in Qualitative Evaluation: A framework for assessing research evidence*, Cabinet Office.

Staples, M. & Niazi, M. (2008), 'Systematic review of organizational motivations for adopting CMM-based SPI', *Information and Software Technology* **50**, 605–620.

Steinmacher, I., Chaves, A. & Gerosa, M. (2013), 'Awareness support in distributed software development: A systematic review and mapping of the literature', *Computer Supported Cooperative Work (CSCW)* **22**(2-3), 113–158.

Straus, S. E., Tetroe, J. & Graham, I. (2009), 'Defining knowledge translation', *Canadian Medical Association Journal* **181**(3-4), 165–168.

Sun, Y., Yang, Y., Zhang, H., Zhang, W. & Wang, Q. (2012), Towards evidence-based ontology for supporting systematic literature review, in *'Proceedings of 16th International Conference on Evaluation and Assessment in Software Engineering (EASE 2012)'*, pp. 171–175.

Tahir, A., Tosi, D. & Morasca, S. (2013), 'A systematic review on the functional testing of semantic web services', *Journal of Systems & Software* **86**, 2877–2889.

Thomas, J. & Harden, A. (2008), 'Methods for the thematic synthesis of qualitative research in systematic reviews', *BMC Medical Research Methodology* **8**(45).

Thorne, S., Jensen, L., Kearney, M. H., Noblit, G. & Sandelowski, M. (2004), 'Qualitative metasynthesis: Reflections on methodological orientation and ideological agenda', *Qualitative Health Research* **14**(10), 1342–1365.

Tichy, W. F. (1998), 'Should Computer Scientists Experiment More?', *IEEE Computer* **31**(5), 32–40.

Tomassetti, F., Rizzo, G., Vetro, A., Ardito, L., Torchiano, M. & Morisio, M. (2011), Linked data approach for selection process automation in systematic reviews, in *'Proceedings of 15th International Conference on Evaluation and Assessment in Software Engineering (EASE 2011)'*, pp. 31–35.

Torres, J., Cruzes, D. S. & do Nascimento Salvador, L. (2012), Automatic results identification in software engineering papers, is it possible?, in *'Proceedings of 12th International Conference on Computational Science and Its Applications (ICCSA 2012)'*, pp. 108–112.

Toye, F., Seers, K., Allcock, N., Briggs, M., Carr, E., Andrews, J. & Barker, K. (2013), 'Trying to pin down jelly - exploring intuitive processes in quality assessment for meta-ethnography', *BMC Medical Research Methodology* **13**(46).

Trendowicz, A. & Münch, J. (2009), Factors influencing software development productivitity—state-of-the-art and industrial experiences, in *'Advances in Computers'*, Vol. 77, Elsevier, pp. 185–241.

Truex, D., Baskerville, R. & Klein, H. (1999), 'Growing systems in emergent organisations', *Communications of the ACM* **42**(8), 117–123.

Tsafnat, G., Glasziou, P., Choong, M., Dunn, A., Galgani, F. & Coiera, E. (2014), 'Systematic review automation technologies', *Systematic Reviews* **3**(1), 74.

Turner, M., Kitchenham, B., Brereton, P., Charters, S. & Budgen, D. (2010), 'Does the technology acceptance model predict actual use? A systematic literature review', *Information and Software Technology* **52**(5), 463–479.

Verner, J., Brereton, O., Kitchenham, B., Turner, M. & Niazi, M. (2012), Systematic literature reviews in global software development: A tertiary study, in 'Evaluation Assessment in Software Engineering (EASE 2012), 16th International Conference on', pp. 2–11.

Verner, J., Brereton, O., Kitchenham, B., Turner, M. & Niazi, M. (2014), 'Risks and risk mitigation in global software development: A tertiary study', *Information and Software Technology* **56**(1), 54–78. Special sections on International Conference on Global Software Engineering – August 2011 and Evaluation and Assessment in Software Engineering – April 2012.

Viechtbauer, W. (2007), 'Accounting for heterogeneity via random-effects models and moderator analyses in meta-analyses', *Journal of Psychology* **215**(2), 104–121.

Viechtbauer, W. (2010), 'Conducting meta-analyses in r with the metafor package', *Journal of Statistical Software* **36**(3).

Walia, G. S. & Carver, J. C. (2009), 'A systematic literature review to identify and classify software requirement errors', *Information and Software Technology* **51**(7), 1087–1109.

WHO (2005), Bridging the "know-do" gap: Meeting on knowledge translation in global health, Technical report, World Health Organisation.

Wieringa, R., Maiden, N., Mead, N. & Rolland, C. (2006), 'Requirements engineering paper classification and evaluation criteria: A proposal and a discussion', *Requirements Engineering* **11**(1), 102–107.

Williams, B. J. & Carver, J. C. (2010), 'Characterizing software architecture changes: A systematic review', *Information & Software Technology* **52**(1), 31–51.

Wohlin, C., Runeson, P., Höst, M., Ohlsson, M. C., Regnell, B. & Wesslen, A. (2012), *Experimentation in Software Engineering*, 2nd edn, Springer.

Yin, R. K. (2014), *Case Study Research: Design & Methods*, 5th edn, Sage Publications Ltd.

Yin, R. K. & Heald, K. A. (1975), 'Using the case survey method to analyze policy studies', *Administrative Science Quarterly* **20**, 371–381.

Zelkowitz, M. V. & Wallace, D. R. (1998), 'Experimental models for validating technology', *IEEE Computer* **31**, 23–31.

Zhang, C. & Budgen, D. (2012), 'What do we know about the effectiveness of software design patterns?', *IEEE Transactions on Software Engineering* **38**(5), 1213–1231.

Zhang, C. & Budgen, D. (2013), 'A survey of experienced user perceptions about design patterns', *Information & Software Technology* **55**(5), 822–835.

Zhang, H. & Babar, M. A. (2013), 'Systematic reviews in software engineering: An empirical investigation', *Information and Software Technology* **55**(7), 1341–1354.

Zhang, H., Babar, M. A. & Tell, P. (2011), 'Identifying relevant studies in software engineering', *Information and Software Technology* **53**(6), 625 – 637.

Zwarenstein, M. & Reeves, S. (2006), 'Knowledge translation and interprofessional collaboration: Where the rubber of evidence-based care hits the road of teamwork', *Journal of Continuing Education in the Health Professions* **26**, 46–54.

Index

Printed and bound by CPI Group (UK) Ltd, Croydon, CR0 4YY

24/10/2024

01778277-0013